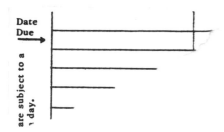

D0065153

SURFACE AND COLLOID SCIENCE

Volume 9

Advisory Board

SURFACE AND COLLOID SCIENCE

Volume 9

Editor: EGON MATIJEVIĆ

Institute of Colloid and Surface Science
Clarkson College of Technology
Potsdam, New York 13676

A WILEY-INTERSCIENCE PUBLICATION

1976

JOHN WILEY & SONS

New York—London—Sydney—Toronto

Library of Congress Catalog Number: 67-29459

ISBN: 0-471-57638-7

Printed in the United States of America

10 9 8 7 6 5 4 3 2 1

Preface

A need for a comprehensive treatise on *surface and colloid science* has been felt for a long time. Our series endeavors to fill this need. Its format has been shaped by the features of this widely roaming science. Since the subjects to be discussed represent such a broad spectrum, no single person could write a critical review on more than a very limited number of topics. Thus, the volumes will consist of chapters written by specialists. We expect this series to represent a treatise by offering texts and critical reviews which will describe theories, systems, and processes, handle these in a rigorous way, and indicate solved problems and problems which still require further research. Purely descriptive colloid chemistry will be limited to a minimum. Qualitative observations of poorly defined systems, which in the past have been so much in evidence, will be avoided. Thus, the chapters are neither supposed to possess the character of *Advances*, nor to represent reviews of authors' own work. Instead, it is hoped that each contribution will treat a subject critically giving the historic development as well as a digest of the newest results. Every effort will be made to include chapters on novel systems and phenomena.

It is impossible to publish a work of this magnitude with all chapters in a logical order. Rather, the contributions will appear as they arrive, as soon as the editor receives sufficient material for a volume. A certain amount of overlap is unavoidable, but will be kept to a minimum. Also, uniform treatment and style cannot be expected in a work that represents the effort of so many. Notwithstanding these anticipated difficulties, the series presented here appears to be the only practical way to accomplish the task of a high level and modern treatise on surface and colloid science.

Some general remarks may be in order. In modern times, few disciplines fluctuated in "popularity" as much as colloid and surface science. However, it seems that these sporadic declines in interest in the science of "neglected dimensions" were only apparent. In reality, there has been a steady increase in research through the years, especially in industrial laboratories. The fluctuations were most noticeable in academic institutions, especially with regard to teaching of specialized courses. It is thus only natural that the university professors with surface and colloid science as their abiding interest were frequently concerned and have repeatedly warned of the need for better and more intensive education, especially on the graduate level.

There are several reasons for the discrepancy between the need of industrial and of academic research laboratories in well trained surface and colloid scientists and the efforts of the academic institutions to provide specialization in these disciplines. Many instructors believe that a good background in the basic principles of chemistry, physics, and mathematics will enable a professional person to engage in research in surface and colloid science. This may be true, but only after much additional professional growth. Indeed, many people active in this area are self-educated. Furthermore, this science deals with an unusually wide range of systems and principles. This makes a uniform treatment of problems in surface and colloid science, not only challenging, but also a very difficult task. As a matter of fact certain branches of colloid science have grown into separate, independent disciplines which only in a broad sense are now considered a part of the "parent" science. Finally, there is often a stigma associated with the name "colloids." To many, the term symbolizes empirically and poorly described, irreproducible, etc., systems to which exact science cannot as yet be applied. The latter impression is in part based upon the fact that a considerable number of papers were and are published which leave much to be desired in regard to the rigorousness of the approach.

Yet, during the first half of this century some of the most acclaimed scientists have occupied themselves with colloid and surface science problems. One needs to mention only a few like Einstein, von Smoluchowski, Debye, Perrin, Loeb, Freundlich, Zsigmondy, Pauli, Langmuir, McBain, Harkins, Donnan, Kruyt, Svedberg, Tiselius, Frumkin, Adam, and Rideal, who have made substantial contributions to the classical foundations of colloid and surface science. This work has led to many fundamental theoretical advances and to a tremendous number of practical applications in a variety of systems such as natural and synthetic polymers, proteins and nucleic acids, ceramics, textiles, coatings, detergents, lubricants, paints, catalysts, fuels, foams, emulsions membranes, pharmaceuticals, ores, composites, soils, air and water pollutants, and many others.

It is therefore our hope that this treatise will be of value to scientists of all descriptions, and that it will provide a stimulating reference work for those who do not need to be convinced of the importance of colloid and surface science in nature and in application.

EGON MATIJEVIĆ

February, 1969

Addendum to the Preface

In order to broaden the circle of specialists whose advice the editor can seek and to still keep the size of the advisory board to a reasonable number of people, the memberships of Professor Stig Claesson, Dr. D. G. Dervichian, and Professor G. Ehrlich have been terminated.

The editor is taking this opportunity to express his deepest appreciation for all the assistance these members of the board have rendered since the inception of the series *Surface and Colloid Science*.

We take great pleasure in welcoming Professor Ervin Wolfram, Lorand Eötvös University, Budapest, Hungary, to the advisory board of the series.

EGON MATIJEVIĆ

October 1975

Contents

SURFACE AND COLLOID SCIENCE

Volume 9

1

The Stability of Emulsions and Mechanisms of Emulsion Breakdown

B. J. CARROLL

Unilever Research Laboratory
Port Sunlight
Merseyside
England

I. INTRODUCTION

A. Preliminary Remarks

Colloid scientists broadly understand what they mean when talking about an emulsion, but semantic differences are evident in the way the generic term is used by different authors and it is desirable to establish a working definition. Becher, discussing this point in his monograph (1), cited nine definitions of varying degrees of meaning, some being unhelpfully nebulous. Becher himself proposed to define an emulsion as

> . . . a heterogeneous system, consisting of at least one immiscible liquid dispersed in another in the form of droplets whose diameters, in general, exceed 0.1 μm. Such systems possess a minimal stability which may be accentuated by such additives as surface active agents, finely-divided solids, etc.

It seems very likely that a necessary condition for "minimal stability" is the presence of trace amounts of material adsorbed at the surface, as there is no likely, general mechanism whereby perfectly clean interfaces bounding droplets can hinder their immediate coalescence. Such surface active material might be inorganic ions or, more commonly, of one of the classes shortly to be described. This point apart, Becher's concept of an emulsion is better than many: It seems to be more explicit than the most recent description (2):

> . . . a dispersion of droplets of one liquid in another with which it is incompletely miscible.

It will be noted that neither this nor the operating definition recognises a foam as an emulsion of air in a liquid. This is in keeping with the several differences in behavior, to be enumerated in later sections, observed between emulsions and foams.

The most fundamental characteristic of an emulsion is its overall thermodynamic instability. A dispersion of droplets of a liquid in another has a high free energy associated with its large interfacial area and will always tend to reduce this free energy via the interfacial area decrease that accompanies droplet coalescence. In emulsions this approach to thermodynamic equilibrium takes place more or less slowly, and the term "stable" is frequently employed, the word being used relatively and in a strictly kinetic sense. Thus, although emulsions may have a half-life of minutes, this time may also be days or years.

Stability towards breakdown or coalescence is generally conferred by the accumulation at the interface of certain types of material that create a large energy barrier in the coalescence "reaction path" (droplet separation–potential energy or V-H curve). Disregarding certain inorganic compounds, which act only feebly (3), the most common stabilizers are (a) amphiphilic organic compounds such as the detergent type RX, where R is hydrophobic and X is hydrophilic and which may be ionic, ionize at the interface, or be nonionic, (b) macromolecules, such as proteins and polyelectrolytes, and (c) solid particles, which are partially wetted by both phases. For none of the above classes can it be said that the nature of the energy barriers existing between the droplets is properly understood, even after a century of research into the problem, although a considerable number of partial correlations exist. As a corollary, the breaking of unwanted emulsions, such as occur in industrial effluents, for example, is still at best a semiempirical process—an embarrassing state of affairs in the present climate of ecological concern.

This review attempts to set out the present state of the theory and, to some extent, the technology of the stabilization and destabilization of emulsions. In Section II is considered the factors controlling the aggregation (coagulation) of emulsion droplets. In the next section the breakdown (coalescence) of aggregated droplets into large droplets is discussed. Finally, in Section IV the available practical means by which the above-mentioned processes can be induced or accelerated are outlined.

B. General Considerations

1. Definitions

Because variations in usage in the literature are evident, it is worthwhile to define the terminology used in this review. IUPAC recommendations (2) have been followed.

An *aggregate* is an ensemble of two or more droplets held together by forces of unspecified magnitude. Quantitizing the interdroplet forces, aggregates comprise *flocculated* droplets when the forces are weak (interaction energy a few times kT) or *coagulated* droplets when the forces are stronger. Flocculates can frequently be redispersed by shaking, while coagulates can not. The process whereby two droplets unite to become one is termed *coalescence*. The term *breaking* of an emulsion covers the two consecutive processes flocculation/coagulation and coalescence. In this review, coalescence is deemed to take over from aggregation at a certain point, as discussed in Section III.A. It covers the last stages of the thinning of the film of continuous phase between the droplets and its eventual rupture. The rupture process is always very rapid (unless extremely viscous liquids are involved) and although a research field in its own right (4–8), the process does not need to be considered here because of its neglibible influence on the overall kinetics of breakdown.

2. Breakdown Mechanisms

Before discussing in detail the individual stages of emulsion breakdown, it is instructive to examine the possible forms these stages may take and the ways in which they are interlinked in the breakdkown process.

Schemes 1 and 2 (Figure 1), respectively, contrast examples of particularly simple and extremely elaborate breakdown mechanisms. In scheme 1 aggregation is irreversible and is followed only by coalescence. Scheme 2 indicates other possibilities. Aggregation, particularly flocculation, may be reversible and multiple aggregation (clumping) of droplets may occur; flocculated droplets and multiplets may coagulate while coagulated doublets and multiplets may either coalesce or form relatively stable conglomerates characterized by black film formation between droplets (Section III.D).

It is relatively easy to calculate the overall rate of breakdown in scheme 1 but very difficult to treat scheme 2 similarly. However, such a treatment is not necessary, as it is most unlikely that the breakdown occurs to a significant extent by more than a few of the many possible routes in a given system. The path actually followed is determined by the droplet concentration, the droplet interaction energy (which determines the rate of aggregation and also the nature of the aggregate), and by the rate of coalescence. The overall rate

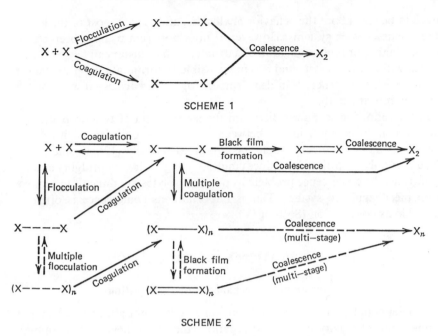

SCHEME 1

SCHEME 2

Fig. 1. Two possible routes to the coalescence of two emulsion droplets (denoted by X).

of breakdown will be determined by the rate constants of the separate stages and by the concentrations of the "reactants" involved in each stage.

In illustration of the foregoing, consider the system of scheme 1. Anticipating future sections, coagulation of two particles is a second order process with rate constant about 10^{-17} m^3/sec while coalescence (of coagulated droplets) is first order with rate constant of order 10^{-4}/sec. The steady state rate of coagulation is about the same as the rate of coalescence at a concentration of 10^{13} droplets per cubic meter. Concentrations higher than this will result in the rate of coagulation predominating over coalescence and extensive clumping of aggregates is likely. Concentrations below 10^{13}/m^3 lead to coagulation-controlled breakdown kinetics.

Complex schemes of type 2 are simplified by the fact that droplet coalescence is a first order process (9): This means that rupture of a given inter-droplet film in a clump does not affect the stability of adjacent films. Therefore coalescence in clumps is no more complicated than coalescence in a doublet and study of the latter is sufficient to establish the behavior of clumped droplets. To a certain extent similar remarks can be made about the aggregation process. The addition of one droplet to a cluster is analogous to the coagulation of the droplet with a single, larger droplet, although perhaps of different mass. So studies of two-droplet coagulation should allow much to be said about the behavior of clusters, and indeed most of the literature discusses such systems. However, it does not seem to have been established whether or not coagulation of a droplet with a cluster can trigger coalescence within the cluster, and it is not possible to formulate the problem for clusters with confidence. Van den Tempel (10, 11) has discussed some of the above points at length.

In the following section is discussed the aggregation of two droplets. Aggregation induced by bridging between droplets by polyelectrolytes is excluded, as a review on this topic has appeared recently (12). Furthermore, the use of polyelectrolytes to induce emulsion coagulation by bridging usually aggravates the coalescence problem as polyelectrolytes are difficult to remove from the coagulated system. This is important, as coalescence is often as desirable as coagulation (Section IV).

II. FLOCCULATION AND COAGULATION

A. Introduction. Mechanisms of Aggregations

Droplet conglomeration in emulsions may be brought about by settling of droplets under gravity or by centrifugation (the phenomenon of creaming). Conglomerates—generally aggregates in the defined sense of the word—may also form in the bulk liquid by collision of droplets brought about by their thermal (Brownian) motions. Aggregation may also be induced by the presence of solids—either the container wall or else a material introduced from outside, either deliberately or accidentally (e.g., dust). The last-mentioned is discussed in Section IV.E and the present section is chiefly concerned with aggregation occurring in the bulk liquid. However, although creaming is sometimes regarded as a trivial phenomenon, a few remarks on the topic do not seem amiss. From the practical point of view creaming is, after all, likely to result in an increased rate of aggregation/coalescence by virtue of increased droplet concentration in the creamed part of the emulsion.

B. Creaming

When an emulsion is placed in a force field that differentiates the dispersed from the continuous phase, creaming occurs. Gravity is the usual motive force, but creaming in electrostatic or magnetostatic fields is also possible (see Section IV.D and IV.F.2). In a gravitational field, if there is a density difference between the dispersed and continuous media, dispersed droplets experience a vertical force that tends to concentrate them in a layer above or below the continuous phase, according to the sign of the density difference. This tendency is opposed in two ways: temporarily by Stokes-type viscous forces and perpetually by the thermal motion of the droplets. In a dispersion of equally sized droplets the result is a Boltzmann distribution of droplet concentration in the vertical direction. The characteristic thickness of this distribution of creamed droplets depends basically on the droplet size and density difference; and calculations indicate that for emulsions it seldom exceeds a few millimeters. An obvious prerequisite for creaming—that the vertical dimension of the containing vessel greatly exceeds this characteristic length—is thus normally fulfilled.

The observed long-term stability of many emulsions towards creaming can be attributed to slowness of the settling speed of the droplets. This, too, is dependent on droplet size and on the density difference, and is of the order a few micrometers per second in typical stable emulsions. Random convection currents in the liquid bulk are frequently sufficient to counteract settling at this rate (13a).

It should be stressed that creaming is not *ipso facto* an indication of the breakdown of a newly prepared emulsion. It is an indication of droplet size, which depends on the emulsification conditions independently of aggregation. (Aggregation is, of course, more likely to occur in the creamed emulsion). Creaming is often, however, symptomatic of aggregation in the bulk phase, being the result of the formation of larger droplets or clumps that settle more rapidly. In a polydisperse emulsion, differences in droplet settling speeds may also give rise to aggregation rate enhancement via orthokinetic aggregation (*vide infra*).

C. Aggregation in the Bulk Phase

Droplet–droplet collision in the bulk is brought about primarily by Brownian motion. It is believed that adhesion to form aggregates depends on the existence of attractive van der Waals (dispersion) forces between droplets, which also tend to enhance the collision frequency because of their relatively large range of action (of order of the droplet radius). These attractive forces are often mitigated by forces associated with the overlap of electrical double layers that tend to reduce both the force of adhesion between aggregated droplets and the collision frequency. These two forces complicate the theoretical approach to aggregation. However, to fix ideas it is convenient to consider aggregation occurring solely as a result of Brownian motion and to discuss actual processes in terms of deviations from this situation. In certain situations it is found that stirring, or other processes bringing about shearing in an emulsion, can enhance the aggregation rate. Orthokinetic aggregation, as the effect is known, is discussed briefly after the following consideration of nonsheared or perikinetic emulsion aggregation. The topic of perikinetic aggregation is discussed at some length, for besides the fact that aggregation can be the rate-determining stage of a coagulation/coalescence process (in dilute emulsions, for example), there is an effect originating in the inevitable droplet size heterogeneity of emulsion. Different aggregated doublets are possible: large droplets with small, large with large, and small with small. It is quite plausible that such doublets coalesce at different rates. The conditions of aggregation can, thus, directly influence the coalescence process.

In considering the aggregation of emulsions, it is necessary, because of the dearth of literature on the aggregation of *droplets*, to look to the case of the aggregation of solid *particles*. In fact, as long as the approaching droplets are a few tens of nanometers apart, it is expected that their behavior will be substantially similar to that of solid particles. It has been established that, when surfactant is present, circulation within droplets is greatly retarded (14) and deformation of the droplets is unlikely to occur until late in the aggregation/coalescence process (see Section III.A).

D. Perikinetic Brownian Aggregation

The Brownian aggregation of a monodisperse suspension of particles, in which each collision results in an aggregate forming, has been discussed by Smoluchowski (15). The collision frequency at the start of the aggregation gives the rate of formation of doublets. It is important to note that in a system where the mean free path of a particle between collisions is less than the order of the particle diameter, the problem is to calculate the diffusive flux of particles towards a given central particle rather than one of calculating the number of particles moving in the right direction with sufficient energy to overcome the energy barrier where one exists. In the absence of such barriers, the two approaches are equivalent, but energy barriers are usually present. The treatment is outlined in ref. 16a. By use of the Einstein expression for the diffusion coefficient of a particle, $D = kT/6\pi\eta$ (17), Smoluchowski obtained the half-life of the process

$$t_{1/2} = \frac{3\eta}{4RTn_0} \equiv \frac{K}{n_0} \tag{1}$$

For continuous phase of viscosity 1 mN sec/m^2, K has the value $2 \times 10^{17}/m^3$ at 25°. The value of n_0 is of order $10^{14}/m^3$ in dilute emulsions and $10^{20}/m^3$ in concentrated emulsions, therefore, no emulsion is expected to be stable for more than 1 hr: usually less. Slow aggregation is believed to be the result of repulsion between the electrical double layers of particles and is treated in the classical DLVO theory outlined in Section II.F. Some extensions to the above "rapid aggregation" theory have been made and are worth mentioning before considering the treatment for interacting particles, because the versatility of the theory as a point of reference for the latter systems is thereby increased.

The most important extension is to allow for the particle size variation always found in practice, particularly in emulsions. The effect is to replace a single collision diameter for the system by a distribution of such diameters. This affects the aggregation kinetics—the collision frequency—in a way determined by the form of the distribution, but the result is always to accelerate aggregation. Müller (18) considered a mixture of the two monodisperse systems as a model of a polydisperse system and derived an expression for enhanced collision rates. This has recently been semiquantitatively confirmed experimentally (19). A more comprehensive treatment of polydisperse systems undergoing Brownian aggregation has been given by Swift and Friedlander (20), who have shown that as aggregation proceeds a reduced form of the particle size distribution function is unchanged or "self-preserving."

E. Orthokinetic Aggregation

In Section II.C above the enhanced collision rate that obtains in sheared solutions as a consequence of velocity gradients was mentioned. If there is relative motion between a particle and continuous phase over and above that due to Brownian motion, an increase in the flux of particles towards the central particle will occur. The flux turns out to depend on the third power of the collision diameter for a pair of particles, and so the effect is strongly size dependent. The Smoluchowski theory predicts that significant effects for the aggregation rate appear, in a briskly stirred emulsion, for particles of size 10 μm or more (13b). The effect is thus important for all but very fine emulsions. Experimental evidence from study of a monodisperse system supporting Smoluchowski's theory (21) has been given by Swift and Friedlander (20). The same authors discussed the extension of the theory to polydisperse systems and found that a self-preserving size distribution function exists, as was the case with perikinetic aggregation. Some experimental evidence was given in support.

In practical situations, in stirred emulsions, orthokinetic aggregation can be made to occur at some 10^4 times the perikinetic rate. Such a large effect renders typical emulsions sometimes quite unsuitable for the study of the theoretical aspects of coagulation/coalescence, as orthokinetic coagulation is often important and difficult to separate from other effects. Even in the absence of mechanical stirring, thermal convection currents are often important; in addition, the highly polydisperse nature of many emulsions gives rise to differential sedimentation (creaming) speeds which results in enhanced collision frequencies (13c). Creaming can, of course, be eliminated by the use of oil of exactly the same density as the aqueous phase, but this is unlikely to occur fortuitously. Differential sedimentation rates should not appear in monodisperse emulsions, but these are extremely difficult to prepare. It will be seen later that much of the most informative work on colloid stability has been performed with monodisperse solid dispersions; the particle size used has been usually smaller than that of most emulsions and such systems are therefore less susceptible to orthokinetic coagulation effects from whatever source. Working with mono-sized systems has additional advantages in that the theoretical interpretation of results is simplified.

F. The DLVO Theory

1. General Theory

According to equation (1) above, no emulsion is expected to be stable with respect to aggregation for more than at most an hour. Yet, in fact, emulsions stable for years can be prepared. This indicates that a factor, which we shall label W, of order 10^5 to 10^{11} for dilute and concentrated suspensions, respectively, should be inserted in the rate equation for rapid coagulation to account for the observed stability. The theoretical derivation of W has been the subject of much of the work on colloidal systems in the last 30 years, most notable being the work of Verwey and Overbeek (16) and of Derjaguin and Landau (22). Because this work will be frequently cited, it is outlined briefly below.

The Smoluchowski theory of Brownian aggregation breaks down because, contrary to the assumptions of the theory, a pair of particles interacts at separations far outside the range of the Born repulsion. Where water is the continuous phase, it is generally true that because of ionization of material at the interface, or the adsorption of free ions, all particles possess an electrical double layer which, overlapping the double layer of similar particles, results in a nett repulsion force. These double layer forces are to be combined with the attractive force due to the van der Waals interaction to give the nett interaction. Both interactions are functions of the particle separation and the result is that the sign of the nett interaction force depends very often on the interparticle separation. Other forces, sometimes held to be significant, are discussed later (Section II.H).

More quantitatively, repulsive (double layer) interactions between two particles whose centers are a distance $r = x + 2a$ apart are given by Derjaguin's approximate expression (23):

$$V_R = \frac{2\pi \times 10^{-2}}{9} \varepsilon\varepsilon_0 a\psi_0^2 \ln\left[1 + \exp\left(-\kappa x\right)\right], \tag{2}$$

which is widely used, although properly so only for low values of ψ_0 and when $\kappa a \gg 1$.

The van der Waals dispersion force between particles is classically calculated by integrating the London expression (24) for the force between two atoms over the whole volume of the particles concerned, on the assumption that pairwise summation is valid (25). An approximation valid for small particle separations is

$$V_A = \frac{-Aa}{12x} \tag{3}$$

If the interparticle medium is not a vacuum, A, the Hamaker constant (26), is replaced by an effective value

$$\begin{aligned} A' &= A_{11} + A_{22} - 2A_{12}, \\ &= A_{11} + A_{22} - 2(A_{11}A_{22})^{1/2} \end{aligned}$$

This, and the derivation of equation 3, is elaborated in ref. 16b. More elaborate expressions have been given (27) and are discussed later.

Qualitatively, the double layer interaction forces become important for particle separations of the order of the double layer parameter $1/\kappa$, while dispersion interactions are, for emulsion-sized particles, generally negligible outside separations of order of the particle radius. It is impossible to discuss analytically the total interaction $V_R + V_A$, due to the large number of variables present. It has been treated graphically and at length in ref. 16b. The basic features are shown in Figure 2, which depicts the quantities V_R (curve A), V_A (curve B) and two possible forms of $V = V_R + V_A$ (curves C and D).

Curves of type C show a nett repulsion between particles over certain particle separations (16b) and so lead to a retardation in aggregation with respect to the reference (Brownian) rate. The retardation is in part dependent on the height E of the potential evergy maximum. Curves of type C appear for systems having a high surface potential and in which the electrolyte concentration is not too high.

2. General Application to Aggregation

Addition of electrolyte progressively reduces the energy barrier E preventing aggregation until the point is reached where no barrier remains. This point is reached before $E = 0$, as will be evident later. When $E = 0$ (curve D) aggregation is, in fact, more rapid than Brownian aggregation owing to nett attractive forces at other values of the particle separation. Aggregation of this type is generally in the primary minimum P.

The value of E that gives rise to significant stabilization is of great interest, both theoretically and experimentally. As has been mentioned, the problem is one of diffusion of particles in a field of force rather than one of passage of separate particles over a potential energy barrier—a subtlety that has been ignored more than once (28, 29). It is not, therefore, possible to say that a value of E of a few kiloteslas is sufficient for stability. Treatment of the problem as one of diffusion is possible using a method of Fuchs (30). This leads to the expression for the factor W, (the stability ratio):

$$W = 2\int_0^\infty \frac{e^{-V/RT}}{H^2}\,dH \qquad (4)$$

The value of W depends mainly on the value of V in the region of the poten-

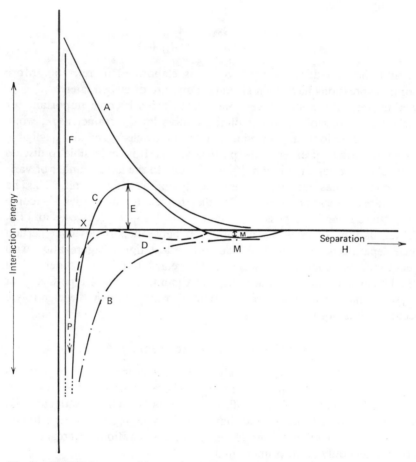

Fig. 2.   Potential energy of interaction between two particles as a function of separation. The net interaction energy depends on the relative magnitudes of the electrostatic (curve A), dispersion (curve B) and Born (curve F) interaction energies at a given separation. Two such curves are C and D. P and M denote primary and secondary minima, respectively. The barrier E to coagulation in the primary minimum on curve C has just disappeared in curve D.

tial energy maximum. This depends on the value of E (Figure 2). Values of E sufficient to ensure stability of even concentrated emulsions, of the order 20 kT, have been suggested (16a).

3. Size Effects

The effect of size on particle interaction energy will be referred to in other contexts and merits discussion. According to ref. 16a, when particles are smaller than the double layer parameter $1/\kappa$ ($\kappa a \ll 1$), stability always increases with increased particle size because the double layer repulsion then operates at distances outside the range of the dispersion forces (approximately the particle radius), and according to equation 2, V_R is proportional to a. However, ref. 16a points out that this size effect can be reversed for sols at concentrations of electrolyte near to the critical flocculation concentration C_e. The relative range of electrical double layer and of dispersion forces are then reversed. Experimental evidence for such size effects should, therefore, provide one test of the theory. Such tests are discussed in Section II.G below.

A feature of curve C not yet discussed is the shallow minimum M at relatively large particle separation. This minimum is generally significant when particle sizes of order 1 μm or larger are involved, when M can be of order of a few kiloteslas. This minimum, therefore, features in many instances with emulsified systems. When M is of order a few kiloteslas, reversible aggregation, that is, flocculation, is possible when particles are at the appropriate separation (16a). (Unlike the case with the quantity E, flocculation into the minimum M can be described by an exponential term as no energy barrier is involved in the process. One can, therefore, say at once of what order M should be for the formation of flocculates.) The phenomenon of flocculation in the secondary minimum is characterised by its easy reversibility—on shaking, for instance. It will always tend to complicate the kinetics of emulsion breakdown. Two droplets so flocculated should be more likely to coalesce than two isolated droplets in the same dispersion. An apparent increase in the coagulation kinetics in such systems is expected, which can mask the true coagulation process (into the primary minimum) (see, e.g., ref. 34).

G. The DLVO Theory and Experimental Evidence

For emulsions and dispersions of solids, experimentally accessible quantities are the factor W, its dependence on electrolyte concentration and, from these two, the concentration C_e of electrolyte at which $W = 1$, that is, at which aggregation proceeds at the Brownian rate. In an approximate analysis, Reerink and Overbeek (31), show that log W/log C plots are roughly linear and of slope

$$\frac{d \log W}{d \log C} = -2.15 \times 10^7 \, a\chi^2/v^2 \tag{5}$$

where $\chi = [\exp{(ve\, \psi_d/2kT)} - 1]/[\exp{(ve\psi_d/2kT)} + 1]$, provided that the potential ψ_d is constant.

The value of C_e many be calculated approximately on the assumption that $V = 0$ when $W = 1$, although it has been seen that $W < 1$ in this case. In practice, however, the Hamaker constant is usually unknown and is estimated from C_e. The surface potential (equation 2) is approximated by the electrokinetic potential ζ which often closely reflects ψ_d (109). This method ignores the possibility that other terms besides V_R and V_A appear in V (see Section II.H).

Early experimental work (13c) gave order-of-magnitude agreement with the above theory but was subject to large uncertainities originating in the experimental technique. In early work troubles originated in polydispersity of the suspensions—size effects are theoretically expected to affect coagulation kinetics (*vide infra*). Recent work by Ottewill and Shaw (32) and by Watillon et al. (33, 38) on near-monodisperse polystyrene dispersions has largely circumvented this problem. Theoretically reasonable C_e values were determined by both groups of workers but no agreement with equation 5 was found. In particular, the size dependence was found to be contrary to theory. Cooper (34) has pointed out that the log W/log C curves of refs. 32 and 33, which, should be drawn for constant ψ_d for equation 5 to apply, were actually plotted for systems in which the ζ potential varied widely: Points allegedly on one curve actually fell one each on a family of log W/log C curves. Analysis of the data in this way is not, therefore, permissible. Similar remarks apply to the data of Mathai and Ottewill on AgI sols (35, 36).

The DLVO theory does, however, make some interesting predictions about the effects of particle size as has been seen (Section II.F.3). Some experimental evidence for these predicted effects has appeared.

The primary size effect (that is, subject to the condition $\kappa a \ll 1$) that an increase in dispersion stability with an increase in particle size is expected, has found a tentative confirmation in the work of Matthews and Rhodes (37) who observed an increase in stability with particle size for monodisperse particles of size 0.74 and 1.4 μm, respectively. However, in their work $1/\kappa$ was often comparable with or smaller than the particle size. The situation $\kappa a \ll 1$ is in fact less frequently realized in systems in which particle size of order tens of micrometers are involved than when the size is less than one micrometer: Concentrations of 1:1 electrolytes considerably below 10^{-3} mole/dm^3 are required for the condition $\kappa a \ll 1$ to hold, and this is much below the usual range (10^{-2} to 10^{-1} mole/dm^3) of C_e. An added complication in such systems is, however, the possible existence of secondary minima of significant depth. It is, unfortunately, not certain from ref. 37 that aggregation was not due to this effect.

A recent paper by Joseph-Petit et al. (38) throws light on the contradictions apparent in the work above. Working with monodisperse selenium hydroxide hydrosols, it was shown that over the particle radius range 45 to 135 nm, stability increased with increasing particle size up to 50 nm, passed through a maximum, and decreased thereafter. Calculations based on the DLVO theory, incorporating corrections for the Vold effect (Section II.H.1) and viscous interactions (Section II.H.5) fitted the experimental data only up to the turning point at 50 nm. The results beyond this point were shown to be described by a secondary minimum coagulation. A good quantitative account of the data could be given for the two types of coagulation.

The reversal in size effects discussed in Section II.F.3 seems to have had experimental confirmation in a recent paper by Frens (39), who observed that gold sols of particle size 5 to 100 nm showed decreasing stability with increasing size.

The existence of secondary minimum flocculation is now well established experimentally. The early work of von Buzagh (40) on the adhesion of glass particles to plane glass sheets as a function of particle size demonstrated a maximum in adhesion: larger particles tended to be removed from the plate by gravitational forces while small ones were in an insufficiently deep secondary minimum. Schenkel and Kitchener (27) have demonstrated the phenomenon in polystyrene dispersions, while Srivastava and Haydon (41) and Lemberger and Mourad (42, 43) found the effect with emulsions stabilized by proteins and by conventional surfactants, respectively. The effect has been conveniently brought into the general theoretical framework by Wiesse and Healy (44). On a graph of surface potential versus electrolyte concentration, or particle size ratio versus concentration, regions or domains are delineated in which particular types of aggregation, or no aggregation at all, occur. The approach was shown to be equivalent to the DLVO scheme. A similar approach has been described by Miller (45).

These authors also suggest that the data of refs. 32 and 33 may in fact refer to secondary minimum flocculation, so that even the making of proper allowance for variation of the surface (ψ_d) potential, as suggested by Cooper (34) may still not allow these data to test the basic predictions of the DLVO theory.

It is apparent that the DLVO theory cannot be regarded as adequately tested with regard to the aggregation kinetics of a dispersion, which is the most important of the fields to which the theory is applicable. It is necessary to work with monodisperse systems with well-characterized surfaces and in which the particles are sufficiently small to preclude the possibility of secondary minimum flocculation at the electrolyte concentration in question (see also Section II.I).

Because the theory is inadequately tested by aggregating systems, the work of Scheludko et al. (46–52) and of Sonntag et al. (53–61) on microscopic thin liquid films is particularly useful as it has provided some quantitative evidence from a system closely related to aggregating droplets. Both schools have studied the properties of liquid films of thickness some 1 μm downwards and diameter of order 0.1 mm. Scheludko et al. have worked principally with films in air, but have demonstrated qualitatively similar behavior of films in liquid (47). Sonntag and collaborators have worked chiefly with liquid–liquid systems. The microscopic films studied by Scheludko et al. were usually nearly plane-parallel but this was not true of Sonntag's system. The latter, however, is arguably a better model for emulsions, while Scheludko's work is strictly more relevant to very fine foams. The last-mentioned work has, however, established beyond doubt the importance of dispersion and electrical double layer forces in such systems and has provided quantitative and semiquantitative verification of the theoretical expressions for these forces, giving one confidence in applying the same approach to the interaction of two emulsion droplets.

Essentially, Scheludko et al. have established (a) that microscopic, circular films drain radially and obey the Reynolds equation for such drainage (62), (b) by working at very low and very high electrolyte concentrations in aqueous films, and also with nonaqueous films, double layer and dispersion interactions could be separately examined and were shown to be quite well described by expressions relevant to flat films, similar to equation 2 above. This was proved in both static and dynamic ways, details being given in Scheludko's very able review of the subject (48). However, Platikanov and Manev (47) found that in liquid-liquid systems the dispersion interaction is best fitted by an expression corrected for retardation effects (see below Section II.H.1).

An expression for the double layer repulsion, essentially a simplification of that derived by Verwey and Overbeek (16c) has been verified experimentally by Lyklema and Mysels (63) for large soap films with not too high electrolyte concentrations or too thin a film. Fair agreement with theory has also been reported by Mysels and Jones using a different experimental technique (64). It might be suggested that thin film work such as the above provides a first-hand model of emulsions. However, a drawback is the very different time scales involved in the thinning of these films and the aggregation of emulsion droplets. One consequence of this may be that changes in the interfacial charge distribution may occur in the former but not in the latter process. This point is discussed in Sections II. H. and III. C.

Nevertheless, the thin film work discussed above has provided much-needed quantification of the two forces involved in the DLVO theory. Unfortunately, the results for very thin (< 10 nm) films do not fit in with the general DLVO picture. They are also, owing to experimental difficulties, not very precise. This fact, together with other discrepancies, found for instance from aggregation work, has resulted in a number of modifications being made to the basic theory, to be discussed in the next section. While many of these amendments are theoretically sound, it has been remarked regretfully that the elegant simplicity of the original DLVO theory has now been largely lost (65).

H. Modified Forms of the DLVO Theory

Several proposed changes to the original theory are discussed below. Because many of these corrections are quite system-specific in their dependence, it is beyond the scope of the present work to attempt to describe their effect when applied simultaneously.

1. Corrections to V_A

The expression for V_A derived by Hamaker should be corrected for relativistic retardation of the dispersion interaction, the effect of which is to reduce the attractive force, increasingly so as the particle separation increases (66). Physically, the effect can be thought of in terms of the time taken for the fluctuation field from one atom to reach the second atom and of the lifetime of the fluctuation. When these two quantities become comparable the interaction decreases. Approximate expressions, appropriate for low and high particle separations, have been given by Schenkel and Kitchener (27). V_A takes a modified form, which depends upon the value of a parameter $p = 2\pi x/\lambda$, where λ is the London wavelength. For the range $0 < p < 3$

$$V_A = \frac{Aa}{12x}\left(\frac{1}{1 + 1.77p}\right) \qquad (6)$$

while for $3 < p < \infty$

$$V_A = \frac{Aa}{12x}\left(-\frac{2.45}{5p} + \frac{2.17}{15p^2} - \frac{0.59}{35p^3}\right) \qquad (7)$$

The correction is expected to be important when the particle separation exceeds the London wavelength of electrons (~ 100 nm). Direct experimental measurements on the interaction between mica surfaces have recently appeared (67), and it is shown that the transition from fully unretarded to fully retarded interaction occurs for plates, in the region 10 to 60 nm. It is stressed here that this system shows some important differences from emulsion-type systems, as is elaborated below.

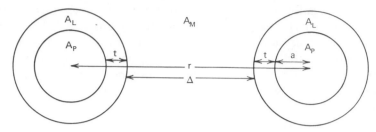

Fig. 3.   Van der Waals dispersion interaction of two identical particles sheathed with a medium of different material.

The effect of adsorbed layers at the interface on the dispersion interaction has been discussed by Vold (68), who showed that appreciable reductions in the attractive force were sometimes to be expected. The effect depends on the relative magnitudes of the Hamaker constants of the several components, on the thickness of the adsorbed layer, t, and on the particle radius. Denoting the Hamaker constants for particle, adsorbed layer and dispersion medium A_P, A_L, and A_M, respectively (cf. Figure 3), the following expression is obtained for identical particles:

$$- 12V_A = (A_M^{1/2} - A_L^{1/2})^2 f(L) + (A_L^{1/2} - A_P^{1/2})^2 f(p)$$
$$+ 2(A_M^{1/2} - A_L^{1/2})(A_L^{1/2} - A_P^{1/2})f(p, L) \qquad (8)$$

where $f(\alpha, \beta) \dfrac{\beta}{\alpha^2 + \alpha\beta + \alpha} + \dfrac{\beta}{\alpha^2 + \alpha\beta + \alpha + \beta} + 2\ln\left(\dfrac{\alpha^2 + \alpha\beta + \alpha}{\alpha^2 + \alpha\beta + \alpha + \beta}\right)$

in which for $f(L) = \Delta/2(a + t)$, $\beta = 1$;
for $\qquad\qquad f(p) = (\Delta + 2t)/2a$, $\beta = 1$;
and for $\qquad f(p, L) = (\Delta + t)/2(a + t)$, $\beta = a/(a + t)$.

V_A may be repulsive if the particles are not identical. A certain sequence of the various Hamaker constants involved in such a situation can bring this about (69).

Vold's treatment has been rescrutinized by Vincent et al. (69, 70), who have considered changes in V_A consequent to the adsorption of a layer of matter under two headings: a Hamaker constant effect, arising from the difference in values of the constant for the several different types of matter present; and a core effect, arising from the change in the spatial distribution of matter. The second effect is apparently the more important, although it was with the first that Vold was primarily concerned. The core effect always brings about a reduction in the van der Waals attraction, but it is only in limiting cases of small particle size or small adsorbed layer thickness that the first-mentioned effect becomes important. An error in the original treatment is noted which requires the sequence of Hamaker constants necessary for the Vold effect to be corrected. Israelachvili and Tabor (67) have shown directly that adsorbed layers of stearic acid on mica plates reduce the dispersion interaction. Work by Joseph-Petit et al. already quoted (38) is best interpreted by a DLVO treatment modified, *inter alia*, for the Vold effect.

A completely fresh approach to the calculation of the van der Waals interaction forces is that of Lifshitz (71, 72), whose treatment involves fewer approximations than does Hamaker's (26) and which expresses the interaction in terms of the macroscopic properties—dielectric dispersion behavior and density—of the media involved. As has been observed, the classical treatment depends on the unproved assumption that pairwise summation over atoms is valid. Furthermore, the effect of the presence of matter between interacting bodies on the force between them is allowed for by simply introducing the permittivity of the intervening medium as a factor in the interaction expressions. Both these assumptions have been questioned (73). Also, of course, there exists no reliable theoretical derivation of the Hamaker constant which is always an empirical quantity. Lifshitz's theory was originally thought to require very detailed (and, largely, not available) data on the dielectric dispersive properties of the media involved, and was believed to be of limited practical use. This was because an integral of the type

$$\int_0^\infty \left[\frac{\varepsilon_1 - \varepsilon_2}{\varepsilon_1 + \varepsilon_2} \right]^2 d\nu$$

over the entire frequency range is involved, the ε's both generally being frequency dependent. Recently, however, Ninham et al. (74–77) have shown that knowledge of the dispersive behavior of the media over the whole spectrum is not, after all, essential, it being adequate to extrapolate from the mean absorption frequencies and the oscillator strengths, properties available in many cases.

An important outcoming of these authors' work has been the prediction that in many systems of interest in colloid science the dispersion properties of the dielectrics in regions other than the ultraviolet can be important, especially at fairly large separations. The typical London frequency of electrons is generally in the uv region and hence attention had hitherto been focused on interactions of this wavelength. However, when the medium between interacting bodies has a similar electron density or, what amounts to the same thing, similar bulk density, contributions to the interaction from the uv region cancel out and lower frequency interaction becomes important. This is often so for emulsion systems. Because, in such systems, the effective London frequency is reduced the retardation effects are expected to be less important (74). Because of the differing density difference between dispersed and continuous phases for emulsions and suspensions, it is to be expected that magnitude of the dispersive interaction and retardation behavior will differ between the two otherwise comparable (size, charge, etc.) systems.

Other important innovations introduced include a predicted significant temperature dependence of the van der Waals interaction in those systems where contributions outside the uv region are important (76). The effect of material added to one of the media, for example, dissolved electrolyte, should also be significant, as the dispersion properties of the medium are thereby changed (75) (cf. the well-known change of refractive index of water with added electrolyte).

The essential soundness of the new approach is demonstrated by its remarkably good prediction (75) of the van der Waals interaction energy between lipid bilayers, as measured by Haydon and Taylor (78). It does not, incidentally, throw any light on the odd behavior of thin films observed by Scheludko and co-workers, who found that water films between oil droplets exhibited retarded van der Waals forces at thicknesses of the order expected while air-water-air systems did not. This observation does not yet seem to have been satisfactorily explained.

In the light of the above a re-appraisal of much of the published data and conclusions based thereon seems to be called for.

The extension of the Lifshitz approach to geometries other than parallel planes is difficult and has only recently been broached (79, 80). In this context an interesting, though not rigorous, treatment has recently appeared (81). The dispersion interactions are treated in terms of image forces according to the well-established electrostatic theory of this subject. This approach frequently results in formulas that are either identical with, or else good approximations to the rigorously derived expressions. It is shown to be possible to embrace a wider range of geometries than does formal theory at present. The effect of dielectric inhomogeneity on the magnitude of the van der Waals interactions has recently been discussed in terms of the Lifshitz theory by Weiss et al. (82), who show that pairwise summation is not a good approximation in those cases where the variation in the dielectric is over a distance comparable with the particle separation. This result makes the application of Vold's corrections, discussed above, a matter for caution.

An interesting attempt to reconcile the differences between the Hamaker-London and the Lifshitz approach has recently appeared (83). This paper corrects for the nonadditivity of atom–atom interactions using recent theory (73). General reviews of the theory of van der Waals interactions have been given by Krupp (72), Gregory (84) and more recently by Visser (85).

2. Modified Forms of V_R

The expression (2) for V_R is valid only (a) when particle dimensions are large compared with the double layer extension, that is, when $\kappa a \gg 1$, and

(b) when $\kappa H \ll 1$. The first condition is usually true for emulsions: $1/\kappa$ has the value 1 μm at (1:1) electrolyte concentration of 10^{-7} mole/dm^3 but the second condition limits the use of equation 2. Most refinements in this respect have been generalized to cover systems heterogeneous in particle type and size and are considered later (Section II.H.4).

A further point is that the derivation of equation 2 presupposes that particle approach is at constant surface charge rather than constant surface potential. Which of the two conditions best represents actuality is currently still an open question: it may, indeed, depend on the system. The duration of a Brownian collision between elastic particles is of the order of 10^{-6} sec, and it is in some cases arguable that changes in the surface charge distribution on one particle under the influence of the field of the approaching particle can occur in this time. Particles charged by fully ionized groups strongly adsorbed on the surface may be compared with oil droplets stabilized by nonionic surfactants—a system charged by the adsorption of highly mobile hydroxyl ions. The former is less likely than the latter to satisfy the constant potential conditions. It seems doubtful whether ideal constant potential collisions are often, if at all, met with, but constancy of neither charge nor potential does seem feasible. Frens and Overbeek (86) regard the constant charge approach to be more justified than that of constant potential, but show that, in either case, V_R is not greatly changed at moderate particle separation ($\kappa H \approx 1$) and changes typically by 10 to 20% at closer separations. Honig and Mul (87) have solved the problem numerically, for both constant charge and constant potential, in a more rigorous treatment and have tabulated solutions. An analytical expression that reproduces these numerical results approximately up to moderately large surface potentials has been described by Gregory (88). The case for the approach of particles under the condition of constant energy of adsorption of potential-determining ions has recently been discussed (89). Ohshima (90, 91) has discussed the case where partial decay of the electrostatic potential occurs inside the particles.

Shilov (92) has discussed the approach at constant charge in the case of conducting spheres. The approach of two particles, on one of which the condition of constant potential and on the other of which that of constant change applies has also been discussed, to the limitations of the Derjaguin integration method and the Debye–Hückel approximation to the solution of the Poisson-Boltzman equation (93). A variational solution of the Poisson-Boltzman equation for the electrical double layer around a spherical particle has recently been described (94), the results being given in analytical form.

Levine and Bell (95) have considered situations in which Stern-type (96) adsorption may be important and have incorporated the discreteness-of-charge effect into the general stability theory. This effect, possibly important at high surface charge densities, is held to be responsible for reduced critical flocculation concentrations in silver iodide systems at high pI (97).

Equation 2, the expression for V_R is based on Gouy–Chapman theory of the electrical double layer and so ultimately on the Poisson-Boltzman equation. The use of the last has been criticized at times (e.g., ref. 98) and a local thermodynamic treatment of the electrical double layer (99) has been extended to embrace corrections to the classical DLVO theory by Sanfeld et al. (100). The influence of molar volume, ion polarizability, and activity corrections on interaction energy were discussed. However, it is debatable whether refinements to double layer theory are worth seriously considering at the present stage, in view of the quite large corrections to the DLVO theory that can arise from other sources.

A start has recently been made by Chen and Levine on the double layer repulsion theory for concentrated systems (101). A notable result is an increased repulsive energy at low particle separation compared with the case for dilute emulsions. The authors suggest that care should be taken to check the particle concentration before ascribing deviations from the DLVO theory to solvent structure effects (see below Section II.H.5).

3. Polydisperse Systems

Size heterogeneity has already been considered in the context of Brownian coagulation (Section I.D). More general treatments, encompassing the DLVO-type particle interactions have only comparatively recently been discussed (34, 102). Previous work ignored either interaction forces (Swift and Friedlander (20)) or else particle size heterogeneity (e.g., Hogg, Healy and Fuerstenau (103)). Cooper (34) postulated an initially Gaussian distribution of particle size and investigated the variation of an overall stability ratio W with the standard deviation of the particle size distribution. It emerges from this study that stability decreases with increased size heterogeneity (as was found to be the case with Brownian aggregation). Increase in the electrical surface potential tends to increase stability (102), while variation in surface potential among particles has the opposite effect. Cooper was able to explain an apparent discrepancy between the data of Ottewill and Shaw on monodisperse suspensions and the DLVO theory, first on the basis of the small polydispersity of the systems actually used, and secondly on the incorrect application of Reerink and Overbeek's treatment (31) (see Section II.G). Reference 102 shows that a self-preserving distribution function for these systems also exists (cf. ref. 20).

4. General Heterogeneity

Besides size heterogeneity, it is possible to have systems composed of mixtures of two or more different types of dispersion. In such cases heterogeneity in both electrical double layer and dispersion interactions is to be expected. Although the classical DLVO theory made no attempt to treat such systems, the general concepts of this theory can be applied. The dispersion interaction between particles of dissimilar material is quite simply treated and the principal problem is to describe the interactions of dissimilar double layers. Derjaguin (104), Devereux and De Bruyn (105), and Miller (45) have given treatments similar to that employed by Verwey and Overbeek for homogeneous systems, but the results of the two treatments are not readily applicable to practical systems and approximate methods have been devised. Hogg, Healy, and Fuerstenau (HHF) (103) have solved the problem approximately and demonstrate agreement with ref. 105 for moderate surface potentials (75–100 mV). Their method was essentially to take a form of the Debye–Hückel approximation to the potential between particles and to use the Derjaguin method (23) to evaluate the interaction energy. Combining this with the expression (3) for dispersion interactions, they concluded that for the coagulation of oppositely charged particles W is almost constant and is close to the value expected for rapid coagulation of uncharged particles. Use of the Derjaguin method limits the analysis to $\kappa a > 1$, and the authors were unable to comment on the disputed observation that, at $\kappa a < 1$, an enhanced coagulation rate for particles of charge of opposite sign but similar magnitude exists (106). The HHF theory has recently been qualitatively confirmed by Harding (107).

The HHF treatment has been extended (44) to cover the case of particles interacting at constant surface charge, and the effects of retardation were included. Following the same lines, but extending the treatment to cover the particle concentration dependence of the coagulation rate, Pugh and Kitchener have developed criteria for selective coagulation of mixed dispersions via control of the ionic strength (108) and have experimentally demonstrated a predicted instance (109). Parsegian and Gingell (110) have discussed the interaction of approaching unlike double layers with respect to conditions of constancy of charge and of potential, for systems in which $\kappa a > 1$. They point out that while like-charged plates always repel, oppositely charged plates need not always attract and plates with potential of the same sign need not repel one another. They describe graphically regions of attraction and repulsion as a function of plate charge ratio. The phenomenon hinges on the movement of sufficient counter charge into the region between plates to over-neutralize the total plate charge. Bell and Peterson (111) and Ohshima (90, 91) have arrived at essentially the same conclusions.

Bell, Levine, and McCartney (112) have developed more general expressions for double layer interaction for particles heterogeneous in size and potential, for large and small interparticle distances. It now seems that reasonably good estimates of the electrical double layer interaction forces are available for most of the possible permutations of size, separation, and charge. It must be noted that Stern adsorption has been largely neglected, and also that particle size differences are not properly treated, in that the diffusion coefficient for a pair of unequal particles is taken as twice that of the individual particles, an approximation only valid for particles of not too disparate size. Effects of the sort dealt with by Müller (18) are not treated. The droplet size distribution in emulsions can be very broad, and this may vitiate the application of the above to such systems.

5. *Viscous Interactions*

The viscosity of the continuous phase enters DLVO theory via the Einstein equation for the diffusion coefficient $D = \kappa T/6\pi\eta a$ and the value of η is always taken as the bulk value. The use of the Einstein expression for D becomes hard to justify when particles are close together for at least two reasons. First, the fluid flow pattern around a particle is no longer that for an isolated sphere in a continuum—the presence of the second particle modifies the flow pattern. Secondly, the viscosity of the medium may at small particle separations differ from the bulk viscosity.

The first point has been discussed recently by Spielman (113), who found that a significant correction is necessary, particularly when $\kappa a \gg 1$, when a tenfold decrease in coagulation rate is predicted. The size of the correction depends on the location (at a particle separation $H \approx 1/\kappa$ (16b)) of the potential energy maximum, being greatest when this is at small separations. This problem has also been discussed recently by Chow and Hermans (114) from a rather different angle of less use in the present context. Lips et al. (115) have measured the absolute coagulation rate of a polystyrene latex and have found it to be of the order 60 to 70% of the Smoluchowski value. This is in good agreement with Spielman's estimate for the effect of viscous interactions, although, of course, this agreement may be fortuitous, other corrections possibly contributing.

The second possibility, that the viscosity of fluid films a few nanometers thick is different from the bulk value, is still a controversial issue. The two interfaces can conceivably promote solvent structure in the common case of water via orientation by increased hydrogen bonding, for example. Derjaguin has long been an exponent of this view (116), but experimental evidence, particularly for liquid—liquid systems, is at present not unequivocally in favour of either viewpoint. Work by Stigter (117) and Carroll (118), respectively, on micellar systems and emulsions containing ionic surfactant shows no evidence for solvent structuring, but Vincent and Lyklema (119) have recently invoked such behavior to interpret data on the potentiometric titration of silver iodide systems. It is possible that the ability to promote solvent structuring varies with the interface involved: for instance, the formation of polywater species is allegedly possible only through the mediation of silica or quartz surfaces (120). Perhaps also two adjacent interfaces are more potent in promoting structure than a single interface (the work in refs. 117 and 118 refers to a single interface). Direct studies on thin liquid films are the most likely to elucidate this problem, but measurements on very thin films are, for technical reasons, not very accurate at present. Some evidence for the existence of anisotropic layers between coalesced emulsion droplets in certain systems has been described (121).

This effect can be viewed in another way, in terms of the potential energy—separation diagram (Figure 2). Enhanced viscosity of the solvent should broaden the maximum and consequently W (equation 4) should increase. Explanations of deviations from the DLVO theory in these terms have long been popular (see, e.g., refs. 29, 122–124). The poor present state of the theory of the structure of liquids in general and of water in particular makes it impossible to quantitize the effects of structure enhancement, and appeals to the effect must be regarded as rather *ad hoc* at the present time. The possibility of other effects being responsible (see, e.g., ref. 101 and Section II, H.3) for the observed deviations must never be discounted (125).

6. Entropic and Related Forces

When particles approach very closely, interaction between the stabilizing surfactant molecules in different particles is possible if the latter protrude greatly into the continuous phase. This is especially the case with W/O type emulsions, in which the long hydrocarbon "tail" of the stabilizer is in the continuous phase and in the case of O/W emulsions stabilized by certain nonionic surfactants (e.g., polyethoxylated ethers C_nE_x) (126, 127) or by macromolecules. In such systems, particles are surrounded by a layer which, depending on the extent of surface coverage, is more or less penetrable. This layer may extend some 5 nm beyond the surface of the particle (more in the case of some macromolecules), and interpenetration becomes possible as the particles approach.

The detailed explanation of what happens when interpenetration occurs is complex and is still under discussion. Repulsive energies of interaction arising from a loss of configurational entropy have been discussed on the basis of a simple model by Mackor and van der Waals (128), and a more comprehensive discussion was given by Hesselink et al. (129). The latter authors applied the volume-restriction treatment of Flory and Huggins to the interpenetration of the chains. Unfortunately, theory and experiment are not yet in good agreement. Evans and Napper (130) stress that the role of the solvent in the interaction process has been ignored by most previous treatments, although it is demonstrably important (131), and in a further paper (132) they discuss this effect in an adaption of an early treatment of Fischer (133).

The topic of steric stabilization is of peripheral rather than central interest in the present review, and its complexity forbids its discussion at length. The interested reader is referred to reviews by Phillips (134), Ottewill (135), and Napper (130) for a more comprehensive treatment.

I. Concluding Remarks

It is noted here that studies of coagulation kinetics do not seem to be a particularly useful way of investigating interparticle forces, as the quantity measured—the stability ratio W—is an integral over all particle separations and is therefore relatively insensitive to the spatial variation of these forces. Knowledge of W does not allow one to say whether the terms V_R and V_A are incorrectly calculated or that the DLVO concept is fundamentally wrong and that other forces, such as are discussed above, are important for determining stability. It is for this reason that more direct methods of investigating interparticle forces are welcome (e.g., refs. 46–61). The conclusion to be drawn from the preceding sections is that the DLVO theory should best apply to aggregation in those systems where the maximum in the interaction energy is located at relatively large particle separations. This maximum is typically at a separation $H \approx 1/\kappa$. Relatively low ionic strength is, therefore, desirable. At higher ionic strength ($\approx 10^{-2}$ mole/dm^3) corrections due to viscous interactions, entropic effects, and so on can become important during the aggregation process. They are in all cases important in the coalescence process that follows aggregation, and it will be found that the DLVO theory does not adequately describe coalescence.

III. THE COALESCENCE PROCESS

A. Introduction

After two particles have approached closer than the separation corresponding to the potential energy maximum, they tend to draw together under the influence of the van der Waals forces until short-range forces (Born repulsive forces) become dominant. The process with two drops is qualitatively different, as at small separations some flattening of the drops occurs (cf. photographs in refs. 3 and 136). A thin circular film of continuous phase separates the two droplets. The small but finite radius of this film reduces the rate of drainage of the continuous phase from between the droplets as compared with the situation for two particles. Coalescence of the drops happens when the dividing film ruptures. As this often occurs in a still-thinning film rather than in one which has ceased to drain, there is no obvious way of defining the end of coagulation and the beginning of coalescence. In what follows coalescence is arbitrarily defined to start when the droplets have passed the point of maximum repulsive energy, for after this point the nett attractive force increases very rapidly with decreasing droplet separation and deformation of the droplets is expected, marking the departure of the system from the analogy with two particles.

Studies of emulsion beakdown have generally been in the following directions:

1. Direct study of a coalescing emulsion via the time variation of droplet concentration, size distribution, and so on.
2. Accelerated aggregation/coalescence by use of an ultracentrifuge.
3. Study of isolated thin films in air and under another liquid.
4. The coalescence of macroscopic (millimeter) drops at phase interfaces.

Theoretical studies of the coalescence process are very few.

Generally, attempts have been made to interpret results in terms of the DLVO theory or of one of its close relatives (e.g. refs. 19, and Section III.A below). Such interpretations have never been wholly successful. It is really necessary to link up generalized theories of coagulation, thin film formation, and thin film rupture. None of these can be said to be developed at present, although theories for the last two processes have been advanced (Section III.D). The only recent study which is a departure from the DLVO type of approach, seems to be the work of Hill and Knight (137) (see Section III.I).

The coalescence of air bubbles (i.e., foam breakdown) has been discussed by Marrucci (138).

B. Direct Studies

Studies on the coalescence process in emulsions can be made if the emulsion is sufficiently concentrated, so that coagulation is relatively very rapid, as has been discussed in Section I.B.2. The direct approach can give information on the life span of the interdroplet film of continuous phase and on factors influencing this life span, such as the effect of added surfactants or electrolyte. In these systems it is not usually possible to determine the film thickness at rupture. The method suffers from the difficulty in obtaining monodisperse emulsions of size homogeneity similar to that achieved with solids. Efforts have, however, been made in this direction, although better results seem to have been obtained with air—liquid systems (aerosols) (139, 140) than with liquid—liquid systems (141, 142); however, further efforts in this direction have recently been made (143, 144). Droplet size homogeneity is desirable as it leads to uniformly sized interdroplet films that are under similar pressures (of double layer and dispersion force origin) at a given film thickness.

Early work by van den Tempel has established that, in a clump of aggregated droplets, the rupture of one interdroplet film has no repercussions for the stability of adjacent films. This important result forms the rationale for the use of isolated thin films as a model for coalescence, an aspect considered in Sections II.G and III.D.

Another consequence of this result is that plots of log (Total number of particles, N) versus time t are linear and that the slope is a measure of the rate constants of 10^{-3}/sec and 10^{-7} sec for characteristic unstable and stable systems, respectively.

An interesting feature of log N–t plots obtained experimentally is that they often show apparent initial rate constants of up two orders of magnitude greater than the long-term constants (11, 29, 145, 147, 151). Van den Tempel attributed this phenomenon to ageing effects at the interface, possibly coupled with the establishment of a particle size distribution that allows close packing. Sherman (151) considered unequal degrees of surfactant adsorption with differing drop size to be responsible. Matthews and Rhodes (19) suggest that the effect originates in the higher interaction force between small particles than between the bigger ones. Another possibility is interfacial turbulence accompanying mass transport across the film leading to rupture. This could occur if one or more solutes were not in partition equilibrium. The possibility also exists that this phenomenon originates in the dissolution of small droplets, which results in the growth of large drops at the expense of the disappearance of the smaller ones. The process is a consequence of the higher internal (Laplace) pressure of the smaller drops, which enhances their solubility. The molecular diffusion process is discussed further in Section III.D. Support for this mechanism of emulsion breakdown comes from work of Hallworth and Carless (152) and of Davis and Smith (153), on the stability of O/W emulsions in which the oil was one of a series of n-alkanes. The lower members of the series were found to give the least stable emulsions; also, increasing the water solubility in the case of other oils studied in general led to a decrease in stability. The stability of these relatively unstable systems could be increased by addition of hexadecane (153), as would be expected if this mechanism of droplet dissolution operates. It should be remembered also that coalescence leads to an increase in the value of the bulk surfactant concentration, as the oil–water interfacial area has decreased. In those cases where stability is proportional to the bulk surfactant concentration, the decrease in coalescence rate with time is to be expected.

Van den Tempel studied emulsions stabilized by both ionic (SDS) and nonionic (aerosol OT) surfactants. Nonionic surfactants of type C_nE_x $(CH_3(CH_2)_{(n-1)}(CH_2CH_2O)_xH)$ and proteins (bovine serum albumin BSA) have also been studied (145–150, 41).

A series of papers by Elworthy et al. (145–150) described the coalescence of emulsions stabilized by nonionic alphol ($R(CH_2CH_2O)_xH$) type surfact-ants. The effect of varying surfactant concentration (in the range $0.1-10\%$) and the chain length x ($x = 3,6,9$, and 22; R was fixed at C_{16}) was investi-gated, together with the effects of the polar or nonpolar nature of the emulsi-fied oil. Concurrent measurements of the electrophoritic mobility of emul-sion droplets provided some information on the changes occurring in the double layer potentials. Following van den Tempel and eariler workers, a photomicrographic drop-counting technique was used.

The results obtained proved to be difficult to interpret. Stability (towards coalescence, as reflected by the rate constants for coalescence) was found to depend both on surfactant concentration and on the value of x. With the lower members of the series, stability increased with increasing surfactant concentration and with increasing x. The system containing $C_{16}E_{22}$ showed a slight decrease in stability with increasing concentration but fitted in with the trend observed for increased stability with increasing x. The calculated electrokinetic potential decreased with surfactant concentration increase in all cases, and also decreased with increasing x.

An attempt was made to correlate the observed stability trends using a DLVO-type model. The value of V_A (equation 3) was calculated for various values of the Hamaker constant, applying Vold's correction (68) for adsorbed material on the dispersion interactions (important because of the length of the stabilizing molecule). The value of V_R was calculated using the electro-kinetic potential in place of ψ_0 in equation 2. Of course, V_R is varying in the wrong direction with surfactant concentration to explain the results and no satisfactory fit could be obtained. Appeal to entropic-type interactions of the nature considered in Section II.H.6 failed to improve the position as these turn out to be relatively insensitive to surfactant concentration (although they help to explain the increased stability with increasing x, as they do increase with this quantity). The failure to correlate the concentration-stability trends led the authors to believe that the adsorbed surfactant tends to move away from the area of contact as two droplets approach each other. This move-ment will be hindered if surface viscosity is appreciable and one might expect that stability would parallel the surface viscosity. Such a correlation was, in fact, demonstrated, and is not out of line with other work (41, 154).

It may be that in some respects the calculations of interparticle forces were incorrect. The systems mainly studied (chlorobenzene and aniline/water) were of near-equal density and it is expected that in such systems equation 3 will break down (see Section II.H.1). Additionally, the use of the electro-kinetic (ζ) potential in equation 2 is open to question. First, ζ was cal-culated without allowance for plane-of-shear effects (118, 155a). Second, the free charge in these systems originates not from the adsorbed surfactant but from extremely mobile hydroxyl ions on the droplet surface. It is likely in this case that the condition of constant charge, which the use of ζ in equation 2 presupposes, is in fact incorrect (Section II.H.2).

It should be stressed that no attempt to interpret absolute coalescence rates was made. In this system, it is expected that steric/entropic effects should play a role in the coalescence process and an already difficult problem is thereby aggravated. Some theories of coalescence are discussed in Section III.D. The work is of interest, however, in that it is probably the most comprehensive study of its kind on systems containing nonionics and also for the fact that the coalescence kinetics were found to follow Hill and Knight's theory (137) quite well.

The use of the Coulter centrifugal photodensitometer in studies of emulsion stability has been described recently by Matsumoto and Kimsey (29). This instrument, still in the development stage, fractionates particles by size in a centrifugal field and measures the particle concentration as a function of size by light absorption. This technique, although more convenient to use than the Coulter counter, shares with the latter the limitation of being unable to differentiate single particles or droplets from clumped aggregates. Lips et al. (115) have recently shown how it is possible to take into account clumping morphology on the angular variation of light-scattering properties of dispersions. It is, however, not practicable to measure such scattering properties in the type of apparatus at present in use. For this reason, Matsumoto and Kimsey's work is not further discussed here.

A study of BSA-stabilized O/W emulsions on the same lines as ref. 11 has been made by Srivastava and Haydon (41). The irreversible nature of the protein adsorption and the difficulty of controlling the degree of adsorption precluded a comprehensive study of the type made by Elworthy et al. The authors were, however, able to demonstrate an interesting parallelism between the surface shear viscosity and elasticity and the coalescence rate constant, all of which quantities varied widely and conjugately with pH. The results showed a good correlation between these parameters and the half-life of a millimeter-size drop at a plane interface in the same system, which makes the macroscopic drop–plane interface an apparently viable model for the coalescence process (but see Section III. E below).

There have been few other studies of this type published, (vindicating the surmise of Lawrence and Mills (156) that the tedious nature of the task of particle counting would make it an unpopular line). The emergence of electronic particle-counting devices in the last decade has been of limited use for coalescence studies as these are not able to distinguish clustered droplets from clusters which have coalesced. Useful information on coagulation kinetics can, however, be obtained in this way (Matthews and Rhodes (157), Bernstein et al. (158).)

C. Ultracentrifuge Techniques

The gravity-driven creaming of an emulsion can be accelerated if the emulsion is centrifuged. The forces between creamed droplets are also increased and it is possible to investigate the coagulation/coalescence process by this means. The advent of commercially available ultracentrifuges capable of producing the high accelerations needed has led to a considerable amount of study of emulsion stability in this way (159–168). A technique of this kind is attractive because it allows the study of the coalescence process unobscured by effects arising from aggregation: the latter process occurs very quickly in the ultracentrifuge. The time scale of such experiments is also generally one or two orders of magnitude less than conventional stability studies.

The technique is basically simple: stability towards coalescence is related to the appearance of oil at the top of the centrifuge tube and this is usually measured as a function of time of centrifugation. (Exactly how the results should be taken to measure stability, e.g., rate of formation of oil at a given time, or the total amount formed, is not settled (166). However, to compare the stability of given system with another it is preferable to start off with two systems of the same size distribution to eliminate effects caused by size variations. This is only possible if the two emulsions are stabilized by the same agent, where the same stock emulsion can be used to prepare both (161, 162). This is another instance of the desirability of developing methods for the preparation of well-characterized, preferably monodisperse, emulsions (cf. Section III.B above).

The potential attractiveness of the technique has led to the closer scrutiny of the method as a tool for studying emulsion stability (164–168). This work has made it evident that the analogy with the usual coalescence process is incomplete. The oil cream that quickly forms at the top of the tube on centrifugation is further compressed to form a tightly packed polyhedral structure, not unlike a biliquid foam (168, 169), (cf. Figure 4). This is most unlike the structure obtained in the absence of centrifugal fields. Drainage patterns are, therefore, expected to be quite different, and thus such factors as the dilational modulus or the shear viscosity of the interface are also expected to act in a different manner, as they are related to the drainage pattern. Another consequence of the tight, gel-like structure of the "foam" is that oil formed from coalescence of droplets must travel to the top of the tube by a relatively tortuous path, and it has yet to be shown whether or not this transport process is itself rate-determining, comparable in rate with the coalescence process itself (162, 164, 166). Indeed, it has not so far been established whether the coalescence process occurs largely in the bulk of the "foam" or at the top, adjacent to the coalesced oil phase (164, 166). However, some cor-

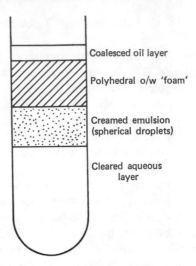

Fig. 4.   State of emulsion during centrifugation. Not all layers are necessarily present: For example, at low centrifugation speeds the uppermost (oil) layer may not appear, while at very high speeds the creamed emulsion layer will disappear quite rapidly.

relations have been shown to exist between the "centrifuge life" and the "self-life" of some emulsions (152, 166, 167) and further work on these lines is not unreasonable, but it seems unlikely that a complete analogy will be established between the ultracentrifuge and "normal" methods.

D. Thin Film Studies

The thinning and rupture of the aqueous film between two coagulated oil droplets has a formal analog in the rupture of planar aqueous films formed under oil, in the same way as free-standing soap films have been studied in connection with the problem of foam stability. The latter systems are a better analog of foam stability than films of water under oil are of emulsions. In the first place, the area of the film between coagulated emulsion droplets is very small, several orders of magnitude smaller than that of a centimeter-scale film and the drainage pattern is actually different in the two films: it is radial between emulsion droplets (170), much less well-defined in larger films (171), which are also often of irregular thickness. Secondly, the time scale may be very different, particularly in the initial thinning stages. The dispersion of the kinetic energy of approaching droplets occurs in a time or order 10^{-6} sec (the life of a Brownian collision), while a free film tends to drain quite slowly (time of order seconds). This affects the relative importance of diffusional exchange between bulk and interface and thereby the effects of surface viscoelasticity; the variation of charge and potential of the surfaces is also different. It also seems very probable that inertial effects, neglected by Reynolds (62), play a significant role in the hydrodynamics of thin film drainage between two emulsion droplets. These criticisms need not apply to films in air used to study foam stability: free-standing aqueous films make good models for foams because they are of size comparable with that of a single foam lamella, and accurate measurements on this system have been made (172). But this argument certainly applies to the thin film formed between a millimeter-size droplet and a plane interface and renders such systems suspect as models of emulsions, in spite of some reported correlations (see Section III.E).

However, *microscopic*, free-standing films have been shown to drain radially (48) and are of uniform thickness up to certain values of the radius. Scheludko's school has intensively investigated these systems. Platikanov and Manev's paper (47) suggests that the work done on free-standing films is at least qualitatively paralleled in liquid–liquid systems. Apparatus used by Sonntag to investigate such systems is shown in Figure 5.

A summary of the main features to emerge from the study of free-standing films is worthwhile. The drainage pattern once estabilshed (Section II.G), the behavior of the film during thinning was studied. This was found to be system-specific, but in general films followed one of three courses:

1. The film ruptured at a statistically well-defined thickness, termed the critical rupture thickness, h_{cr};
2. An "equilibrium" film was formed, in which border suction was just balanced by the nett dispersion/double layer force; or

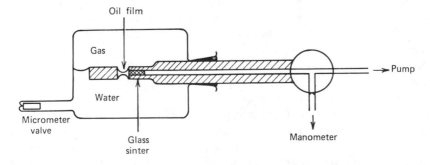

Fig. 5. Apparatus for investigation of disjoining present in thin films as a function of film thickness (61). The glass sinter allows the film thickness to be varied at constant film diameter. Disjoining pressures of order 1 atm can be measured. Film thickness is determined interferometrically. The system shown would correspond to flotation of the disperse phase of a W/O emulsion. (Courtesy Dr. Dietrich Steinkopff Verlag.)

3. The film thickness changed locally and suddenly as shown by the appearance of black spots. The entire film eventually attained this new thickness and was characteristically extremely stable. Two types of this black film have been described (171): first black films, whose thickness is dependent on electrolyte concentration, and second black films, whose thickness is concentration independent and is usually only a few water molecule diameters thicker than twice the surfactant molecule length. The stability of "equilibrium" films (above) is better understood than that of black films, which, although much thinner, are more stable. Black film formation is naturally of great importance for foam and emulsion stability.

An extremely interesting point was that there was very little tendency for a given system to form black films below a certain bulk surfactant concentration C_{b1} (173). This concentration is apparently linked in some way with the attainment of a certain closeness of packing of the adsorbed surfactant molecules and is usually somewhat below the cmc. Apparently, emulsions in which the surfactant concentration is less than C_{b1} should be quite unstable. This has not yet been reported for the few systems for which C_{b1} has been established, however.

The demonstration of a well-defined critical rupture thickness (49) has led to renewed theoretical interest in the nature of this instability. In an early series of papers (174–176) de Vries considered the formation of a small hole in a thin liquid film. Because of the finite thickness of the film, hole formation is accompanied by an initial increase in surface area. This becomes a decrease once the hole grown sufficiently large, but initiation of a hole should thus require an activation energy. This was calculated to be on the order of kiloteslas for film thickness of order nanometers, but to be much higher for films of order tens of nanometers thickness. Since films commonly rupture at the latter rather than the former thickness, rupture must be initiated by an alternative process involving a lower activation energy.

Scheludko (178) pointed out that a localized thinning of a film, perhaps of thermal origin, will tend to be amplified by dispersion interactions with neighboring film elements once a critical thickness is reached. Vrij has formulated this concept quantitatively (179, 180). The problem has been discussed comprehensively taking explicit account of finite values of the interfacial dilational modulus and of variation of this quantity by Lucassen et al. (181, 182). The problem is essentially a calculation of the film thickness below which a randomly occurring wave motion is amplified spontaneously. Once this thickness condition is satisfied a particular wavelength, the value of which is determined by the interplay of capillary and viscous forces, is amplified preferentially. In practice, a film is thinning continuously and the wavelength of optimum growth changes with the film thickness, making the point of rupture later than the "static" rupture thickness. This point can be treated graphically in an approximate way (179).

The agreement of theory with experiment is incompletely investigated at present. The Vrij predictions have not been quantitatively confirmed (48). Ivanov et al. (52) have redeveloped the Vrij theory, eliminating many of his approximations, and have tested the modified theory experimentally. The expected dependence of h_{cr} on the film radius R was not found, experimental values always being too low. The discrepancy increased with increasing surfactant concentration and this has led to the suggestion that an additional repulsive force is operative at small film thicknesses (a few tens of nanometers), which would have the effect of increasing the drainage time before rupture, lowering h_{cr}. The concentration dependence of the phenomenon was proposed to be explained in terms of concomitant changes in surfactant adsorption; it was observed that an increase in the viscosity of the solvent at low film thickness is not ruled out, but that such an effect is not essential to explain the observations. This interpretation has recently been reviewed in the light of the probable nonuniformity of the film thickness in the systems studied (183). There appears to be strong indications that such inhomogenity in film thickness may have a large influence on the observed value of the critical rupture thickness and that the "ideal" h_{cr} ought to be obtained by extrapolation of the observed values at various film radii to a value at an extremely low radius.

Graham and Phillips (184) have considered a similar interpretation of their data on the thinning of protein-stabilized water films in oil.

However, the possibility exists that the Vrij theory has some fundamental flaw. Sonntag, in fact, does not consider there to be even qualitative confirmation of the Vrij theory and has suggested that a better interpretation of h_{cr} is in terms of the thickness at which the potential energy-separation curve passes through the abscissa (point X, Figure 2) (56). This proposal was not, however, accompanied by a mechanism for hole formation and cannot be regarded as theoretically viable at present.

The theory of Lucassen et al., probably the most comprehensive, has yet to be tested experimentally. This theory predicts an important influence of the dilational modulus ε_D of the interface on stability. As ε_D is increased, an abrupt increase in stability, as measured by the damping coefficient of symmetrical waves in the film, was predicted at a definite value of the modulus. There is now some experimental basis for such a change in stability (183). The Vrij treatment was shown to be a limiting case of this theory. The theory is derived for films of constant thickness and will require some modification before it can be applied to the rupture of thinning films.

The extreme stability of black films in spite of their diminutive thickness strongly suggests that different factors are controlling the stability of these systems. Some information about their structure can be obtained from the contact angle between the film and bulk phase (54, 60, 78, 185) and from their dielectric and related properties (e.g., refs. 18, 187). The mechanism of rupture of these films is little studied, either theoretically or experimentally. A two dimensional adaptation of a cavitation theory of liquids has appeared (188), but it has no experimental basis. It is certain that new or refined experimental techniques must be developed before satisfactory progress can be made into this problem.

E. Coalescence of a Drop at a Plane Interface

One of the earliest models for emulsion stability to be studied was the coalescence of a single drop with a plane interface between bulk continuous and dispersed phase liquid. The droplet was generally of millimeter size and the effects of variation in physicochemical parameters on the time spent by the drop at the interface before coalescence occurs—the rest time—was investigated.

Early work (189–193) has in retrospect served to establish the difficulty of obtaining reliable data in these systems, for much of the later work has produced results quite different from those of the early workers. Thus both direct and inverse variation of drop rest times with drop size has been reported (191, 195), and many different equations have been proposed to describe the experimental data on the thinning of the thin film between the drop and interface (190–192, 195). However, certain facts do emerge from this early work. Rigorous control of general cleanliness, material purity and temperature is essential and the early efforts made in these directions were not adequate (197). This is not surprising as the principles for clean working were incompletely known at that time. Reference 191, for example, indicates a polythene–water interface in the apparatus. This is a well-known source of contamination (198). The paper of Charles and Mason (192), perhaps the most comprehensive of the early work, also established the importance of solid impurities (e.g., dust), of partition disequilibrium between bulk phases, and of applied electric fields, all of which tend to decrease rest times. This paper also provided an early indication of the true complexity of the drainage of the film between drop and interface as equations of the Reynolds (62) genre, the most widely used at that time, were incompatible with their data.

Modern work (194–197, 199–204) has investigated the effect of surfactant concentration and added electrolyte on drop rest times. Figure 6 illustrates apparatus used by Hodgson et al. (197). The drop rest time has been found to depend on the former in an intricate way that has been linked with changes in the drainage pattern with surfactant concentration (197, 200, 203). Direct studies of the radial thickness variation (200, 204) have in fact led to no fewer

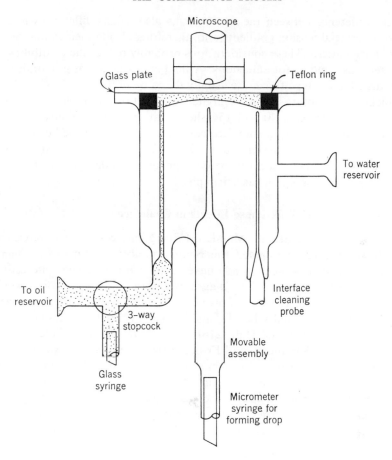

Fig. 6. Apparatus for study coalescence of a droplet with a plane interface (200). Droplets are formed by raising the central tube tip into the oil phase, removing a small pellet of oil, and subsequently releasing this pellet on lowering the tube. Preliminary cleaning of the interface is possible *in situ* and the thickness of the droplet–interface film can be measured interferometrically. (Courtesy Academic Press.)

than five distinct drainage patterns being described, of which two are asymmetrical with respect to the drop center (204).

This complexity in the drainage pattern in millimeter-scale systems raises serious doubts as to the validity of the analogy between these systems and emulsion coalescence. The comparatively low curvature of the interfaces in these systems also means that the interfaces are far more deformable than those in emulsions. The deformed area is, therefore, large and usually the two interfaces are not parallel, dimpling of the central area occurring (205).

The size difference between the two systems also means differences in the scale of interfacial tension gradients and, therefore, also in their influence on the drainage pattern. These considerations probably reduce the credibility of the systems as a model for emulsions. A final criticism is that gravitational, rather than DLVO-type forces are usually the cause of thinning (154) and the film thickness at rupture is usually higher. Correlations have been shown to exist; for example, refs. 40 and 154 show that emulsion stability and the half-life of a drop at a plane interface correlate with viscoelastic behavior of the interface in an O/W system stabilized by bovine plasma albumen, but there does not at present seem to be convincing theoretical justification for believing that the correlation has a firm physical foundation.

F. Bulk Phase Effects in Coalescence

Over the last decade Ekwall, Mandell, Friberg, and co-workers in Sweden (121, 206–213) have investigated the phase equilibria of three-component water–oil–surfactant systems and have demonstrated some interesting parallels between features of these diagrams and the stability of corresponding emulsified systems. The typical properties of such systems have also been reviewed by Lachampt and Vila (214) and by Bierre (215). Other systems have been described by Chen and Hall (216) and Balmbra et al. (217) (cationic–anionic–water) and Kislalioglu and Friberg (218) (nonionic–oil–water).

Two-component systems of water and surfactant usually form micelles above a certain surfactant concentration. These micelles, depending on size, are spherical or ellipsoidal (219). Increasing the surfactant concentration further may lead to the formation of long cylindrical micelles (myelin cylinders) and at still higher concentrations lamellae of surfactant alternating with

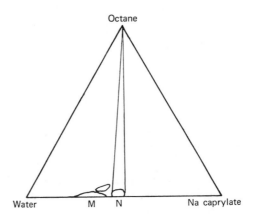

Fig. 7.   Phase diagram for the system: sodium caprylate/water/n-octane. (After ref. 208.)

solvent may form. The micellar, middle and neat phases, as they are respectively known, are capable of solubilizing oil to varying degrees and so the three-component phase diagram has regions of one isotropic phase, together with others of two and sometimes three phases (214). Figure 7 illustrates the phase diagram for system octane/H_2O/Na caprylate (from ref. 208). M and N denote the middle and neat phases respectively. In the system shown in Figure 8 the regions M and N are larger. This type of behavior is found in systems in which oil is polar (e.g., a higher alcohol, in this case decanol). In both the examples given, certain regions outside the drawn one-phase regions M and N, through which the tie lines ending on M and N pass will contain, respectively, phase M and N, in equilibrium with another phase. The point of interest of these diagrams for the problem of emulsion stability lies in the observation that passing from a system containing no neat phase to one containing the neat phase was paralleled by a very significant increase in the stability of the emulsified systems (207). Thus system P in figure 8, containing no N, gives a much less stable emulsion than does the system represented by point Q, which does contain N. Balmbra et al. (217) have also observed a number of parallels between the behavior of black films and the lamellar mesomorphous phase in a mixed cationic–anionic system in the absence of oil.

The ultimate origin of this enhanced stability is not yet understood. Formation of the lamellae is accompanied by an increase in the bulk viscosity, which could retard coalescence. It also seems possible, however, that multilayer formation in the interdroplet medium in some way increases mechanical resistance of the film towards rupture. Another possibility is that the van der Waals forces between the droplets are reduced. Friberg (212) used Vold's

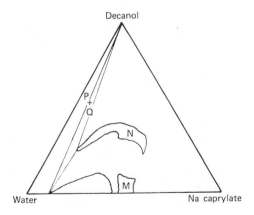

Fig. 8.   Phase diagram for the system: sodium caprylate/water/n-dodecanol. (Afet ref. 208.)

treatment (68) to investigate this possibility and decided it was not likely to be so. However, a calculation using Lifshitz theory is very appropriate in these systems and might show interesting differences when compared with Friberg's calculation.

G. Temperature Dependence of Coalescence

Beyond the fact that boiling or freezing an emulsion (i.e., inducing phase changes) can bring about breakdown (220), little has been established until recently about temperature effects on emulsion stability. Increasing the temperature brings about changes in most of the factors that could affect stability: The bulk viscosity, interfacial tension, and adsorption at the interface all decrease while the potential energy yardstick kT and the Brownian velocity both increase. Double layer potentials can either increase or decrease, according to the sign of the surface potential. The Gouy theory gives the expression:

$$\frac{d\psi_0}{dT} = \frac{\psi_0}{2T} - \frac{R}{e}\tanh\left(\frac{e\psi_0}{2kT}\right) \tag{9}$$

in which the first term on the right is larger than the second term. Dispersion interactions are not usually regarded as temperature dependent, although recent work indicates that this is not true of certain systems (76) in which infrared contributions are appreciable. The overall effect of an increase in temperature is that emulsions usually become decreasingly stable.

In a study of the temperature dependence of the stability of emulsions Shinoda and Saito (221) found such a decrease in stability; they took as measures of stability the mean droplet size as a function of time. They also established the fact that stability decreased very rapidly as the phase inversion temperature (PIT) was approached, and suggested that emulsion stability was related to the difference between the system temperature and the PIT. This idea has received independent support (222). That such a correlation is reasonable is readily seen. Any mechanism for phase inversion must, for topological reasons, involve a coalescence process at one stage (223, 224), so that even in the absence of detailed knoweldge of the process, it is possible to say that inversion is in one respect analoguous to coalescence. Inversion in an unstirred solution can be envisaged as the rapid coalescence of a cluster of coagulated droplets, which leads (for an O/W emulsion) to the formation of a water droplet in an oil-continuous phase (figure 9). The six coalescence processes in Figure 9 must be rapid, to ensure the "encapsulation" of the water droplet and so bring about inversion. Alternatively, in such a system rupture of one interdroplet film may trigger rupture at neighboring sites, even though this is apparently not the case in more stable systems (9). It is at the

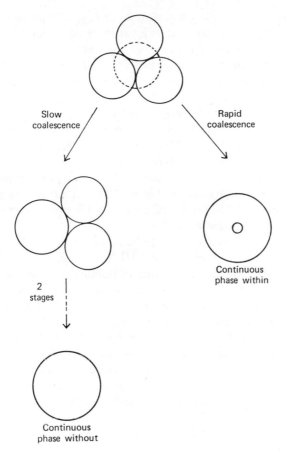

Slow coalescence

Rapid coalescence

2 stages

Continuous phase within

Continuous phase without

Fig. 9. Inversion of an emulsion. Inversion only takes place if the rapid coalescence mechanism can operate.

PIT that this condition of sufficiently rapid coalescence obtains making this temperature in principle predictable, although this has yet to be done.

In sheared systems somewhat different considerations hold, as now both O/W and W/O emulsions are being simultaneously produced and the type of emulsion formed in the long term depends on the relative rates of coalescence of the two types. Davies has discussed this situation (28, 225). Coalescence is still important, but other factors, related to emulsification or to orthokinetic coagulation effects, will also influence the process. One would not expect a sheared system to have quite the same PIT as an unsheared one.

The empirical rule of Shinoda and Saito has thus at least a qualitative theoretical background. This rule can now be turned around: The PIT of a system, easily determined unless it is close to or beyond a phase transition of one of the components, can be used as a yardstick of stability. This is much more conveniently used than particle-counting methods although it depends on the same factors—the coalescence process is common to both. It is also a more quantitative measure than the "visual" methods that sometimes appear in the literature.

Correlations have been made between the PIT method for the prediction/ assessment of emulsion stability and the more traditional method of the hydrophilic-hydrophobic balance (HLB) (1b, 165, 215, 226). Parkinson et al. (222) found that in a series of emulsions stabilized by surfactants of varying HLB, systems of maximum PIT also had an HLB corresponding to maximum stability. Shinoda and Saito claimed that the PIT method should give a more sensitive measure of emulsion stability, although this trait does not stand out from the results of Parkinson et al.

Exceptions to the PIT rule have yet to be reported and are not really to be expected. This makes it superior to the HLB method, which of its nature cannot account for the observed variation of stability with emulsifier concentration or for the effect of minor added components (165). It seems likely to become an increasingly popular measure of emulsion stability. It should be remarked, however, that use of the method has less clear theoretical justification in systems of mixed emulsions undergoing heterocoalescence. (Section IV.C.2). In such systems more than one coalescence mechanism is possible— encapsulation of one droplet by another involves rupture of only one interface, for example—and there is no guarantee that the same mechanism is operating both at the storage temperature and at the inversion temperature.

H. Stability of W/O Emulsions

In the foregoing it has usually been assumed that an oil-in-water emulsion was under discussion, although much of what has been written applies equally to W/O systems (e.g., Brownian coagulation, temperature effects). Water-in-oil systems are, however, sufficiently different to merit separate, if brief, treatment. Their stability towards coagulation and coalescence depends qualitatively on rather different factors for two reasons:

(a) The double layer interaction between droplets is low, because of the low permittivity of the oil medium. This is in spite of the fact that appreciable electrokinetic potentials have been reported in such systems (227). Two effects operate: first, the double layer thickness parameter $1/\kappa$ is of the order of the droplet separation in a fairly concentrated system (i.e., $\sim$ micrometers). Thus two noncoagulated droplets already possess an appreciable repulsive energy with respect to infinite separation and the effective barrier to coagulation is lowered. Second, the double layer interactions of other droplets affects the interaction of any given pair, which therefore cannot be treated in isolation. This serves to reduce the energy barrier to coagulation (228). The coagulation can, in fact, be controlled by dispersion interactions until other forces (Section II.H.4) become appreciable. Likewise, the coalescence process is driven by these forces, double layer interpenetration still being a minor factor.

(b) Forces of the type referred to in Section II.H.6 originate in the interaction of the long hydrocarbon chains of molecules adsorbed on both droplet interfaces, and one way in which they are believed to contribute to stability is in the same way as do alkylated aromatic compounds stabilize graphite dispersions in hydrocarbons, via steric and entropic repulsion (128). This repulsion is high for relatively little interpenetration of adsorbed layers: the limiting thickness of black films in water is about twice the length of the adsorbed molecule (186). Interaction of this type is more important in W/O systems because the hydrocarbon chain of a surfactant is usually much longer than the hydrophilic portion (exceptions to this rule occur with some nonionic surfactants, e.g., C_nE_x). In extreme cases this repulsion can be operative at droplet separations at which the dispersion interaction energy is of the order of 1 kT and so prevent even coagulation (229). In more commonly encountered chain lengths the effect will be to restrict the limiting thickness of the black films, possibly with the result that their stability is increased thereby.

However, black film formation is only one possible fate of a coagulated pair of water droplets and corresponds to maximum stability. As with O/W systems, formation of these black films is dependent on surfactant concentration (53), premature rupture of the film occurring below c_{b1}. The reasons for this are as uncertain as they are for O/W systems.

I. Discussion

In the preceding sections a number of factors that may contribute to emulsion stability have emerged, but no all-embracing principle to explain the phenomenon seems evident and it is still very true to say that the problem remains one of the *bêtes noires* of colloid chemistry. This quandary arises in part from the difficulty of studying any of the various factors to the exclusion of the others in real systems and also from the complexity of the theory needed to estimate threshold values for the several factors, that is, values below which a particular factor is probably negligible.

A good example of the difficulty of unequivocally establishing the relevance of a given factor for stability is afforded by the case of the surface dilational modulus and shear viscosity. The former modulus is defined by $\varepsilon_D = d\gamma/d\ln A$ and is a measure of the resistance of the interface to dilation; the latter is the "surface viscosity" as usually measured (155b). A quantity related to ε_D is the Gibbs elasticity G, equal to $2\varepsilon_D$ in sufficiently thick films (230). G is a measure of the film's resistance to localized thinning. Such thinning results in a rise in interfacial tension, the magnitude of which rise depends on G, and the resulting gradient in interfacial tension is the origin of a movement of adjacent surface and subjacent fluid towards the attenuated region (the Marangoni effect (231).

A larger value of the Gibbs modulus should thus confer on a film a resistance to both drainage and rupture via localized thinning. The surface shear viscosity can also retard these processes, although probably not to the same extent as the elastic modulus. The resistance of a thin film to thinning and rupture has been described theoretically by Marrucci (138) and by Lucassen et al. (181), both authors considering only the effect of the dilational modulus.

Experimental evidence, such as it is, is difficult to interpret. The overall impression given is that both factors can make contributions to stability, the dilational elasticity being the more important, but that other factors, such as the formation of black films or multilayers, are frequently dominant. Thus, Jones et al. (232) quote an instance of a film of low surface viscosity having a lifetime comparable with films of high surface viscosity, but the ultimate reason for stability was not given. Prins et al. (233–235) have concluded that the Gibbs elasticity is only imortant in determining the stability of relatively thick films. These authors give theoretical and experimental reasons, in the form of the effect of minor components in producing large changes in G, in support of these views. A recent paper (236) supports this thesis. It is interesting to note that Van den Tempel (11) observed no effect of minor added components on the stability of actual emulsions, perhaps because the additives were soluble in both phases (*vide infra* Section IV.C). This underlines the incompleteness of the analogy between foam and emulsion films. Vold and Mittal (165) do report such an effect, but their paper tacitly stresses the difficulty of unambiguous interpretation by attributing the effect to closer packing of the adsorbed layer in the presence of a minor component. Even this conclusion is not in accord with the close correspondence between emulsion stability and the surface dilational and shear moduli reported by Haydon et al. (40, 154). The resistance of a film towards displacement normal to its plane has been considered a factor in determining stability by Sonntag et al. (55), who have thus added another possible factor to the list of probable ones.

Hill and Knight's (137) proposal that, in systems where electrostatic forces are unimportant, the probability of coalescence of two drops coming into contact is proportional to the integral $\int$ (Pressure x area) dt (the mechanical impulsive force) is of interest, chiefly because it has some experimental foundation. Although it neglects explicit treatment of the several factors enumerated in this section, it is surprisingly successful. It predicts a linear decrease in the reciprocal of the specific surface area of the emulsion with time, whereas the Smoluchowski theory is of reciprocal cube form. The data of Lawrence and Mills (156) and later work by Elworthy et al. (145) are in good agreement with the Hill and Knight theory. It is unfortunate that the physical basis of the theory is rather obscure, as successful theories of coalescence lie very thinly on the ground.

IV. THE BREAKDOWN OF EMULSIONS

As was remarked in the introduction, incomplete understanding of the factors controlling emulsion stability is unavoidably paralleled by a certain empiricism in the approach to the technically important problem of emulsion breaking. A number of established techniques are reviewed below and in addition a few more esoteric proposals are discussed.

A. Coagulation Aspects

In certain technically important systems relatively dilute emulsions are involved in which coagulation presents a problem. The best way to deal with such systems is apparently by electrical methods that concentrate and sometimes coalescence the emulsions. Electrophoresis of O/W emulsions has been described recently in this context (237). The droplets must carry an appreciable charge, but this is usually the case or can be arranged, via pH changes, for example. A novel variant, whereby the electrical field is produced internally in a column packed with two dissimilar metals (238) is apparently very efficient and often combines coagulation with coalescence (see Section IV.D). Success with W/O systems was also claimed. The process was found to work very well with dilute emulsions. The metal granules do not become deactivated by oil films after a time, and it is apparently not an instance of coalescence induced by third bodies in the usual sense. The fact that granules of only one metal can be made to produce the same effect, in the presence of added electrolyte (239), has led to the suggestion that coagulation is in fact brought about by the high concentration of electrolytically produced electrolyte around the "electrodes" in the two-metal systems.

Emulsions of water-in-oil can be concentrated by high fields without concurrent electrolysis at the electrodes (240). If the droplets carry only a low charge, it is sometimes possible to induce charges via the applied field. Migration of induced dipoles in a divergent field—dielectrophoresis—can also be effected (241). The theory of this process has been discussed by Pickard (242). Concentration of emulsions may also be achieved via filters, centrifugation, and magnetic means. These are discussed later (Section IV.F).

Concentrated emulsions, once obtained, can be coalesced by classical methods, for example, addition of electrolyte, preferably of high counter-ion valence (cf. the Schulze–Hardy rule). Occasionally, complications are met with when complexes are found or Stern-type adsorption occurs. Calcium soaps, for example, can bring about inversion of the emulsion.

Coagulation may also be accelerated by the presence of oppositely charged particles, which interact in ways already discussed (Section II.H.4). This point is discussed further in Section IVC.2.

B. Coalescence: Introduction

The major problem in practice is more often the coalescence of a coagulated oil. The oil may be the desired product, as it is in the Kunerator process for vegetable oil extraction, or it may be an effluent contaminant. In the latter case coalescence can mean that the oil can be recirculated or sold rather than become another disposal problem.

It should be asked what clues the previous sections hold for the practical solution of the coalescence problem. Having obtained a reasonably concentrated coagulate, the ideal is to bring about (*a*) an increase in the attractive force between droplets, (*b*) the reduction or annihilation of known or suspected stabilizing factors (dilational elasticity, for example), (*c*) change in the mechanism of coalescence by introduction of a third body of large surface area: droplet coalescence then occurs via the third body and spreading. Most of the following methods utilize one or more of these principles.

C. Physicochemical Methods

1. Effect of Added Electrolyte

The addition of electrolyte will, by its suppression of double layer repulsion, increase the overall coalescence rate via an increased rate of thinning. The effect of a given electrolyte concentration is usually greater, the higher the charge of the ion of the same sign as the counter-ions of the double layer (cf. the Schulze–Hardy rule for colloid flocculation). Multivalent counterions can be particularly effective in breaking emulsions stabilized by ionic surfactants, in which systems high interfacial potentials are often present, making a major contribution to the overall stability. Addition of electrolyte will also reduce the rate of deaggregation of coagulated droplets (42, 43). It has been mentioned above (Section II.H.1), however, that electrolyte can reduce dispersion interactions in some systems, thereby reducing the overall effect of addition of electrolyte. It is possible that salting out effects (243) also operate.

2. Heterocoalescence

Unlike the situation with homocoalescence, the coalescence of two different drops need not be thermodynamically expected: When the two droplets are miscible the interfacial tensions of the coalesced and uncoalesced droplets may be such as to render the decreased surface area of the coalesced droplet inadequate compensation for the changed interfacial tension. This point has been discussed by Sonntag et al. (57) who also discuss the various possibilities (e.g., encapsulation) that arise when immiscible drops coalesce. Such systems have also been discussed by Torza and Mason (244, 245). In the following the droplets are assumed to be miscible and the various interfacial tensions to be such that the overall process is thermodynamically favorable.

Addition of an emulsion, the droplets of which carry a charge of opposite sign to the original emulsion, can accelerate coalescence as well as coagulation provided that conditions are chosen with care. If both emulsions are stabilized by surfactants insoluble in the continuous (that is, aqueous) phase, coagulation and coalescence of AB pairs are favored for purely electrostatic reasons. More commonly, however, ionic emulsifiers are appreciably soluble in water and mutual neutralization of charge on the emulsion droplets is likely, possibly accompanied by complex formation between the surfactants. Electroneutral droplets are obviously more likely to *coagulate* than are droplets carrying like charge, but *coalescence* is not necessarily accelerated; it may even be retarded. If the electroneutral combination is oil soluble (e.g., dodecyl trimethylammonium dodecyl sulphate) coalescence is likely to be favored (*a*) because of reduced double layer interaction, (*b*) because the surface concentration of surfactant will be reduced, and (*c*) because having surfactant in the bulk (droplet) phase rather than in the film will, due to greatly increased short-circuiting of interfacial tension changes by diffusion, reduce drastically the Gibbs elasticity of the film (*vide infra*), but (*d*) complex formation may increase the mechanical rigidity of the film, resulting in increased stability (cf. Section IV.C.4 below).

A technique related to the above genre has been employed by Skrylov et al. (246, 247) for the extraction of charged colloids from suspensions. A surfactant is added to the suspension and is adsorbed, presumably with polar head towards the particles, making them oleophilic. The suspension is then shaken with oil, forming an emulsion that is often quite easily broken.

If the two coalescing drops are immiscible, however, encapsulation of one drop by the other is possible given suitable combinations of drop size and of the interfacial tensions of the three interfaces: oil A/water, oil B/water and oil A/oil B. For equally sized drops, the condition for encapsulation of A by B is

$$\gamma_{AW} - \gamma_{AB} > (2^{2/3} - 1) \gamma_{BW}, \tag{10}$$

a condition that can often be arranged. This equation gives no information about the kinetics of the process, of course, but encapsulation may in the right circumstances be of use for demulsification—if for example, an emulsion of A is difficult to coalesce but drops of A encapsulated in B are not. Little attention has been given to this point to date. The engulfment of polymer particles by a solidifying liquid melt has been studied experimentally by Neumann et al. (248).

Another aspect of heterocoalescence is the coalescence of gaseous and liquid droplets with solids. The former is an important stage in the flotation of minerals, and some careful work on model systems has recently appeared. The coalescence of liquid drops with solids is of especial interest for demulsification for its relevance to the performance of filters. Further discussion appears in Section IV.E.

3. Reduction of the Gibbs Modulus

The principle mentioned in Section IV.B, whereby the Gibbs elasticity of the interdroplet film is reduced, although potentially useful, is in practice difficult to apply: There are usually technical problems in getting the surfactant into the dispersed phase if it is not already there or formed *in situ* as in Section IV.C.2. (In contrast, it is relatively easy to get, for example, ether into a foam to break it (249).) This effect in emulsions could be one of the reasons for the obeying of Bancroft's rule by many systems: any emulsion formed with the emulsifier in the dispersed phase will coalesce quite quickly.

The Gibbs modulus of the film is reduced if the moduli of the individual surfaces are somehow reduced other than by diffusion. This is not often found to be possible, but in systems stabilized by a mixture of surfactants the selective extraction of the minor component should reduce G. In protein-stabilized systems it has been found that the addition of cationic surfactant to bovine plasma albumin (154) and of certain nonionics to β-casein stabilized films lowers the elastic modulus, as also does a change of pH.

4. Mass Transport Effects

Distinct from the above is the exploitation of mass transport *across* the interdroplet film rather than to the bounding interfaces. This can occur when a component is dissolved to different extents in two coagulated droplets. Diffusion of the component across the film results in a local lowering of the interfacial tension in the region. Consequently, there is a tendency for surfactant to move radially outwards (Figure 10), and the film is thinned via the

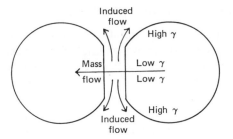

Fig. 10   Mass transport across film separating two coagulated drops. The right-hand drop has a higher concentration of some third component. The diffusion of this component across the intervening phase can upset the distribution of interfacial tension and result in Marangoni flow of the dispersion medium.

Marangoni effect. This phenomenon could be brought about by mixing the emulsion with a similar emulsion containing the extra component in the disperse phase. It could also perhaps be achieved by partial extraction of a component soluble in the disperse phase prior to coagulation. The extraction rate would presumably be proportional to the surface area of the droplet, while the rate of decrease of concentration is inversely as the volume. On this basis the concentration of extracted component would decrease more rapidly in small droplets. It is likely that this effect is responsible for the initially high rate of coalescence frequently observed (see Section III.B), and it almost certainly operated in the systems studied by Lin (250–253).

The nature of the adsorbed interfacial layer can influence mass transfer profoundly, as recent work by Booij (254) exemplifies. Booij studied the extraction of viscous oil from an O/W emulsion shaken with a bulk oil phase (petrol ether). Mixed anionic/cationic emulsifiers stabilized the systems studied. Chemical similarity with compounds known to form p-coacervates (255) led to the postulate that a complex with the hydrocarbon phase was formed. Booij's interpretation of his results was that a film of p-coacervate surrounding each droplet acted as a semipermeable barrier that allowed transfer of short chain (petrol ether) but not long chain (viscous oil) molecules. The droplets grew in size, as observed, as a result of this one-way transport until a point was reached where the available p-coacervate was unable to accommodate further expansion of the interface. The droplets then burst, their contents passing into the bulk oil phase. Other examples of p-coacervates have been described (255), but they form a relatively small class of stabilizers, and mass transport across interfaces will only occasionally be retarded by such compounds. It is, however, interesting to note that added ethanol accelerated the extraction process, apparently by its action on the interfacial film. Similarly ethanol accelerates coalescence in altogether different systems (29).

5. Surfactant Concentration Control

Reduction of the concentration of stabilizing surfactant is usually effective in bringing about coalescence. It can, depending on the concentration, involve the removal of a mesomorphous phase from the system (Section III.F), the reduction of concentration below C_{b1} (Section III.D), or reduction of the dilational elastic modulus that passes through a maximum at some concentration below the cmc (256). Such changes in concentration are usually hard to achieve in practice, but can be effected by chromatography (257), hyperfiltration through cellulose acetate membranes (258), or by the formation of surface-inactive compounds (169, 259). Simple dilution is unlikely to be practical in most cases because of its adverse effect on coagulation kinetics. Methods involving the replacement of the surfactant by large amounts of weakly surface-active material (e.g., silicones) have been discussed by Sonntag and Strenge (54) and, in the context of foam breaking, in ref. 3.

Destruction of the surfactant by chemical, biochemical, or electrolytical means is sometimes possible (Cooke (260)). By conventional chemical reactions it is possible to reduce or, sometimes, annihilate the surface activity of certain surfactants. Biological destruction, via the growth of microorganisms at the expense of the surfactant, is an increasingly important technique of use in many systems. It has been reviewed recently (261, 262). The process has the disadvantage of being slow—usually on the order of days. Electrochemical methods are discussed in the next section.

D. Coalescence by Electrical Methods

Emulsion droplets are generally charged to some degree, so that the use of an electrical field to induce coalescence seems obvious. Yet this has not been extensively investigated or employed for the breakdown of O/W emulsions, although it has been for W/O systems. This situation is changing as effluent control becomes more urgent (237, 263–265). Electrical methods are becoming increasingly investigated as it becomes realized that electricity is the most promising energy commodity for the future. There is the further advantage that problems from scaling-up seldom arise, in contrast with most chemical engineering experience (264).

In electrophoresis the migration velocity of charged particles depends linearly on their electrokinetic potential, which is related, but not always very closely, to charge (see, e.g., ref. 118); if $\kappa a \gg 1$

$$u = \frac{\varepsilon \varepsilon_0 X}{\eta} \zeta. \qquad (10)$$

(If $\kappa a < 1$, correction functions have been tabulated (266) that reduce the calculated velocity by up to two-thirds.) Typical migration velocities are of the order 100 μm/sec for field strengths of a few hundred volts per meter. (Higher field strengths are accompanied by polarization problems with even carefully chosen electrodes.) Coalescence can be achieved by inserting a semi-permeable barrier across the migration path of the droplets, upon which the droplets concentrate and coalesce either directly with themselves or via spreading over the barrier surface. In both cases coalescence is assisted by the Maxwell pressure due to the external field. A practical cell based on the above has been described by Bier (237). Kuhn (263) has mentioned cells with swept electrodes and notes the advantage—in stirred systems—of reversing the field direction at intervals as a means of cleaning electrodes and of reducing polarization effects. In systems without semipermeable barriers, coalescence can also be brought about by a type of flotation process in which gas bubbles (the product of electrolysis) capture and entrain oil droplets as they rise to the surface. Special cells exploiting this effect have been described (264). The novel mixed metals technique of Fowkes et al. (238), described previously, probably falls into a different category (239).

Use of the redox properties of the electrodes is, of course, widely established in metal recovery, but breakdown of emulsifiers at the electrodes can also occur. Carboxylic acid salts have been successfully removed via the Kolbé reaction, leading to destabilization of the emulsion (Surfleet (267)). Because carboxylates are a quite common class of stabilizer this technique would be of interest in spite of the high initial cost of the treatment plant. However, decreased efficiency of the process with decreasing emulsion concentration was found by Surfleet. This is in agreement with the findings of the detailed pre-war study of the reaction by Glasstone and Hickling (268). Decreasing the carboxylate concentration was shown by these authors to result in an increased rate of formation of products other than the expected hydrocarbon; some of these products are likely to be surfactants themselves. The mechanism of this complex reaction is still rather uncertain, but it appears that the breakdown is peculiar to carboxylates and that similar reactions do not occur with analogous ionic surfactants, such as sulphonates (268).

The bulk of research on electrically induced coalescence, however, has focused on water-in-oil systems, a reflection of the importance of these emulsions to the oil industry, where they occur during extraction of crude oil. High field strengths can be used in such systems because of the low conductivity of the continuous phase. The transport phenomena are not usually of electrophoretic nature (241); that is, they do not obey the electrophoretic equations. Droplets are often charged and retain their charge if the oil is of low conductivity. Coalescence can occur by collision of particles as they travel between electrodes at different rates (as a consequence of differences in droplet charge and size). A device (Figure 11) to increase the rate of droplet

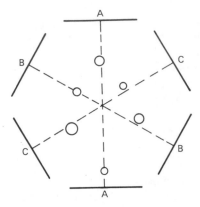

Fig. 11.   Three electrode pair systems for droplet coagulation. The application of an electric field between plates A, B, and C in succession brings about contact of droplets that would never meet were only one electrode pair used. As a result of the finite droplet size, the probability of contact is increased by a factor considerably greater than 3. (After Babalyan et al. (265).)

collision by alternate use of three pairs of electrodes has been described (265). The motion of drops also can result in their distortion and so to nonuniform distribution of the adsorbed interfacial material. This factor may itself promote coalescence, a point that may also be relevant for ultracentrifuge techniques (Section III.C). The high fields are also active in inducing coalescence in these systems and this aspect has been the subject of several studies (269–271). A critical field strength, beyond which coalescence was very rapid, has been reported (271), and is of the order expected from the dielectric strength of oil; that is, rapid coalescence is probably brought about by a discharge across the oil film.

It was mentioned in Section IV.B that effluent emuslions are often very dilute. This raises economic as well as technical problems, as the handling and electrolysis of large quantities of emulsions is never cheap. It is, therefore, essential to make the electrical process as efficient as possible. Aspects of this have been discussed above and in refs. 264, 272.

E. Coalescence Induced by a Third Body. Filters

A solid introduced into an emulsion, if wetted to some degree by the disperse phase, can act as a site for the coagulation and coalescence of the droplets and also as a channel for the removal of coalesced liquid. The emulsion bulk surface and walls of the containing vessel have been thought of in this context (273). If provision for the removal of coalesced material is made, such a coalescing arrangement can be self-regenerating and suitable for continuous rather than batchwise operation. This factor, together with the relatively small amount of maintenance required and the low initial cost makes emulsion breaking by solids attractive, for example, in the field of effluent treatment. However, the principles of the design of filters are still not well understood, and much empiricism is to be found in all but some recent work (274).

Some of the principles of solid–liquid coagulation/coalescence have been discussed previously. DLVO-type forces are expected to dominate the interaction of the solid and the approaching droplet, although the inhomogeneity of the situation makes the description of these forces quite complicated. The presence of the solid surface may also influence the solvent structure near the surface: there are many indications that effects of this sort are more pronounced with solid–liquid than with liquid–liquid interfaces. The solid interface also alters the flow pattern for drainage of the continuous phase from between the approaching bodies, halving the drainage rate (275).

The topic of solid–liquid coalescence has not been studied as extensively as has homocoalescence. In part this is due to the difficulty of obtaining reproducible surfaces: Surface roughness and inadequate cleaning techniques have limited experimental work to a few reports (276–279). So far as filtration is concerned, most of the fundamental work has been theoretical and is further discussed shortly. The solid may be in plate (280) or in cylindrical form (281). Beds of granular material are less often used for emulsion breakdown although they are widely used in other contexts, such as sewage treatment. The reason seems to be one of convenience more than anything else. Filtration by filaments is potentially more efficient than by plates on a volume basis because of the greater surface area available and such filters are more flexible in design potential than are granular beds. In this section are discussed some aspects of filtration by packed fiber filters or meshes. Depending on the fiber spacing and coarseness, such filters range from open meshworks down to filter paper.

Basically, when an emulsion flows through, a fiber must capture a droplet and retain it until a large number of such droplets have been captured and coalesced to form one large drop which can easily be dealt with—by sedimentation, for example. This drop should then be released to reduce back pressure caused by partial blockage of the filter by the drop's presence. The capture, retention, and release are all complex processes that can conceivably depend upon a large number of system parameters.

The capture process resembles coagulation of drops, in that it is controlled largely by Brownian-type collisions augmented by double layer and dispersion forces. In addition, however, the particle has velocity relative to the filament because of the fluid flow. The DLVO forces both depend upon the radii of both particle and filament, as does the drainage of the continuous phase from between the two. In practice, however, the filament radius is seldom less than the particle radius (of order micrometers) and will often considerably exceed it. To this approximation the problem has recently been discussed by Spielman and Goren (282) neglecting double layer forces, and by Spielman and Cuckor (283), taking the latter into account.

The former authors discussed their theoretical results in terms of a single fiber efficiency F, the ratio of the number of captured droplets in the presence and absence of interaction forces. In the absence of retardation to the dispersion forces, F was given by

$$F = \left(\frac{3\pi}{4} \cdot N_{Ad_F} \right)^{1/3} \tag{11}$$

where the dimensionless number N_{Ad_F} is given by

$$N_{Ad_F} = \frac{A a_F^2}{\pi A_F a^4}$$

where a_F and a are, respectively, the filament and droplet radii. The hydrodynamic flow factor A_F characterizes the fluid flow around a cylinder. The ratio of filament to particle radius is clearly important for the filter efficiency. Spielman and Cuckor in their more complete, computed solution used the treatment of Hogg et al. (103) for DLVO forces and distinguished regions of coagulation in the primary and secondary minima.

It should be noted, however, that F is not always a very good measure of the collecting ability of an array of filaments. Coalesced oil can increase the effective filament diameter and alter both disperison and double layer interaction forces. Also, when the grid spacing becomes comparable with the droplet diameter a large increase in droplet capture is expected on purely steric grounds.

The retention and release of droplets are related problems and are not easily analyzed because of the multiplicity of possible conformations a droplet may adopt on an individual fiber. Also, depending on its size, it may be in contact with one or many filaments. Some information about droplet behavior in the former case has recently been reported (284, 285). Some tentative extrapolations from single to multiple droplet–filament contacts can be made.

If a droplet initially spreads on a single filament on impact, forming a more or less uniform coating on the filament surface, it will tend to recover droplet form owing to the operation of the Rayleigh instability of fluid cylinders (284). This reformed droplet is always symmetrically located on the filament immediately after formation (Figure 12a) but, subject to critical combinations of values of the contact angle, drop dimensions and filament radius, this

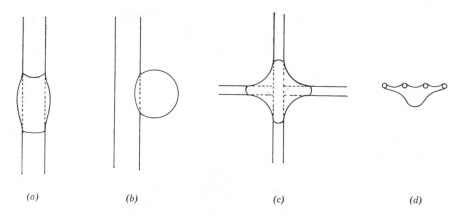

(a)   (b)   (c)   (d)

Fig. 12.   The wetting of filaments by drops. (a) Single filament wetted by drop (zero contact angle). (b) Single filament, high contact angle. (c) Drop at crossover point (zero contact angle). (d) Large drop on a filament matrix (cross section).

drop may move to the side of the filament, becoming asymmetric with respect to the filament exist (285).

The last-mentioned phenomenon is of importance in the present context. The adhesion of oil droplets to a filament may be expressed in terms of the work of adhesion W_A (286).

$$W_A = O\,(\gamma_{LW} - \gamma_{LS}) \tag{12}$$

where O = contact area between drop and filament and L, W, and S denote the oil, water, and solid phases, respectively. Tending to remove the droplet from the filament is the drag of the liquid flowing through the filter, which is proportional to the flow speed, fluid viscosity, and the "effective" radius of the adhering droplet—the last a measure of the projection of the droplet into the liquid flow (287, 288).

The movement of the droplet to the side of the filament tends, first, to reduce the contact area O and thus also W_A and, second, to increase the effective radius of the droplet and so also the force of detachment. The asymmetrical droplet is plainly more easy to remove. In a filter it is desirable for the captured droplet to be retained until it has grown relatively large, but not so large that blockage of the filter becomes serious. The residence time can be controlled via the flow speed, contact angle, and filament diameter, and will also depend on the uncontrollable input droplet diameter. Sareen et al. (289) have demonstrated that an optimum contact angle exists, but conclusions about filament size effects cannot be drawn from their work as this quantity was not systematically varied. A summary of currently used filtration media has appeared recently (290).

It is, however, a fact that droplet adhesion at filament crossover points and also over several cells of a meshwork are important (Figures 12c and d), although no properly formulated experimental study has appeared on this topic. However, it is easy to demonstrate experimentally that such points retain larger droplets than do single filaments (291). The cell-bridging phenomenon is most likely to occur when the interfilament distance is of the order of the droplet diameter and such filters may have too high a resistance to flow, by reason of their tight packing, to be practically useful. Detachment from single filaments and from crossover points may, therefore, be the most important mechanisms. A drop at a crossover point, being in contact with two filaments, has roughly twice the contact area of a comparable drop on a single filament, while the effective radius is not expected to be greatly different except at low droplet/filament diameter ratio. It is to be expected that these points provide sites for the growing droplets in the filter, growth of the droplet occurring via drainage from adjacent filament and by direct capture from the emulsion. Some of these principles have found experimental confirmation recently (292). The authors also describe the cell-bridging phenomenon as it occurs with millimeter-size droplets in settling tanks (Section IV.F.2), but the extrapolation to emulsion droplets is uncertain. A filter incorporating some of the present considerations has recently been described (293). No instance of a filter being used as an electrode for electrophoretic coalescence, or surrounding such an electrode, has come to the present author's attention although such a system has attractive possibilities. A metal foil electrode would be expected to acquire a layer of oil on its outermost parts after a short time of operation. Unlike the case with a plane electrode, this will not appreciably reduce the field strength across the emulsion until the foil electrode is completely saturated with oil.

It seems likely that a number of innovations are possible in the field of filter design. Recent work, for example, has indicated that mixed packing materials gives improved performance (294). The reason for this has not been established, although it is presumably a wetting phenomenon. Size inhomogeneity of the filaments should lead to preferential movement of captured droplets towards regions of low capillary pressure; that is, towards regions where large filaments predominate. This could be a useful way of bringing about controlled release of the coalesced oil.

F. Miscellaneous Methods

1. Magnetic Field-induced Coalescence

Oil-soluble ferro-fluids with magnetic properties are now available commercially and the magnetically induced coalescence of emulsion droplets containing a ferro-fluid has been demonstrated (295). The practical weakness of this novel method lies in the difficulty of introducing the ferro-fluid into preformed emulsion droplets: It has been remarked already that the disperse phase in emulsions is relatively inaccessible.

2. Settling

Actually a creaming process, settling is often used on a moderate-scale basis in industrial processes. Settling tank design has been discussed recently by Mizrahi and Barnea (296) and by Mumford and Thomas (269), who describe the use of meshes in settling tanks as an aid to coalescence.

The use of a magnetic field as an aid to settling does not seem to have been considered, yet may be technically feasible. There is a fairly large difference in the diamagnetic susceptibilities of most oils and water and application of a strong, inhomogeneous magnetic field should result in creaming of the oil phase at a rate some 100 times that of gravitational creaming for low density differences. Added inorganic paramagnetic salts may enhance this effect. The process is obviously capital intensive, however.

3. Centrifugal Separation

Separation by centrifugation has been described (297) (cf. also Section III.C). The method has been used successfully to separate emulsions. The method is rather costly, but is useful where removal of traces of the disperse phase is important: for example, in the separation of water from aircraft fuel.

4. Dissolution

If the dispersed phase is slightly soluble in the continuous phase, the latter acts as a connection between any pair of drops. Because of their lower internal (Laplace) pressure, large droplets tend to grow at the expense of the small.

The solubility $S(a)$ of a drop of radius a can be written

$$S(a) = S(\infty) \exp\left(\frac{K}{a}\right) \tag{13}$$

where $S(\infty)$ refers to the bulk solubility and K is a constant proportional to the interfacial tension. By considering a mixture containing two different droplet sizes, Higuchi and Misra (298) derived an equation for the rate of change with time of the radius of the smaller particle (which tends to dissolve). This rate depends mainly on the values of $S(\infty)$ and of the droplet radius a:

$$\frac{da_2}{dt} = \frac{DKS(\infty)}{a_2^2} \cdot \frac{n_2(a_2 - a_1)}{n_1(n_1 a_1 - n_2 a_2)} \cdot \tag{14}$$

where D is the diffusion coefficient of droplet phase in the dispersion medium and n_1, n_2 are the number densities of the larger and smaller droplets, respectively. For values of $S(\infty)$ of the order of that of n-decane (50 μg/dm^3, extrapolated from ref. 299) the disappearance of micron and submicron droplets from emulsions containing larger droplets can readily occur by this mechanism.

5. Inhibition

Rather than deal with the problem of emulsion breakdown, it is better to prevent the initial formation of emulsions if possible (143). This may be achieved by avoiding as far as possible the use of processes involving shearing forces, turbulent mixing, and so on. Naturally, this kind of thinking is often wishful as the unwanted emulsions are frequently desirable at some earlier stage of a process, for example in extraction processes or chemical reactions between immiscible liquids.

SYMBOLS

Latin

A	Hamaker constant
A_F	Hydrodynamic factor (cf. equation 7)
a_F	Filament radius
a	Droplet radius
C_{b1}	Critical concentration for black film formation
C_e	Critical coagulation concentration
E	Height of maximum in V-H curve
e	Electronic charge
F	Filter efficiency function
G	Gibbs elasticity of thin film
H	Reduced particle separation(r/a)
h_{cr}	Critical rupture thickness of film
k	Boltzman constant
M	Depth of secondary minimum in V-H curve
N_{Ad_r}	Dimensionless adhesion number
n_0	Initial droplet concentration
O	Area of contact
r	Particle–particle separation
P	Depth of primary minimum in V-H curve

R	Gas constant
T	Temperature
$t_{1/2}$	Half-life for coagulation
u	Electrophoretic velocity
V	Droplet–droplet interaction energy
W	Stability constant of dispersion (equation 4)
W_A	Work of adhesion
X	Electrostatic field strength
x	r-$2a$

Greek

γ	Interfacial tension
$\varepsilon, \varepsilon_0$	Permittivity; permittivity of free space
ε_D	Dilational modulus of interface
ζ	Electrokinetic potential
η	Viscosity
$1/\kappa$	Double layer length parameter
ν	Frequency
ψ_x	Electrical potential at position x

REFERENCES

1. P. Becher, *Emulsions Theory and Practice*, Reinhold, New York, 1965, (a) Chapter 1, (b) Chapter 6.

2. *IUPAC Manual of Definitions, Terminology, and Symbols in Colloid and Surface Chemistry*, IUPAC Secretariat, Oxford, 1970.

3. J.A. Kitchener and P.R. Musselwhite, in *Emulsion Science*, P. Sherman, Ed., Academic, London, 1968, Chapter 2.

4. W. McEntee and K.J. Mysels, *J. Phys. Chem.*, **73**, 3018 (1969).

5. S.P. Frankel and K.J. Mysels, *J. Phys. Chem.*, **73**, 3028 (1969).

6. K.J. Mysels and J.A. Strikeleather, *J. Colloid Interface Sci.*, **35**, 159 (1971).

7. G. Frens, K.J. Mysels, and B.R. Vijayendran, *Spec. Discuss. Faraday Soc.*, **1**, 12 (1970)

8. A.T. Florence and G. Frens, *J. Phys. Chem.*, **76**, 3024 (1972).

9. M. van den Tempel, *Rec. Trav. Chim.*, **72**, 442 (1953).

10. M. van den Tempel, *Rec. Trav. Chim.*, **72**, 433 (1953).

11. M. van den Tempel, *Proc. 2nd Int. Congr. Surface Activ.*, **1**, 439 (1957).

12. J.A. Kitchener, *Brit. Polym. J.*, **4**, 217 (1972).

13. J.Th.G. Overbeek, in *Colloid Science*, Vol. 1, H.R. Kruyt, Ed., Elsevier, Amsterdam, 1952, (a) Chapter 2, (b) Chapter 7, (c) Chapter 8.

14. M. Linton and K. Sutherland, *Proc. 2nd Int. Congr. Surface Activ.*, **1**, 494 (1957).

15. M. Smoluchowski, *Phys. Z.*, **17**, 557 (1917).; *Z. Phys. Chem.*, **92**, 129 (1917).

16. E.J.W. Verwey and J.Th.G. Overbeek, *Theory of Stability of Lyophobic Colloids*, Elsevier, Amsterdam, 1948, (a) Chapter 12, (b) Chapter 11, (c) Chapter 5.

17. A. Einstein, *Ann. Phys. (Leipzig)*, **17**, 549 (1905).

18. J. Müller, *Kolloid-Z.*, **38**, 1 (1926).

19. B.A. Matthews and C.T. Rhodes, *J. Colloid Interface Sci.*, **32**, 333 (1970).

20. D.L. Swift and S.K. Friedlander, *J. Colloid Sci.*, **19**, 621 (1964).

21. M. Smoluchowski, *Z. Phys. Chem.*, **92**, 155 (1917).

22. B.V. Derjaguin and L.D. Landau, *Acta Physicochim. U.R.S.S.*, **14**, 633 (1941).

23. B.V. Derjaguin, *Trans. Faraday Soc.*, **36**, 203 (1940).

24. F. London, *Z. Phys.*, **63**, 245 (1930).

25. H. Hellman, *Einführung in die Quantenchemie*, Leipzig und Wein, 1937, p. 189.

26. H.C. Hamaker, *Physica IV*, (10), 1058 (1937).

27. J.H. Schenkel and J.A. Kitchener, *Trans. Faraday Soc.*, **56**, 161 (1960).

28. J.T. Davies, *Proc. 2nd Int. Congr. Surface Activ.*, **1**, 426 (1957).

29. A.S.C. Lawrence and O.S. Mills, *Discuss. Faraday Soc.*, **18**, 98 (1954).

30. N. Fuchs, *Z. Phys.*, **89**, 736 (1934).

31. H. Reerink and J.T.G. Overbeek, *Discuss. Faraday Soc.*, **18**, 74 (1954).

32. R.H. Ottewill and J.N. Shaw, *Discuss. Faraday Soc.*, **42**, 154 (1966).

33. A. Watillon and A.M. Joseph-Petit, *Discuss. Faraday Soc.*, **42**, 143 (1966).

34. W.D. Cooper, *Kolloid-Z.*, **250**, 38 (1972).

35. K.G. Mathai and R.H. Ottewill, *Trans. Faraday Soc.*, **62**, 750 (1966).

36. K.G. Mathai and R.H. Ottewill, *Trans. Faraday Soc.*, **62**, 759 (1966).

37. B.A. Matthews and C.T. Rhodes, *J. Pharm. Sci.*, **57**, 557 (1968).

38. A.M. Joseph-Petit, F. Dumont, and A. Watillon, *J. Colloid Interface Sci.*, **43**, 649 (1973).

39. G. Frens, *Kolloid-Z.*, **250**, 736 (1972).

40. A. von Buzagh, *Kolloid-Z.*, **47**, 370 (1929).

41. S.W. Srivastava and D.A. Haydon, *Proc. 4th Int. Congr. Surface Activ.*, **2**, 1221 (1964).

42. A.P. Lemberger and N. Mourad, *J. Pharm. Sci.*, **54**, 229 (1965).

43. A.P. Lemberger and N. Mourad, *J. Pharm. Sci.*, **54**, 233 (1965).

44. G.R. Wiese and D.W. Healy, *Trans. Faraday Soc.*, **66**, 490 (1970).

45. T.P. Miller, *Computer Control Abstr.*, **6**, 3470 (1971).

46. A. Scheludko and R. Tschernev, *Proc. 4th Int. Congr. Surface Activ.*, **2**, 1109 (1964).

47. D. Platikanov and E. Manev, *Proc. 4th Int. Congr. Surface Activ.*, **2**, 1189 (1964).

48. A. Scheludko, *Advan. Colloid Interface Sci.*, **1**, 391 (1967).

49. A. Scheludko and E. Manev, *Trans. Faraday Soc.*, **64**, 1123 (1968).

50. A. Scheludko, B. Radoev, and T. Kolarov, *Trans. Faraday Soc.*, **64**, 2213 (1968).

51. T. Kolarov, A. Scheludko, and D. Exerowa, *Trans. Faraday Soc.*, **64**, 2264 (1968).

52. V.T. Ivanov, B. Radoev, E. Manev, and A. Scheludko, *Trans. Faraday Soc.*, **66**, 1262 (1970).

53. H. Sonntag, *Proc. 4th Int. Congr. Surface Activity*, **2**, 1089 (1964).

54. H. Sonntag and K. Strenge, *Koagulation und Stabilität disperser Systeme*, Deutscher Verlag, Berlin, 1970.

55. H. Sonntag, J. Netzel and B. Unterberger, *Spec. Discuss. Faraday Soc.*, **1**, 57 (1970).

56. H. Sonntag, *Kolloid Zh.*, **33**, 529 (1971).

57. H. Sonntag and N. Buske, *Kolloid-Z.*, **246**, 700 (1971).

58. H. Sonntag, N. Buske, and B. Unterberger, *Kolloid-Z.*, **248**, 1016 (1971).

59. N. Buske and H. Sonntag, *Kolloid-Z.*, **249**, 1133 (1971).

60. H. Sonntag and J. Netzel, *Z. Phys. Chem. (Leipzig)*, **250**, 119 (1972).

61. H. Sonntag, N. Buske, and H. Furhner, *Kolloid-Z.*, **250**, 330 (1972).

62. O. Reynolds, *Phil. Trans. Roy. Soc., London*, **177**, 157 (1866).

63. J. Lyklema and K.J. Mysels, *J. Amer. Chem. Soc.*, **87**, 2539 (1965).

64. K.J. Mysels and M.N. Jones, *Discuss. Faraday Soc.*, **42**, 42 (1966).

65. B.A. Pethica, *Spec. Discuss. Faraday Soc.*, **1**, 1 (1970).

66. H.B.G. Casimir and D. Polder, *Phys. Rev.*, **73**, 360 (1948).

67. J.N. Israelachvili and D. Tabor, *Proc. Roy. Soc., London*, **A311**, 19 (1972).

68. M.J. Vold, *J. Colloid Sci.*, **16**, 1 (1961).

69. D.W.J. Osmond, B. Vincent, and F. Waite, *J. Colloid Interface Sci.*, **42**, 262 (1973).

70. B. Vincent, *J. Colloid Interface Sci.*, **42**, 270 (1973).

71. E.M. Lifshitz, *Sov. Phys.—JETP*, **2**, 73 (1956).

72. H. Krupp, *Advan. Colloid Interface Sci.*, **1**, 111 (1967).

73. I.E. Dzyaloshinski, E.M. Lifshitz, and L.P. Pitaevski, *Sov. Phys.—JETP*, **10**, 161 (1960).

74. B.W. Ninham and V.A. Parsegian, *Biophys. J.*, **10**, 646 (1970).

75. V.A. Parsegian and B.W. Ninham, *Nature*, **224**, 1197 (1969).

76. V.A. Parsegian and B.W. Ninham, *Biophys. J.*, **10**, 664 (1970).

77. V.A. Parsegian and B.W. Ninham, *J. Colloid Interface Sci.*, **37**, 332 (1971).

78. D.A. Haydon and J.L. Taylor, *Nature*, **217**, 739 (1968).

79. D.J. Mitchell and B.W. Ninham, *J. Chem. Phys.*, **56**, 1117 (1972).

80. B. Davies, B.W. Ninham, and P. Richmond, *J. Chem. Phys.*, **58**, 744 (1973).

81. J.N. Israelachvili, *Proc. Roy. Soc., London*, **A331**, 39 (1972).

82. G.H. Weiss, J.E. Kiefer, and V.A. Parsegian, *J. Colloid Interface Sci.*, **45**, 615 (1973).

83. D. Bargeman and F. van Voorst Vader, *J. Electroanal. Chem.*, **37**, 45 (1972).

84. J. Gregory, *Advan. Colloid Interface Sci.*, **2**, 396 (1970).

85. J. Visser, *Advan. Colloid Interface Sci.*, **3**, 331 (1972).

86. G. Frens and J.Th.G. Overbeek, *J. Colloid Interface Sci.*, **38**, 376 (1972).

87. E.P. Honig and P.M. Mul, *J. Colloid Interface Sci.*, **36**, 258 (1971).

88. J. Gregory, *J. Chem. Soc. Faraday Trans., II*, **69**, 1723 (1969).

89. V.L. Sigel and A.M. Alekseeno, *Kolloid Zh.*, **33**, 737 (1971).

90. H. Ohshima, *Colloid Polym. Sci.*, **252**, 158 (1974).

91. H. Ohshima, *Colloid Polym. Sci.*, **252**, 257 (1974).

92. V.N. Shilov, *Kolloid-Zh.*, **34**, 147 (1972).

93.　G. Kar, S. Chander, and T.S. Mika, *J. Colloid Interface Sci.*, **44**, 374 (1973).

94.　S.L. Brenner and R.E. Roberts, *J. Phys. Chem.*, **77**, 2367 (1973).

95.　S. Levine and G.M. Bell, *J. Colloid Sci.*, **17**, 838 (1962).

96.　O. Stern, *Z. Electrochem.*, **30**, 508 (1924).

97.　B. Težak, *Discuss. Faraday Soc.*, **42**, 175 (1966).

98.　J. MacDonald and C.A. Barlow, in *Electrochemistry*, Proc. 1st Australian Conference, Sydney 1963, p. 199.

99.　I. Prigogine, P. Mazur, and R. Defay, *J. Chim. Phys.*, **50**, 146 (1953).

100.　A. Sanfeld, C. Devillez, and P. Terlinck, *J. Colloid Interface Sci.*, **32**, 33 (1970).

101.　D. Chen and S. Levine, *J. Chem. Soc. Faraday Trans. II*, **68**, 1497 (1972).

102.　A. Suzuki, N.F.H. Ho, and W.I. Higuchi, *J. Colloid Interface Sci.*, **29**, 552 (1969).

103.　R. Hogg, T.W. Healy, and D.W. Fuerstenau, *Trans. Faraday Soc.*, **62**, 1638 (1966).

104.　B.V. Derjaguin, *Discuss. Faraday Soc.*, **18**, 85 (1954).

105.　D.F. Devereux and P.L. de Bruyn, *Interaction of Plane-parallel Double Layers*, MIT Press, Cambridge, Mass., 1963.

106.　H.R. Kruyt and S.A. Troelstra, *Kolloid-Beih.*, **54**, 277, 284 (1943).

107.　R.D. Harding, *J. Colloid Interface Sci.*, **40**, 165 (1972).

108.　R.J. Pugh and J.A. Kitchener, *J. Colloid Interface Sci.*, **35**, 656 (1971).

109.　R.J. Pugh and J.A. Kitchener, *J. Colloid Interface Sci.*, **38**, 656 (1972).

110.　V.A. Parsegian and D. Gingell, *Biophys. J.*, **12**, 1192 (1972).

111.　G.M. Bell and G.C. Peterson, *J. Colloid Interface Sci.*, **41**, 542 (1972).

112.　G.M. Bell, S. Levine, and L.N. McCartney, *J. Colloid Interface Sci.*, **33**, 335 (1970).

113.　L.A. Spielman, *J. Colloid Interface Sci.*, **33**, 562 (1970).

114.　T.S. Chow and J.J. Hermans, *Kolloid-Z.*, **250**, 404 (1972).

115.　A. Lips, C. Smart, and E. Willis, *Trans. Faraday Soc.*, **67**, 2979 (1971).

116.　B.V. Derjaguin, *Pure Appl. Chem.*, **24**, 95 (1970).

117.　D. Stigter, *Proc. 4th Int. Congr. Surface Activ.*, **2**, 507 (1964).

118.　B.J. Carroll, Thesis. Cambridge, England, 1970.

119.　B. Vincent and J. Lyklema *Spec. Discuss. Faraday Soc.*, **1**, 148 (1970)

120.　B.V. Derjaguin *Sci. Amer.*, **223**, 52 (1970).

121.　S. Friberg and L. Rydhag, *Kolloid-Z.*, **244**, 233 (1971).

122.　W.I. Higuchi, T.O. Rhee, and D.R. Flanagan, *J. Pharm. Sci.*, **54**, 510 (1965).

123.　G.A. Johnson, J. Goldfarb, and B.A. Pethica, *Trans. Faraday Soc.*, **61**, 2321 (1965).

124.　G.A. Johnson, S.M.A. Lecchini, E.G. Smith, J. Clifford, and B.A. Pethica, *Discuss. Faraday Soc.*, **42**, 120 (1966).

125.　A. Holtzer and M.F. Emerson, *J. Phys. Chem.*, **73**, 26 (1969).

126.　P. Becher, S.E. Trifilleti, and Y. Machida, *Theory and Practice of Emulsion Technology*, S.C.I. Symposium, London, Academic Press, 1976.

127.　R.H. Ottewill and T. Walker, *J. Chem. Soc. Faraday Trans.* I, **70**, 917 (1974).

128.　E.L. Mackor and J.D. van der Waals, *J. Colloid Sci.*, **7**, 535 (1952).

129.　F.T. Hesselink, A. Vrij, and J.T.G. Overbeek, *J. Phys. Chem.*, **75**, 2094 (1971).

130.　R. Evans and D.H. Napper, *Kolloid-Z.*, **251**, 409 (1973).

131.　D.H. Napper, *Ind. Eng. Chem. Prod. R & D.*, **9**, 467 (1970).

132. R. Evans and D.H. Napper, *Kolloid-Z*, **251**, 329 (1973).

133. E.W. Fischer, *Kolloid-Z.*, **160**, 120 (1958).

134. M.C. Phillips, in *Water: A Comprehensive Treatise*, F. Franks, Ed., Plenum Press, New York (1974).

135. R.H. Ottewill, in *Nonionic Surfactants*, Vol. 1, M. Schick, Ed., Marcel Dekker, New York, 1969, Chapter 19.

136. P. Sherman, *J. Colloid Interface Sci.*, **27**, 282 (1968).

137. R.A.W. Hill and J.T. Knight, *Trans. Faraday Soc.*, **61**, 170 (1965).

138. G. Marrucci, *Chem. Eng. Sci.*, **24**, 975 (1969).

139. J.H. Burgoyne and L. Cohen, *J. Colloid Sci.*, **8**, 364 (1953).

140. L. Ström, *Rev. Sci. Instrum.*, **40**, 778 (1969).

141. M.A. Nawab and S.G. Mason, *Trans. Faraday Soc.*, **54**, 1712 (1958).

142. R.E. Wachtel and V.K. La Mer, *J. Colloid Sci.*, **17**, 531 (1962).

143. B.J. Carroll and J. Lucassen, *Theory and Practice of Emulsion Technology*, S.C.I. Symposium, London, Academic Press, 1976.

144. E.J. Clayfield and D.G. Wharton, *Theory and Practice of Emulsion Technology*, S.C.I. Symposium, London, Academic Press, 1976.

145. P.H. Elworthy and A.T. Florence, *J. Pharm. Pharmacol.*, **19**, 140S (1967).

146. P.H. Elworthy and A.T. Florence, *J. Pharm. Pharmacol.*, **21**, 72 (1969).

147. P.H. Elowrthy and A.T. Florence, *J. Pharm. Pharmacol.*, **21**, 70S (1969).

148. P.H. Elworthy and A.T. Florence, *J. Pharm. Pharmacol.*, **21**, 79S (1969).

149. P.H. Elworthy, A.T. Florence, and J.A. Rogers, *J. Colloid Interface Sci.*, **35**, 1199 (1964).

150. P.H. Elworthy, A.T. Florence, and J.A. Rogers, *J. Colloid Interface Sci.*, **35**, 34 (1971).

151. P. Sherman, *Proc. 4th Int. Congr. Surface Activ.*, **2**, 1199 (1964).

152. G.W. Hallworth and J.E. Carless, *Theory and Practice of Emulsion Technology*, S.C.I. Symposium, London, Academic Press, 1976.

153. S.S. Davies and A. Smith, *Theory and Practice of Emulsion Technology*, S.C.I. Symposium, London, Academic Press, 1976.

154. B. Biswas and D.A. Haydon, *Kolloid-Z.*, **185**, 31 (1962).

155. J.T. Davies and E.K. Rideal, *Interfacial Phenomena*, Academic Press, New York, 1960, (a) Chapter 3, (b) Chapter 5.

156. A.S.C. Lawrence and O.S. Mills, *Discuss. Faraday Soc.*, **18**, 98 (1954).

157. B.A. Matthews and C.T. Rhodes, *J. Colloid Interface Sci.*, **32**, 339 (1970).

158. D.F. Bernstein, W.I. Higuchi, and N.F.H. Ho, *J. Pharm. Sci.*, **60**, 690 (1971).

159. E.R. Garrett, *J. Pharm. Sci.*, **51**, 35 (1962).

160. S.J. Rehfeld, *J. Phys. Chem.*, **66**, 1966 (1962).

161. R.D. Vold and R.C. Groot, *J. Phys. Chem.*, **66**, 1969 (1962).

162. R.D. Vold and R.C. Groot, *J. Phys. Chem.*, **68**, 3477 (1964).

163. R.D. Vold and R.C. Groot, *J. Colloid Sci.*, **19**, 384 (1964).

164. E.R. Garrett, *J. Soc. Cosmet. Chem.*, **21**, 393 (1970).

165. R.D. Vold and K.L. Mittal, *J. Colloid Interface Sci..* **38**, 451 (1972).

166. R. D. Vold and K. L. Mittal, *J. Soc. Cosmet. Chem.*, **23**, 171 (1972).

167.   S. J. Rehfeld, *J. Colloid Interface Sci.*, **46**, 448 (1974).

168.   A. L. Smith and D. P. Mitchell, *Theory and Practice of Emulsion Technology* S. C. I. Symposium, London, Academic Press, 1976.

169.   F. Sebba, *Nature*, **197**, 1195 (1963).

170.   M. van den Tempel, *J. Colloid Sci.*, **13**, 125 (1958).

171.   K. J. Mysels, K. Shinoda, and S. P. Frankel, *Soap Films*, Pergamon, London, 1959.

172.   J. Lyklema and K. J. Mysels, *J. Amer. Chem. Soc.*, **87**, 2539 (1965).

173.   D. Exerowa and A. Scheludko, *Proc. 4th Int. Congr. Surface Activ.*, **2**, 1097 (1964).

174.   A. J. de Vries, *Rec. Trav. Chim.*, **77**, 81 (1958).

175.   A. J. de Vries, *Rec. Trav. Chim.*, **77**, 209 (1958).

176.   A. J. de Vries, *Rec. Trav. Chim.*, **77**, 283 (1958).

177.   A. J. de Vries, *Rec. Trav. Chim.*, **77**, 383 (1958).

178.   A. Scheludko, *Proc. Kon. Ned. Akad. Wetensch.*, **B65**, 86 (1962).

179.   A. Vrij, *Discuss. Faraday Soc.*, **42**, 23 (1966).

180.   A. Vrij and J. Th. G. Overbeek, *J. Amer. Chem. Soc.*, **90**, 3074 (1968).

181.   J. Lucassen, M. van den Tempel, A. Vrij, and F. T. Hesselink, *Proc. Kon. Ned. Akad, Wetensch.*, **B73**, 109 (1970).

182.   A. Vrij, F. T. Hesselink, J. Lucassen, and M. van den Tempel, *Proc. Kon. Ned. Akad. Wetensch.*, **B73**, 124 (1970).

183.   E. Manev, A. Scheludko, and D. Exerowa, *Colloid Polym. Sci.*, **252**, 586 (1974).

184.   D. E. Graham and M. C. Phillips, *Theory and Practice of Emulsion Technology*, S. C. I. Symposium, London, Academic Press, 1976.

185.   K. J. Mysels, F. Huisman, and R. I. Razouk, *J. Phys. Chem.*, **70**, 1339 (1966).

186.   J. L. Taylor and D. A. Haydon, *Discuss. Faraday Soc.*, **42**, 51 (1966).

187.   D. M. Andrews, E. Manev, and D. A. Haydon, *Spec. Discuss. Faraday Soc.*, **1**, 46 (1970).

188.   B. V. Derjaguin and Y. V. Gutop, *Kolloid-Zh.*, **24**, 431 (1962).

189.   E. G. Cockbain and T. S. McRoberts, *J. Colloid Sci.*, **8**, 440 (1953).

190.   T. Gillespie and E. K. Rideal, *Trans. Faraday Soc.*, **52**, 173 (1956).

191.   G. A. H. Elton and R. G. Picknett, *Proc. 2nd Int. Congr. Surface Activ.*, **1**, 288 (1957).

192.   G. E. Charles and S. G. Mason, *J. Colloid Sci.*, **15**, 236 (1960).

193.   G. V. Jeffreys and J. L. Hawksley, *J. Appl. Chem.*, **12**, 329 (1962).

194.   S. Hartland, *Trans. Inst. Chem. Eng.*, **45**, T97 (1967).

195.   S. Hartland, *Trans. Inst. Chem. Eng.*, **45**, T102 (1967).

196.   S. Hartland, *Trans. Inst. Chem. Eng.*, **45**, T109 (1967).

197.   T. D. Hodgson and J. C. Lee, *J. Colloid Interface Sci.*, **30**, 94 (1969).

198.   A. D. Bangham and M. W. Hill, *Nature*, **237**, 408 (1972).

199.   D. Robinson and S. Hartland, *Tenside*, **9**, 301 (1972).

200.   T. D. Hodgson and D. R. Woods, *J. Colloid Interface Sci.*, **30**, 429 (1969).

201.   K. A. Burrill and D. R. Woods, *J. Colloid Interface Sci.*, **30**, 511 (1969).

202.   K. A. Burrill and D. R. Woods, *J. Colloid Interface Sci.*, **42**, 15 (1969).

203.   K. A. Burrill and D. R. Woods, *J. Colloid Interface Sci.*, **42**, 35 (1973).

204.   D. R. Woods and K. A. Burrill, *J. Electroanal. Chem.*, **37**, 191 (1972).

205. S. P. Frankel and K. J. Mysels, *J. Phys. Chem.*, **66**, 190 (1962).

206. L. Mandell and P. Ekwall, *Proc. 4th Int. Congr. Surface Activ.*, **2**, 659 (1964).

207. S. Friberg, L. Mandell, and M. Larsson, *J. Colloid Interface Sci.*, **29**, 155 (1969).

208. L. Mandell and P. Ekwall, *Proc. 5th Int. Congr. Surface Activ.*, **2**, 943 (1969).

209. S. Friberg and L. Mandell, *J. Pharm. Sci.*, **59**, 1001 (1970).

210. S. Friberg and I. Wilton, *Amer. Perfum. Cosmet.*, **85**, 27 (1970).

211. S. Friberg and L. Mandell, *J. Amer. Oil Chem. Soc.*, **47**, 149 (1970).

212. S. Friberg, *J. Colloid Interface Sci.*, **37**, 291 (1971).

213. S. Friberg, *Kolloid-Z.*, **244**, 333 (1971).

214. F. Lachampt and R. M. Vila, *Amer. Perfum. Cosmet.*, **82**, 29 (1967).

215. M. Bierre, *Soap Perfum. Cosmet.*, Oct. 1971, p. 623.

216. D. Chen and D. G. Hall, *Kolloid-Z.*, **251**, 41 (1973).

217. R. R. Balmbra, J. S. Clunie, J. F. Goodman, and B. T. Ingram, *J. Colloid Interface Sci.*, **42**, 226 (1973).

218. S. Kislalioglu and S. Friberg, *Theory and Practice of Emulsion Technology*, S. C. I. Symposium, London, Academic Press, 1976.

219. D. Stigter, *J. Colloid Interface Sci.*, **23**, 379 (1967).

220. A. W. Adamson, *The Physical Chemistry of Surfaces*, Interscience, New York, 1960, Chapter IX.

221. K. Shinoda and H. Saito, *J. Colloid Interface Sci.*, **30**, 258 (1969).

222. C. J. Parkinson and P. Sherman, *J. Colloid Interface Sci.*, **41**, 328 (1972).

223. G. H. Clowes, *J. Phys. Chem.*, **20**, 407 (1916).

224. J. Schulman and E. G. Cockbain, *Trans. Faraday Soc.*, **36**, 661 (1940).

225. J. T. Davies, *Proc. 3rd Int. Congr. Surface Activ.*, **2**, 585 (1960).

226. W. C. Griffin, *J. Soc. Cosmet. Chem.*, **1**, 311 (1949). *Encyclopaedia of Chemical Technology*, Vol. 8, Wiley, New York, 1965, p. 117.

227. W. Albers and J. Th. G. Overbeek, *J. Colloid Sci.*, **14**, 501 (1959).

228. W. Albers and J. Th. G. Overbeek, *J. Colloid Sci.*, **14**, 510 (1959).

229. E. H. Lucassen-Reynders, *Kolloid-Z.*, **197**, 201 (1964).

230. J. W. Gibbs, *Collected Works*, Vol. 1, Dover, New York 1961, p. 301.

231. L. E. Scriven and C. V. Sternling, *Nature*, **187**, 186 (1960).

232. T. G. Jones, K. Durham, W. P. Evans, and M. Camp, *Proc. 2nd Int. Congr. Surface Activ.*, **1**, 225 (1957).

233. A. Prins and M. van den Tempel, *Proc. 4th Int. Congr. Surface Activ.*, **2**, 1119 (1964).

234. A. Prins, C. Arcuri, and M. van den Tempel, *J. Colloid Interface Sci.*, **28**, 84 (1967).

235. A Prins and M. van den Tempel, *J. Phys. Chem.*, **73**, 2828 (1969).

236. H. Princen and E. D. Goddard, *J. Colloid Interface Sci.*, **38**, 523 (1972).

237. M. Bier, *A.I.Ch.E. Symposium Series*, **68**, 84 (1971).

238. F. M. Fowkes, F. W. Anderson, and J. Berger, *Sci. Technol.*, **4**, 510 (1970).

239. S. Liberman, M. Inque, and S. G. Mason, *J. Colloid Interface Sci.*, **48**, 175 (1974).

240. S. E. Sadek and C. D. Hendricks, *Ind. Chem. Eng. Fundam.*, **13**, 139 (1974).

241. L. C. Waterman, *Chem. Eng. Progr.*, **61** (10), 51 (1965).

242. W. F. Pickard, *Cruft Laboratory Technical Report*, **423**, Cambridge, Mass., 1963.
243. B. E. Conway, in *Modern Aspects of Electrochemistry*, J. D. Bockris, Ed., Butterworths, London, 1954, Chapter 1.
244. S. Torza and S. G. Mason, *Science*, **163**, 813 (1969).
245. S. Torza and S. G. Mason, *J. Colloid Interface Sci.*, **33**, 67 (1970).
246. V. I. Borisikhina, L. D. Skrylov, and S. G. Mokrushin, *Kolloid-Zh.*, **24**, 256 (1962).
247. L. D. Skrylov and Y. P. Kosorskaya, *Kolloid-Zh.*, **33**, 746 (1971).
248. A. W. Neumann, J. Szekely, and E. J. Rabenda, *J. Colloid Interface Sci.*, **43**, 727 (1973).
249. W. E. Ewers and K. L. Sutherland, *Aust. J. Sci. Res.*, **5**, 697 (1952).
250. T. J. Lin, *J. Soc. Cosmet. Chem.*, **19**, 683 (1968).
251. T.J. Lin and J.C. Lambrechts, *J. Soc. Cosmet. Chem.*, **20**, 185 (1969).
252. T.J. Lin and J.C. Lambrechts, *J. Soc. Cosmet. Chem.*, **20**, 627 (1969).
253. T.J. Lin, *J. Soc. Cosmet. Chem.*, **21**, 365 (1970).
254. H.L. Booij, *J. Colloid Interface Sci.*, **29**, 365 (1969).
254. H.L. Booij and H.G. Bungenberg de Jong, *Biocolloids and their Interactions, Protoplasmatoligia*, Vol. 1, Springer-Verlag, Vienna, 1956.
256. E.H. Lucassen-Reynders and J. Lucassen, *Advan. Colliod Interface Sci.*, **2**, 347 (1969).
257. T. Green, R.P. Harker, and F.O. Howitt, *Nature*, **174**, 659 (1954).
258. J. Kepinski and N. Chulubek, *Int. Chem. Eng.*, **12**, 499 (1972).
259. S. Saito and T. Taniguchi, *J. Amer. Oil Chem. Soc.*, **50**, 276 (1973).
260. B. Cooke, *Process Eng.*, 1974, (4), 64.
261. B.A. Harris, in *Industrial Pollution Control*, K. Teale, Ed., Business Books Ltd., London, 1973, Chapter 9.
262. P.L. Busch and W. Stumm, *Environ. Sci Technol.*, **2**, 49 (1968).
263. A.T. Kuhn, *Chem. Ind.*, 946 (1971).
264. A.T. Kuhn, in *Electrochemistry of Cleaner Environments*, J.D. Bockris, Ed., Plenum, New York, 1972, Chapter 4.
265. G.A. Babalyan and M. Kh. Akhonadeev, *Dokl. Akad. Nauk SSSR*, **206**, 406 (1972).
266. P.H. Wiersema, A.L. Loeb, and J.Th.G. Overbeek, *J. Colliod Interface Sci.*, **22**, 78 (1966).
267. R. Surfleet, *Electricity Council (UK) Report*, ECRC R/165.
268. S. Glasstone and A. Hickling, *Trans. Electrochem. Soc.*, **75**, 333 (1939).
269. R.S. Allan and S.G. Mason, *Trans. Faraday Soc.*, **57**, 2027 (1961).
270. R.S. Allan and S.G. Mason, *J. Colloid Sci.*, **17**, 383 (1962).
271. A.M. Brown and C. Hanson, *Trans. Faraday Soc.*, **61**, 1754 (1965).
272. Committee report. *J. Amer. Works, Assoc.*, Feb. 1971, p. 99.
273. S. Ross, E.S. Chen, P. Becher, and H.J. Rananto, *J. Phys. Chem.*, **63**, 1681 (1959).
274. G.V. Jordan, *Chem. Eng. Progr.*, (10), 64 (1965).
275. A. Scheludko and D. Platikanov, *Kolloid Z.*, **175**, 150 (1961).
276. B.V. Derjaguin and Z.M. Zorin, *Zh. Fiz. Khim.*, **29**, 1010 (1955).
277. A.D. Read and J.A. Kitchener, *J. Colliod Interface Sci.*, **30**, 391 (1969).
278. T.D. Blake and J.A. Kitchener, *J. Chem. Soc. Faraday Trans. I*, **68**, 1435 (1972).

279. J.F. Padday, *Spec. Discuss. Faraday Soc.*, **1**, 64 (1970).

280. R. Francois, in *Industrial Waste Water*, B. Göransson, Ed., Butterworths, London, 1972, p. 163.

281. L.A. Spielman and S.L. Goren, *Ind. Eng. Chem.*, **62** (10), 10 (1970).

282. L.A. Spielman and S.L. Goren, *Environ. Sci. Technal.*, **4**, 135 (1970); **5**, 254 (1971).

283. L.A. Spielman and P.H. Cuckor, *J. Colloid Interface Sci.*, **43**, 51 (1973).

284. B.J. Carroll and J. Lucassen, *J. Chem. Soc. Faraday Trans. I*, **70**, 1228 (1974).

285. J. Lucassen and B.J. Carroll, to be published.

286. T.E. Belk, Chem. *Eng. Progr.*, **61** (10), 72 (1965).

287. M.E. O'Neill, *Chem. Eng. Sci.*, **23**, 1293 (1968).

288. A.D. Zimon, *Kolloid-Zh.*, **34**, 197 (1972).

288. S.S. Sareen, P.M. Rose, R.C. Gudeson, and R.C. Kinter, *AIChE. J.* **12**, 1045 (1966).

290. L. McLain, *Process Eng.*, (1), 80 (1973).

291. H. Schott, in *Detergency Theory and Test Methods*, W.G. Culter and R.C. Davies, Eds., Arnold, 1972, Chapter 5.

292. C.J. Mumford and R.J. Thomas, *Process Eng.*, (12), 54 (1972).

293. W.M. Langdon, P.P. Naik, and D.T. Wasan, *Environ. Sci. Technol.*, **6**, 905 (1972).

294. G.A. Davies, G.V. Jeffreys, F. Ali, and M. Afzal, *The Chem. Eng.*, October 1972, p. 392.

295. R. Kaiser, C.K. Colton, G. Miskolszy and L. Mir, *AIChE Symposium Series*, **68**, (124) 115 (1972).

296. J. Mizrahi and E. Barnea, *Process Eng.*, (1), 60 (1973).

297. D.M. Landis, *Chem. Eng. Progr.*, **61**, (10) 58 (1965).

298. W.I. Higuchi and J. Misra, *J. Pharm. Sci.*, **51**, 459 (1962).

299. C. McAuliffe, *J. Phys. Chem.*, **70**, 1267 (1966).

2

Nuclear Magnetic Resonance of Surfactant Solutions

TOSHIO NAKAGAWA

Shionogi Research Laboratory
Shionogi & Co., Ltd., Fukushima-ku, Osaka, Japan

AND

FUMIKATSU TOKIWA

Tokyo Research Laboratories
Kao Soap Co., Ltd., Sumida-ku, Tokyo, Japan

69

I. INTRODUCTION

The nuclear magnetic resonance (nMr) technique developed after the initial discovery of the nMr absorption phenomenon by Purcell (1) and Bloch (2) in 1946 is now recognized as a powerful tool for the investigation of various chemical and physical properties of substances. At an early stage this technique threw new light on many difficult problems of organic chemistry and was then rapidly applied in various fields.

In colloid science and related disciplines, the initial application of nMr was made before high resolution spectrometers were commercially available. Probably, the earliest work in this area is that of Spence et al. (3) in 1953, who used broad line technique, on the mesomorphic or liquid crystalline state of single substances such as *p*-azoxy anisole and its homologs. The work of McDonald (4) is also worthy of note as the first nMr work on surfactant systems; he studied the mesomorphic phases in aqueous solutions of surfactant and/or amphiphile with a view of investigating molecular arrangement and hydrogen bonding in these phases. The application of nMr was then extended to colloid-chemical studies of micellar solutions of surfactants, emulsions, or suspensions, as well as phase transitions of surfactants in solid and liquid states, and interactions of adsorbed molecules with substrates such as proteins. The nMr technique is inherently a useful tool for studying micellar solutions of surfactants, because protons of a surfactant molecule in different environments (in this case, micellar and bulk water phases) generally give different nMr resonance signals at different positions. The measurements in surfactant solutions usually consist of

1. The shift of a resonance signal caused by, for example, micelle formation.
2. The split of a signal caused by, for example, solubilization of organic materials.
3. A change in the half height width of a signal.
4. The area and shape of a signal.
5. The spin–lattice and/or spin–spin relaxation times of particular group(s) of a surfactant and a solvent, from which one can speculate on various micellar and colloidal properties of surfactants in solutions, and also on the molecular motion and interaction of surfactants in isotropic micellar and mesomorphic phases.

Recently, pulse Fourier transform ^{13}C nMr spectrometers, which can measure accurately spin–lattice and spin–spin relaxation times, have become commercially available (5). Although the principle is different from that of conventional spectrometers, quite similar spectrograms are obtained after the Fourier transformation of the original response has been automatically carried out. Fourier transform ^{13}C spectrometers will promote the studies of the colloid-chemical behavior of surfactants because they give individual carbon resonances and also direct information about the segmental motion of carbon chains of surfactant molecules, not obtainable with conventional ^{1}H spectrometers (6–10).

A literature survey shows that it would be impossible to cover the entire area of application of nMr in the field of colloid and surface science in such a limited number of pages. Therefore, in this chapter, topics will be restricted to the application of nMr to surfactant solutions and especially to micellar solutions, colloidal dispersions, solubilization behaviors, mesomorphic phases, and interactions of surfactants in mixed micelles or with other materials. The topics and problems that have not been fully discussed will be compiled in the last section. Emphasis has been put, as far as possible, on new knowledge and on information obtained by nMr that is not easily obtainable in any other way. In this chapter the description of nMr spectroscopy is brief but readers unfamiliar with this principle and technique can refer, for example, to the books of Pople et al. (11), Slichter (12), Abragam (13), and

Carrington and McLachlan (14), which will provide the elementary knowledge necessary to understand the succeeding sections. Further, there are several review articles on nMr theory which discuss chemical shifts, relaxation times, and so on (15–18).

II. MICELLAR PROPERTIES OF SURFACTANT SOLUTIONS

A. Micelle Formation and Micelle Structure

1. Chemical Shifts in Aqueous Solutions

a. Chemical Shift of Water Protons in Ionic Surfactant Solutions

The effect of various ions on the chemical shift of water protons has usually been discussed in terms of two factors: (a) hydration of the ion, with a consequent breaking of hydrogen bonds in the water structure nearby and a shift to higher applied field for the water protons, and (b) polarization of the O–H bond in the water molecules by the electrostatic field of the ion, resulting in a shift to lower applied field (19). For most single-charged ions the structure-breaking effect is dominant, giving a net shift to high applied field, while for other ions the polarization effect is considered more important.

As is well known, surfactant molecules in aqueous solution associate to form micelles at a definite concentration, the so-called critical micelle concentration (CMC). Above the CMC, micelles are equilibrated with single surfactant molecules, the concentration of which usually remains equal to the CMC. The effect of surfactant ions, which have a long hydrocarbon chain, on the chemical shift of water protons is considered to be different from the effect of ordinary electrolyte ions. This effect will also be different below and above the CMC, since some of the surfactant is in the form of micelles above the CMC. The chemical shift of water protons in solutions of surfactants is very significant in the consideration of forces leading to the formation of micelles.

Clifford and Pethica (20) have measured the chemical shift of water protons in aqueous solutions of ionic surfactants, sodium alkyl sulfates, and have discussed the effect of micelle formation on this shift and the change of hydrogen bonding in the water structure. When an alkyl sulfate is dissolved in water, the water protons give a single, sharp peak which shifts to higher magnetic field with increasing concentration. The result is shown in Figure 1, in which the shift is expressed in reference to the pure water peak. For the C_2 and C_4 sulfates, which do not form micelles, the chemical shift $\Delta\nu_{H_2O}$ against concentration C curves are linear over the concentration range shown. For the C_6, C_8, and C_{12} sulfates the $\Delta\nu_{H_2O}$ versus C curves have two linear

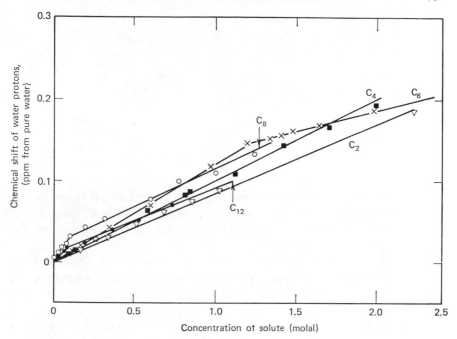

Fig. 1. The chemical shift of water protons in solution of sodium alkyl sulfates at 34°C: ▽, C_2; ■, C_4; ×, C_6; ○, C_8; ●, C_{12} sulfate. (From ref. 20.)

parts that meet at a concentration corresponding approximately to the CMC for each surfactant. It is assumed that below the CMC only single ions exist, whereas above the CMC the concentration of single ions remains equal to the CMC and all the excess of alkyl sulfate is in the form of micelles. Then the slope of the $\Delta\nu_{H_2O}$–C curve below the CMC is a measure of the effect of the nonmicellized alkyl sulfate and sodium ions on the water, and the slope above the CMC is a measure of the effect of the micelles on the water (Table 1).

Below the CMC, the observed shift would be considered to be due to (*a*) the effect of sodium ions, (*b*) the effect of the $-SO_4^-$ groups, and (*c*) the effect of the hydrocarbon chains. The first two cause a shift to higher field, but they would be independent of the number of carbon atoms in the chain. The effect of the hydrocarbon chains cannot be simply explained. The influence of the first few CH_2 groups (next to the ionic group of $-SO_4^-$) on water structure would be small because they are in contact with the water whose hydrogen bonds have already been broken up by the charge of the ion. The effect of the shorter alkyl chains on the water structure is, therefore, likely to be different from the effect of the longer chains. Indeed, if the chemical shifts for water protons are plotted against the number of carbon atoms

TABLE 1

Effect on the chemical shift of water protons[a]

Alkyl Sulfate	As Single Ions (ppm/mole/1000 g of water)	As Micelles
C_2	0.09	—
C_4	0.10	—
C_6	0.13	0.05
C_8	0.20	0.09
C_{12}	(0.34)	0.09

[a]From ref. 20.

in the chain, it is found that the water proton chemical shift is smaller for the shorter chains and becomes significant at C_6. Between C_6 and C_{12} the effect of chain length is approximately linear, each methylene group causing a molal shift of about 0.035 ppm.

Table 1 shows that the $\Delta\nu_{H_2O}$–C curve has a smaller positive slope in the presence of micelles than for single ions. Micellization removes the hydrocarbon chains from contact with water. In addition, the $-SO_4^-$ head groups come closer together, and a high proportion of the Na^+ ions are held near them. However, as estimated by Clifford and Pethica (20), the effects of the head groups and counter-ions on the water chemical shift are not significant. From differences in the slopes of the curves for single ions and micelles, given in Table 1, the process of micellization results in a shift of the water protons of -0.08, -0.11, and -0.25 ppm for the C_6, C_8, and C_{12} sulfates, respectively. As described already, the first four CH_2 groups next to the $-SO_4^-$ part of the ion behave as if they have little effect on the aqueous solvent. Thus, there are 2, 4, and 8 effective CH_2 groups for the C_6, C_8, and C_{12} ions, resectively, and consequently there is a molal shift of the water protons of -0.04, -0.028, and -0.031 per CH_2 group when the alkyl chains are removed into the micelle. These values compare reasonably well with that of $+0.035$ ppm per CH_2 group found when the alkyl chains are dissolved in water.

The above results would seem to show that on dissolution of alkyl chains, hydrogen bonds in water around these chains are broken (or polarization of the O–H bond in the water molecules by the electrostatic field is reduced) to produce the shift of water protons to higher applied field. However, one could instead attribute the effect of hydrocarbon chains on water to an increase in the covalent character of the hydrogen bonds in water.

b. Chemical Shift of Water Protons in Nonionic Surfactant Solutions

Corkill et al. (21) have studied the behavior of nMr signals of water protons in aqueous micellar solutions of nonionic surfactants, alkyl polyoxy-ethylene ethers (abbreviated to $C_n(EO)_p$), and their derivatives (abbreviated to $C_n(EO)_pC_1$), in a range of concentrations higher than the CMC. (In general, the CMC values of this type of surfactants are much smaller than those of ionic surfactants, and therefore the change in the chemical shift of water protons in passing through the CMC cannot be detected.)

In Figure 2 the shift of water proton signals for $C_8(EO)_3$, $C_8(EO)_6$, and $C_8(EO)_6C_1$ are shown as a function of the mole fraction of the solute species. The water proton shift shows a significant change, whereas the alkyl chain shifts are essentially independent of concentration. The introduction of solute molecules into water causes the disruption of water–water hydrogen bonds and the establishment of solute–solvent interaction, thus resulting usually in an upfield shift of the water proton signal. In addition, for a solute in which chemical exchange of protons can take place with the solvent, the chemical shift will be further affected by this exchange.

In the $C_n(EO)_pC_1$ system, where there is no solute–solvent proton ex-

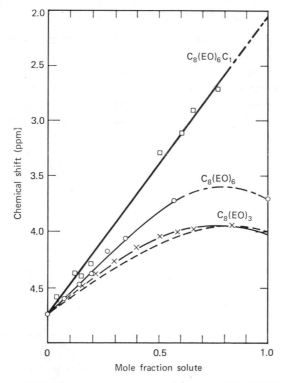

Fig. 2. The chemical shift of water protons in solutions of □, $C_8(EO)_6C_1$; ○, $C_8(EO)_6$; ×, $C_8(EO)_3$ at 32°C. (From ref. 21.)

change, an approximately linear upfield shift of the water signal with increasing solute concentration is observed (Figure 2). The curve of the chemical shift versus mole fraction of $C_n(EO)_p$ shows a maximum (Figure 2); in the latter case chemical exchange occurs between the terminal O–H proton and the water which has already undergone hydrogen bonding changes due to the presence of the solute. The broken curve in Figure 2 was calculated assuming a rapid chemical exchange between water and the solute $C_8(EO)_3$, with the assignment of appropriate values to the solute and water chemical shifts in the absence of exchange.

An increase in the upfield shift of water protons in passing from the $(EO)_3$ to the $(EO)_6$ series reflects a greater solvent perturbation due to the longer polyoxyethylene chain. The replacement of the hydroxyl by the methoxyl group ($-OH$ to $-OCH_3$) also leads to an increase in the water proton shift; in this case, the removal of the hydroxyl group, with its associated downfield shift effect, is more significant for the upfield shift of water protons than any intrinsic disruptive effect of the methyl group.

c. Chemical Shift of Surfactant Protons

Analyzing the signal behavior of particular protons of a surfactant molecule, such as the phenyl protons of a surfactant having an ω-phenylalkyl group, one can obtain some information about micelle formation from the nMr behavior of the surfactant. In principle, if the surfactant molecules are exchanged very slowly between the water and micelle phases, two separate signals of the phenyl protons will be observed at different positions; one corresponds to the surfactant molecules in the water environment and the other to the molecules in the micellar environment (ν_m and ν_M), as shown in Figure 3. On the contrary, if the exchange is sufficiently rapid, one sharp

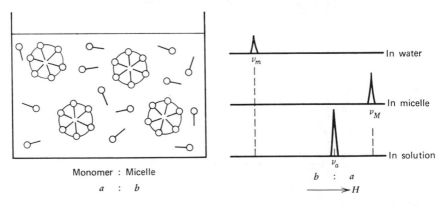

Monomer : Micelle
a : b

ν_m
ν_M
ν_a
b : a
$\longrightarrow H$

In water
In micelle
In solution

Fig. 3.   A schematic illustration of the nMr signal of phenyl protons in aqueous solution $\phi C_n TAB$ above the CMC.

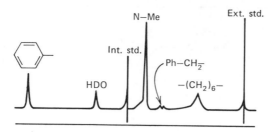

Fig. 4. Nuclear magnetic resonance spectrum of $\phi C_8 TAB$.

signal will be observed at a weight-averaged position (ν_a). At a moderate rate of exchange, the signal shape becomes broader and takes forms varying with the rate (see ref. 11, p. 222).

Figure 4 represents a typical spectrum of ω-phenylalkyltrimethylammonium bromide* $\phi-(CH_2)_n-N(CH_3)_3$ Br (abbreviated as $\phi C_n TAB$) (22). The sharp peak at the lowest field is due to the phenyl protons. Although the *ortho-*, *meta-*, and *para-*protons are not strictly equivalent, they give in effect a sharp singlet signal. With an increase in surfactant concentration above the CMC, this peak shifts to higher magnetic field, owing to the ring current arising from the other phenyl groups in the same micelle, while its width shows no detectable broadening. The fact that above the CMC only one sharp signal is observed, which shifts to higher magnetic field with increasing surfactant concentration, suggests a very rapid exchange of the surfactant molecules between the micelles and the bulk water phase (the residence time of the surfactant molecule in the micelle is probably in the order of 10^{-4} sec or less (22)).

In Figure 5, the signal position of the phenyl protons for a series of $\phi C_n TAB$ ($n = 4 - 12$), with reference to 1,4-dioxane (internal standard), is plotted against the concentration (24). The signal begins to shift abruptly to higher field at the concentration listed in Table 2. As expected, this concentration is nearly equal to the CMC obtained by other methods. The butyl homolog does not exhibit an abrupt shift, suggesting that the micelles are not formed.

Let the following idealized model be assumed for micelle formation: For concentrations higher than the CMC, where the micelles coexist with the monomeric surfactant, each micelle is composed of a definite number of the surfactant molecules. With an increase in concentration, only the number of

---

*From a solubilization study of benzene by sodium dodecyl sulfate, Nakagawa and Tori (23) have found that the ring current arising from benzene molecules in the micelle plays an important role in the signal shift; this led them to study micelle formation of this type of surfactant.

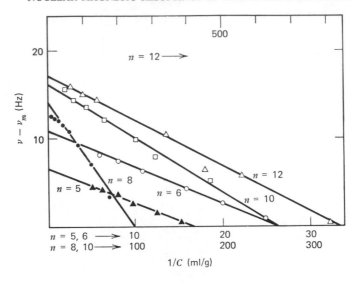

Fig. 5.   The shift of the phenyl-proton signal of $\phi C_n TAB$ as a function of concentration. (From ref. 24.)

the micelles increases, but their structure as well as the concentration of the monomeric surfactant remain invariant. Under these conditions, assuming that the exchange of the surfactant molecules is sufficiently rapid (as in the present case) and neglecting the volume effect of the monomeric surfactant in the solution, the shift measured from the signal position of the monomeric surfactant ν_m is given by

$$\nu - \nu_m = (\nu_M - \nu_m)\left[1 - \frac{(CMC)}{C}\right] \qquad C \geq CMC \qquad (1a)$$
$$= 0 \qquad\qquad\qquad\qquad C \leq CMC \qquad (1b)$$

TABLE 2

Values of CMC, ν_m, ν_M, and $\bar{\nu}$ for $\phi C_n TAB^a$

	CMC $\times 10^2$ (g/ml)		ν_m (Hz)	ν_M (Hz)		$\bar{\nu}$ (ml/g)
$n = 5$	6.0	$(6.1)^b$	-214.7	-208.5	$(-208.5)^b$	0.838
6	3.8	(3.8)	-214.4	-203.8	(-203.9)	0.856
8	1.1	(1.0)	-213.8	-200.1	(-199.8)	0.885
10		(0.38)			(-197.6)	
12		(0.12)			(-196.4)	

^{a}From refs. 22, 24.
^{b}Values in parentheses were obtained from the plot of $(\nu - \nu_m)$ against $1/C$ according to equation 1a corrected for the volume effect of monomeric surfactant.

where ν is the observed signal position, ν_M is the signal position of the micellar surfactant, and C is the concentration of the surfactant. If the volume effect of the monomeric surfactant is taken into account, the right-hand side of equation 1a must be multiplied by a correction factor $1/[1 - \bar{v}(\text{CMC})]$, where the CMC and the partial specific volume of surfactant $\bar{v}$ are expressed in g/ml and ml/g, respectively (22).

For $\phi C_n\text{TAB}$ with n of 5, 6, or 8, the values of CMC and ν_m can be determined from the break point in Figure 5, as shown in Table 2. If, in addition, an appropriate value is assumed for ν_M as given in Table 2 and inserted into equation 1 a together with the values of ν_m and CMC, the curves thus calculated and experimentally obtained are in a fairly good agreement in the range of relatively low concentrations. A small deviation at higher concentrations is probably due to changes of micellar structure.

Equation 1a also enables one to determine the CMC and ν_M simultaneously. The plot of $(\nu - \nu_m)$ against $1/C$ should give a straight line. The intercept on the abscissa is $1/\text{CMC}$, and the intercept on the ordinate is $(\nu_M - \nu_m)$; from these intercepts the CMC and ν_M can be determined. As shown in Figure 6, such plots give straight lines. The values of CMC and ν_M thus obtained are listed in Table 2. Good agreement is seen between the values obtained from Figures 5 and 6.

These results strongly support the view that the position of the phenyl proton signal varies with the ratio of monomeric and micellar surfactants; the signal appears at a weight-averaged position of the signals of both

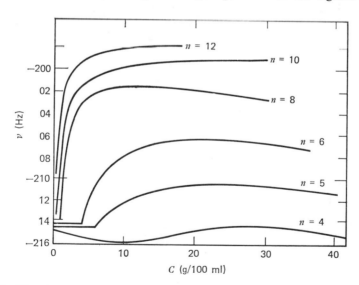

Fig. 6. The plot of $(\nu - \nu_m)$ against $1/C$ (From ref. 24.)

species. It may also be argued that the proposed model of micelle formation is adequate at least in the range of relatively low concentrations.

2. Chemical Shifts in Nonaqueous Solutions

A reversed or inverted micelle formed in a nonaqueous solution is composed of the charged hydrophilic groups in the interior and of the hydrophobic hydrocarbon chains forming the outer layer in contact with nonpolar solvent (25, 26). Although nMr investigations in this field are relatively limited, the studies of micelle formation made by Fendler et al. (27–31) on alkylammonium carboxylates in nonaqueous solvents are interesting. In this section we will describe nMr studies of micelle formation and micellar properties of a series of alkylammonium propionates $H(CH_2)_nNH_3^+ \text{-} O_2CCH_2CH_3$ (abbreviated to C_nAP) in benzene and in carbon tetrachloride.

As in aqueous micellar systems, the nMr spectra of C_nAP in these solvents exhibit single weight-averaged resonance signals for the protons of the monomeric and aggregated surfactants, indicating that the establishment of the monomer–micelle equilibrium is rapid on the nMr time scale. The chemical shift versus concentration relations for C_nAP in benzene are shown in Figure 7, where the shift of the methyl protons of the propionate ion ($CH_3CH_2CO_2^-$) to higher magnetic fields is plotted as a function of surfactant

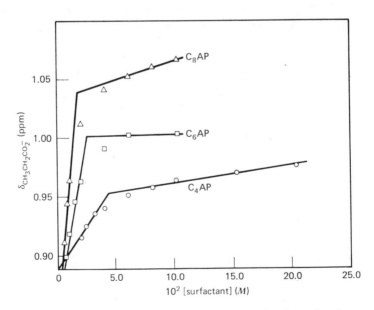

Fig. 7.   Chemical shifts of the $CH_3CH_2CO_2^-$ protons as functions of surfactant concentration in benzene at 33°C. (From ref. 27.)

TABLE 3

Micellar parameters in benzene and carbon tetrachloride[a]

Sample	In C₆H₆			In CCl₄		
	CMC[b], $10^2 M$	m	K, M^{1-m}	CMC[b], $10^2 M$	m	K, M^{1-m}
C_4AP	4.5–5.5	4	1×10^4	2.3–2.6	3	9×10^2
C_6AP	2.2–3.2	7	5×10^{12}	2.1–2.4	7	7×10^{11}
C_8AP	1.5–1.7	5	1×10^8	2.6–3.1	5	5×10^7
$C_{10}AP$	0.8–1.0			2.2–2.7	5	1×10^7
$C_{12}AP$	0.3–0.7			2.1–2.5	4	5×10^4

[a]From ref. 27.
[b]Probably the first figures were obtained from the ν vs. C plot, and the second figures from the ν vs. $1/C$ plot, although this is not clear in the original paper.

concentration. Similar relations were obtained for these surfactants in carbon tetrachloride. For each surfactant there is a break in the chemical shift at a concentration which corresponds to the CMC. The CMC values obtained from these and similar plots are tabulated in Table 3.

The observed chemical shift at a given concentration, ν, is given by rewriting equation la in the form

$$\nu = \nu_M + (\nu_m - \nu_M)\left[\frac{(CMC)}{C}\right] \tag{2}$$

The plots of ν against $1/C$ yield straight lines with intersections at the CMC's, and therefore these intersections also give the values of CMC (Table 3). (The CMC values obtained for C_8AP and $C_{12}AP$ in benzene are compared with the values 0.008 and 0.002 M, respectively, determined at 26°C by solubilization of water (26).) The CMC values in benzene decrease with increasing hydrocarbon chain length, n, whereas those in carbon tetrachloride are nearly independent of n and constant.

The plots of ν against $1/C$ also allow the estimation of the micellar shifts ν_M from their intercepts on the ordinate; these are given in Table 4 together with the monomeric shifts ν_m. For the methyl protons of the propionate ion, it is apparent that both the monomer and micelle resonances shift slightly to lower magnetic field as a function of increasing chain length in carbon tetrachloride; however, in benzene the resonances of the micellar protons move upfield while those of the monomer shift downfield. It is highly probable that the observed differences in shift of the other protons also involve both monomeric and micellar contributions that may or may not differ in sign depending upon the interactions involved.

Alkylammonium propionates in benzene and carbon tetrachloride form rather small micelles with their polar head-groups located in the interior. Available information on other alkylammonium carboxylate micelles in

TABLE 4

Chemical shifts of the monomeric and micellar propionic methyl
protons of C_nAP in ppm[a,b]

Sample	In C_6H_6		In CCl_4	
	ν_m	ν_M	ν_m	ν_M
C_4AP	0.868	0.987	1.547	1.498
C_6AP	0.830	1.027	1.543	1.494
C_8AP	0.826	1.079	1.530	1.492
$C_{10}AP$			1.525	1.478
$C_{12}AP$			1.504	1.456

[a] At 100 HMz and 33°C
[b] From ref. 27.

nonpolar solvents is in agreement with this structure; the aggregation number
of the micelle, m, is in the range 3 to 10 (26). It is interesting to note that the
values of m for a given surfactant are identical in benzene and in carbon
tetrachloride (Table 3). In contrast, the equilibrium constants, K, for micelle
formation for C_6AP or C_8AP having the same m are smaller in carbon
tetrachloride than in benzene (Table 3). Interpretation of the role of nonpolar
solvents in altering these micellar parameters will require additional information.

Fendler et al. have also made nMr studies of micellar properties of octylammonium carboxylates (from propionate to tetradecanoate homologs) in
benzene and carbon tetrachloride to see the effect of the carboxylate structure
(28). In benzene, the CMC values increase with increasing hydrocarbon
chain length of the surfactant carboxyl group, while in carbon tetrachloride
they are relatively independent of the hydrocarbon chain length. Values
of m for these surfactants are in the range 3 to 7, which is comparable to m
for alkylammonium propionates C_nAP (from butyl to dodecyl homologs).
Also, they have studied the effect of nonaqueous solvents having different
polarities on micelle formation of C_nAP with n of 4 to 12 (29). The CMC
values in each solvent decrease with increasing number of carbon atoms in
the alkyl chain, with the exception of carbon tetrachloride as seen in Table 3,
and for each surfactant they increase with increasing solvent polarity. There
are good relationships between log CMC and both the reciprocal dielectric
constant and solvent polarity parameter E_T.

Several structural features of these reversed micelles have been discussed
on the basis of the monomeric and micellar shifts, ν_m and ν_M, in different
solvents (29). Below the CMC, interactions exist between the alkyl groups of
the surfactant and the solvent, and between the polar head-groups of the
surfactant and the solvent. On aggregation, the polar groups pull together,
forming a cavity from which the solvent molecules are essentially excluded,

while the hydrocarbon tails remain in contact with the solvent. The extent of solvent penetration through the hydrocarbon chains of the micelle is governed by factors such as the volume of the micelle, the length of the hydrocarbon chain, and the properties of the solvent. The bulkier the solvent and the longer the hydrocarbon chain, the shorter the distance that the solvent can penetrate.

B. nMr Relaxation Times in Micellar Solutions

1. Changes in T_1 Values of Surfactant Protons through CMC

Studies of the spin–lattice and spin–spin relaxation times, T_1 and T_2, of species present in aqueous micellar solutions can provide information about the molecular motions of the species in these systems (21, 32–38). For example, Lawson and Flautt (35) have determined the T_1 values of various proton moieties of a nonionic surfactant, octyldimethylamine oxide $C_8H_{17}N(O)(CH_3)_2$ (abbreviated to C_8DAO), in D_2O by the progressive saturation method (39). The experimental T_1 values are given in Figure 8 as a func-

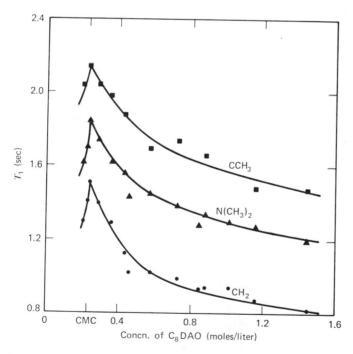

Fig. 8. T_1 values of C_8DAO as a function of concentration at 32°C. (From ref. 35.)

tion of the concentration of C_8DAO. All of the T_1 values show an apparent maximum near the CMC of the surfactant (40). The magnitudes of T_1 are in the order $T_{1,\,CH_2} < T_{1,\,N(CH_3)_2} < T_{1,\,CCH_3}$ at any given concentration.

The increase in T_1 (decrease in relaxation rate) with increasing concentration below the CMC is not fully understood. Lawson et al. explained the result as being due to changes in physical configuration of the hydrocarbon chains of the C_8DAO molecules in the monomer state (41). In contrast to this result, the measurement of T_1 by Clifford (33) for the alkyl CH_2 protons of sodium alkyl sulfates in D_2O showed that the T_1 values decrease with increasing concentration even below the CMC, as will be described later (cf. Figure 9). At concentrations greater than the CMC, C_8DAO exists in two

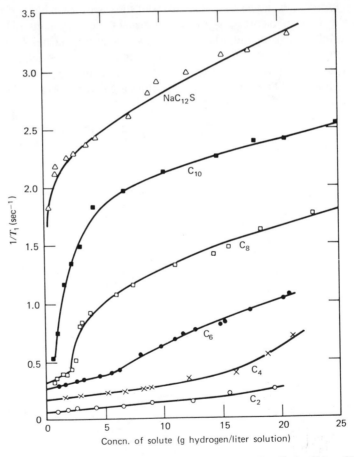

Fig. 9. $1/T_1$ values of CH_2 protons in D_2O solutions of C_2–C_{12} alkyl sulfates. (From ref. 33.)

TABLE 5

Values of $T_{1,m}$ and $T_{1,M}$ extracted from plots of C/T_1 vs.
$[C - (CMC)]^a$

Moiety	$T_{1,m}$ (sec)	$T_{1,M}$ (sec)
CH_2	2.1	0.75
$N(CH_3)_2$	2.3	1.1
CCH_3	2.4	1.4

^{a}From ref. 35.

forms, as monomers and micelles. The surfactant molecules are interchanging between these two environments at some relatively rapid rate. In such a system an average T_1 will be observed if $t_i \ll T_{1,i}$ and $t_j \ll T_{1,j}$, where t is the average residence time of a nucleus or molecule and the subscripts i and j refer to the states i and j (42). If the above conditions are fulfilled, then the observed T_1 is expressed in terms of the relaxation times of the monomeric and micellar species by the equation.

$$\left(\frac{1}{T_1}\right)_c = \frac{(CMC)/C}{T_{1,m}} + \frac{[C - (CMC)]/C}{T_{1,M}} \tag{3}$$

where C is the total concentration of the surfactant.

If equation 3 adequately describes the system and both $T_{1,m}$ and $T_{1,M}$ are independent of C, plots of C/T_1 against $[C - (CMC)]$ should be straight lines with intercepts of $CMC/T_{1,m}$ and slopes of $1/T_{1,M}$. In fact, plots of this type gave straight lines for all of the proton moieties. The values of $T_{1,m}$ and $T_{1,M}$ extracted from the plots are summarized in Table 5. The micellar values $T_{1,M}$ are about 1 sec less than the monomeric values $T_{1,m}$. The decrease in T_1 with micelle formation may reflect the restraint placed upon the molecules when incorporated into micelles. It is expected that internal chain motions (translational and rotational) would be hindered in the micellar species relative to the corresponding motions in the monomeric form. Another factor to be considered is the change in environment that a molecule undergoes during micelle formation. The intermolecular relaxation would be mainly caused by proton–deuteron interactions for the monomer whereas proton-proton interactions would occur mainly for the micelle, and this would lead to a relative decrease in T_1 for the micellar form.

2. T_1 Values of Water and Hydrocarbon Chain Protons in Ionic Micellar Solutions

In order to discuss the interaction between long chain ions and water, Clifford and Pethica (32) have measured the T_1 values of water protons in aqueous solutions of sodium alkyl sulfates $H(CH_2)_nOSO_3Na$ by means of the adiabatic rapid passage method (43).

TABLE 6

Effect of alkyl sulfates on $1/T_1$ of water protons at $32°C^a$

Chain Length	As Single Ions (sec^{-1}/mole/1000 g water)	As Micelles
C_2	0.04	—
C_4	0.07	—
C_6	0.10	0.09
C_8	0.14	0.11
C_{12}	—	0.16

^{a}From ref. 32.

For the solutions of the C_2–C_4 sulfates the relaxation rate $1/T_1$ of water protons varies linearly with concentration. For the solutions of the higher alkyl sulfates a change of the slope of the $1/T_1$ versus concentration curve is observed near the CMC. The slope of the curve below the CMC is a measure of the effect of the nonmicellized alkyl sulfate and sodium ions on water, and the slope above the CMC is a measure of the effect of the micelles on water. Table 6 lists the effect on the $1/T_1$ of water protons (per second per unit molality), calculated from the slopes of the curves.

The data for the monomeric alkyl sulfates given in Table 6 indicate that 1 mole of CH_2 increases the $1/T_1$ value of 1000 g of water by approximately 0.015/sec. The observed relaxation rate of water protons is assumed to be an average of the relaxation rates of protons of water molecules remote from and near to solute molecules:

$$\frac{1}{T_1} = \frac{1}{T_{1a}} \frac{N_a}{N_a + N_b} + \frac{1}{T_{1b}} \frac{N_b}{N_a + N_b} \qquad (4)$$

where T_{1a} and T_{1b} are the relaxation times, and N_a and N_b the number of moles of water protons near to and far from the solute molecules. If $1/T_{1b}$ is taken as being equal to the relaxation rate for protons in pure water (0.244/ sec at 32°C), then $1/T_1$ will be 0.259 (= 0.244 + 0.015)/sec for 1 mole of CH_2 groups in 1000g of water. By assigning reasonable numbers to N_a and N_b (32), $1/T_{1a}$ is calculated to be 0.58/sec, compared with $1/T_1 = 0.244$ for pure water. The alkyl chains, therefore, more than double the relaxation rate of protons in water molecules adjacent to them.

A hydrocarbon chain in aqueous solution probably affects the spin–lattice relaxation of neighboring water molecules in the following two ways: (1) Through its effect on hydrogen bonding in the water it could change the rates of relative motion of protons in water molecules. An increase in hydrogen bonding would result in an increase in $1/T_1$ of the water protons. (2) There could be direct dipole–dipole interaction between alkyl chain protons and water protons. The protons of a water molecule may approach more closely the protons in an alkyl chain, as compared to the distance between the protons of adjacent water molecules. The protons in an alkyl chain seem to be more effective for the spin–lattice relaxation than protons in neighboring water molecules. Thus, it is probable that an alkyl chain dissolved in water increases the relaxation rate of the water protons both by increasing the amount of structure in the water and, also, by direct dipole–dipole interaction, although the relative importance of these two effects cannot be estimated.

In the process of micelle formation there is a reduction of the effect of the alkylsulfate ions on the water molecules. However, a micelle still increases the relaxation rate of water molecules more than an equivalent number of isolated head groups. This is probably because of some electrostriction of the water molecules near the micelle. The double layer at the micelle surface affects the motion of adjacent water molecules so as to increase the relaxation rate of the water protons.

Clifford (33) has also measured the T_1 of CH_2 protons in the alkyl sulfates dissolved in D_2O and discussed the behavior of the protons in the hydrocarbon chains. Solutions in D_2O rather than in H_2O were used because the much lower magnetic moment of the deuteron as compared with the proton ensures that the dipole–dipole interaction between the solvent and the solute is far less in D_2O than in H_2O solutions, and hence the interaction between protons in the alkyl chain molecules is not obscured.

The $1/T_1$ values of CH_2 protons in the C_2–C_{12} sulfates are shown in Figure 9 as a function of concentration in D_2O, in which the concentration is expressed as gram hydrogen/liter solution. For the lower alkyl sulfates, the $1/T_1$ versus concentration curves are linear at low concentrations below the CMC. The results are given in Table 7 in terms of the slope and of the intercept on the $1/T_1$ axis. These data represent the behavior of single ions in

TABLE 7

Values of $1/T_1$ of single alkyl sulfates and alcohols at infinite dilution, and variation of $1/T_1$ with concentration[a]

	Sulfate (S) or Alcohol (A)	C_2	C_4	C_5	C_6	C_8	C_{10}	C_{12}
	S in D_2O	0.07	0.17	—	0.27	0.32	—	—
$1/T_1$, infinite	A in D_2O	0.07	0.13	0.17	0.25	—	—	—
dilution (sec^{-1})	A in CCl_4	(0.24)[b]	0.13	0.16	0.24	0.34	0.45	0.59
Variation of $1/T_1$	S in D_2O	0.008	0.012	—	0.026	0.05	—	—
with concn.	A in D_2O	0.002	0.005	0.008	—	—	—	—
(snec^{-1}/ (g hydrogen/l))	A in CCl_4	(−0.01)[b]	0.005	0.012	0.012	0.012	0.011	0.010

[a]From ref. 33.
[b]Association.

dilute solution. Similar data from CH_2 protons for alcohols dissolved in D_2O and CCl_4 are also given in Table 7 for comparison.

The observed relaxation rate may be made up from the sum of the intra- and intermolecular relaxation rates. The $1/T_1$ at infinite dilution will be $(1/T)_{intra}$ and will depend on the rotational Brownian motion and the internal motion of the molecule. As seen in Table 7, $(1/T)_{intra}$ for a given chain length is about the same for the sulfates in D_2O and for the alcohols in D_2O or CCl_4. This implies that the rates of molecular rotation and internal motion of the chains are much the same for long-chain molecules in aqueous and non-aqueous solutions.

In dilute solutions it can be assumed that $(1/T)_{intra}$ does not vary much with concentration; the slope of the $1/T$ versus concentration curve shown in Figure 9 is, therefore, due to the effect of concentration on $(1/T)_{inter}$. Thus, differences in the slopes of the curves for different solutes can be ascribed to differences in their translational molecular motion in solutions (44). Clifford (33) has pointed out the following possibility: The fact that the slopes for the C_6 and C_8 sulfates in D_2O (0.026 and 0.05 units) are much greater than those for the other solutes in D_2O or CCl_4 (at most 0.012 unit) could be due to an increase in the water structure near the alkyl chains with a consequent increase of the effective viscosity of the liquid through which they must move. This interpretation seems, however, inconsistent with the fact that the C_5 alcohol shows a smaller slope in D_2O than in CCl_4; the viscosity of D_2O is somewhat larger than the viscosity of CCl_4 even when the change of water structure caused by the alkyl chains is disregarded.

The formation of a micelle results in a change from a D_2O environment to a hydrocarbon environment for CH_2 protons in the alkyl chains, leading to a large increase in $1/T_1$. Clifford (33) has assumed that the relaxation rate for monomers $(1/T_1)_m$ varies with concentration in the same way below and above the CMC, and calculated the relaxation rates for micelles $(1/T_1)_M$ of the alkyl sulfates at various concentrations according to equation 3. The $(1/T_1)_M$ of CH_2 protons increases with concentration and also with chain length, probably due to changes in micellar structure (rather than changes in dipole–dipole interaction between micelles) as the concentration is increased.

On the other hand, Henriksson and Ödberg (45) have assumed that $(1/T_1)_m$ is equal to $1/T_1$ at the CMC and independent of concentration. The result for heptafluorobutyric acid in aqueous solutions, obtained in terms of equation 3 under this assumption, was that [19]F $(1/T_1)_M$ is independent of concentration (in contrast to the result obtained by Clifford); this implies that the micellar structure does not change with increasing concentration. A reexamination of the relaxation data of Clifford has been made by the present authors to calculate $(1/T_1)_M$ under the assumption $(1/T_1)_m = (1/T_1)_{CMC}$. The result thus obtained was similar in tendency to his result. This would support his interpretation, which is that the structure of the micelles of the alkyl sulfates changes with concentration. It should be noted here that Lawson and Flautt (35) have evaluated $(1/T_1)_M$ from a plot of C/T_1 versus $[C - (CMC)]$, assuming that both $(1/T_1)_m$ and $(1/T_1)_M$ are independent of concentration, as described already.

3. T_1 Values of Alkyl and Ethoxyl Chain Protons in Nonionic Micellar Solutions

The relaxation behaviors of both hydrophilic and hydrophobic moieties of nonionic micelles have been reported by Clemett (37) for n-decyl pentaoxyethylene ether $C_{10}H_{21}O(C_2H_4O)_5H$ (abbreviated as $C_{10}(EO)_5$), including a comparative sample of the monomethyl ether of polyethylene glycol. Figure 10 shows the T_1 values of the ethoxyl and alkyl methylene protons of 10% $C_{10}(EO)_5$ in D_2O and/or H_2O solution as a function of temperature. The slopes of these plots yield the activation energy $E = 5.3$ kcal/mole for the

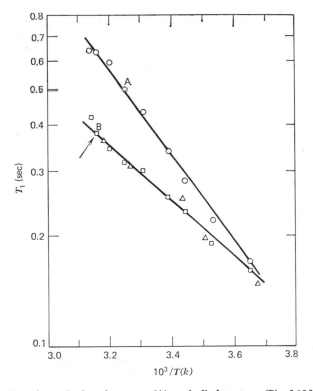

Fig. 10. T_1 values of ethoxyl protons (A) and alkyl protons (B) of 10% w/v $C_{10}(EO)_5$; $\bigcirc$, $\square$ in D_2O; $\triangle$ in H_2O. The arrow indicates the cloud point. (From ref. 37.)

ethoxyl chain, compared with 3.5 kcal/mole for the alkyl chain. (In this case E derived from the plot has no real physical meaning, but it is useful for making qualitative comparisons.)

It is interesting that T_1 of the alkyl protons is independent of whether the solvent is H_2O or D_2O. This suggests that the micellar core is not penetrated to any great extent by water molecules; otherwise the lower magnetogyric ratio of deuterium compared with hydrogen would increase T_1 in D_2O solution. In contrast, it has been found that water penetrates the micelles of sodium alkyl sulfates to some extent (33). The magnitude of T_1 of the alkyl chains in the micelle of $C_{10}(EO)_5$ is 0.32 sec at 34°C, which compares with 0.52 sec for n-dodecane at the same temperature. Thus chain mobility in the micelle core is slightly more restricted than in a pure hydrocarbon of similar chain length. The value of E for the alkyl chains of $C_{10}(EO)_5$ is 3.5 kcal/mole, which compares with 3.0 to 3.5 kcal/mole for sodium dodecyl sulfate (33), depending on concentration. Thus conditions in the cores of these particular nonionic and ionic micelles are similar, and both behave essentially as slightly viscous liquids.

Although we do not have the value of T_1 for the ethoxyl chains in H_2O, the behavior of the ethoxyl chains in the micelle may be discussed from data on the T_1 values of water protons in the solutions of $C_{10}(EO)_5$ and of the monomethyl ether of polyethylene glycol. These data, though not presented here, are consistent with the following model: At low temperatures the ethoxyl chains in the micelle are tightly coiled or tangled, with water molecules trapped among them. As the temperature rises, the chains become extended, progressively allowing the release of the water molecules. As seen in Figure 10, raising the temperature causes T_1 of ethoxyl chains in D_2O to increase much more rapidly than that of the alkyl chains. This can also be explained if the ethoxyl chains extend with increasing temperature, since the extension would increase the average distance between protons in the chains, and would produce an increase in T_1 in addition to that resulting from the greater chain mobility.

Corkill et al. (21) have measured T_1 of water protons in aqueous solutions of dodecyl polyoxyethylene ether $C_{12}(EO)_6$; T_1 decreases with increasing solute concentration at a given temperature, and it increases with increasing temperature at a given concentration. These results may be related to the behavior described above for the alkyl and ethoxyl chains in the nonionic surfactant. Recently Podo et al. (34) also studied the structure and hydration of polyoxyethylene-type nonionic micelles by measuring T_1 values of the surfactant and water protons. Their conclusions regarding the lack of penetration of water into the micelle core and about the temperature dependence of T_1 for the ethoxyl chain agree completely with the results of Clemett (37) described above.

4. T_2 Values in Micellar Solutions

Nuclear magnetic resonance relaxation times are related to correlation times τ_c for molecular motions, as shown in Figure 11 (11). The changes in T_1 and T_2 as τ_c increases from the value found for liquids ($\tau_c \simeq 10^{-11}$ sec) to values typical of solids ($\tau_c > 10^{-4}$ sec) have been calculated for both isotropic (46) and anisotropic (47) motions. With $\tau_c \simeq 10^{-11}$ sec for normal liquids, both T_1 and T_2 have substantially the same value. As τ_c increases up to $1/(2\pi\nu_0)$ (ν_0 = nMr frequency), T_1 and T_2 decrease, with T_1 becoming larger than T_2 at τ_c near $1/(2\pi\nu_0)$. For τ_c values above $1/(2\pi\nu_0)$, T_1 becomes larger and T_2 continues to decrease. These changes in T_1 and T_2 can be used to detect changes in molecular motions.

In normal micellar solutions, which are usually not viscous, the τ_c values are relatively small and hence the T_1 and T_2 values are virtually equal to each other, as are those for normal liquids. In mesomorphic phases appearing in surfactant-containing systems, however, τ_c values are often rather large compared with τ_c for micellar solutions, and T_1 generally differs from T_2. In such cases useful information regarding molecular motions will be obtained from simultaneous measurements of T_1 and T_2, becuase the two relaxation times reflect different kinds of molecular motions. For instance, Tiddy (48)

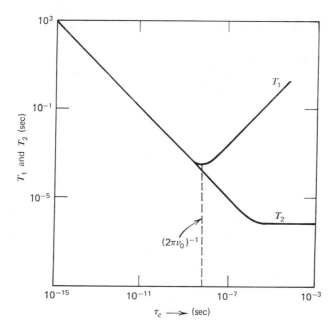

Fig. 11.   Relaxation times, T_1 and T_2, as a function of correlation time τ_c. (From ref. 11.)

has examined T_1 and T_2 values in the lamellar phases of the system sodium caprylate/decanol/water and has discussed the molecular motions of both the amphiphilic and the water molecules. A few good examples of studies about T_1 and T_2 values in mesomorphic phases are described in Section III.B.

Recently, viscoelastic liquids have been found in rather dilute aqueous solutions ($> 90\%$ water) of mixtures of sodium dodecyl sulfate $NaC_{12}S$ and octyltrimethylammonium bromide C_8TAB (49), and of $NaC_{12}S$ and a zwitterion surfactant (50). For instance, in the $NaC_{12}S/C_8TAB$/water system containing approximately 5% surfactants ($NaC_{12}S + C_8TAB$), one liquid, two liquid, and one liquid plus one liquid–crystal phases are observed, depending on the mole ratio of the two surfactants; liquids with different viscoelastic properties are found in the $NaC_{12}S$-rich region. The proton T_1 values obtained from the one-liquid D_2O solutions were 0.5 to 1.0 sec, whereas T_2 values were 0.5 to 0.025 sec and shortest in the composition range 66 to 77 mole $\%$ $NaC_{12}S$ where high viscoelasticity was observed.

The measurement of the signal widths of high resolution resonances is effective when T_2 varies widely, as in the case here, because proton resonances from different groups can be distinguished with this technique. The T_2 measured by spin echo is related to the signal width of resonance, $\Delta\nu_{1/2}$, by the relation $\Delta\nu_{1/2} = \pi T_2^{-1}$. Figure 12 shows the variation of nMr signal widths with $NaC_{12}S/C_8TAB$ ratio in the region where differential broadening of resonance was observed; it can be seen that a dramatic change in signal widths occurs at ca. 77 mole $\%$ $NaC_{12}S$. The change is greater at the head group than at the terminal methyl group, which indicates that in the restricted region shown in Figure 12 the head group is more immobilized than the interior of the micelle.

In the normal micellar liquids, $NaC_{12}S$ and C_8TAB micelles have $T_1 \simeq T_2 \simeq 1$ sec, showing fast molecular motion. If the addition of C_8TAB to $NaC_{12}S$ micelles caused a gradual reduction in molecular motion, then T_1 and T_2 would be reduced by similar amounts until T_1 passed its minimum values of about 20 m sec. This is not the case here, since T_1 is always in the range 0.5 to 1.0 sec while T_2 changes over the range 0.5 to 0.025 sec. It can be concluded that there is a discontinuous change in motion at composition $\leq$ ca. 77 mole $\%$ $NaC_{12}S$. There results and the structure of the viscoelastic solutions can be best explained by the occurrence of both cylindrical (in which $T_2 < T_1$) and spherical (in which $T_2 \approx T_1$) micelles, although the possibility that the viscoelasticity is due to the presence of a microemulsion cannot be excluded.

Similar viscoelastic behavior was found in the system of $NaC_{12}S$/hexadecyldimethylammoniopropane sulfonate $C_{16}H_{33}N^+(CH_3)_2 - C_3H_6OSO_2^-$/water (50). The structure of the viscoelastic solutions and the results for the relaxation times T_1 and T_2 were also interpreted by assuming the existence

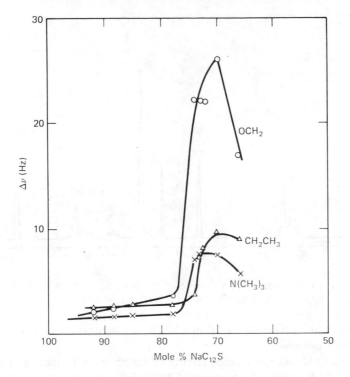

Fig. 12.   Variation of nMr signal-width with $NaC_{12}S/C_8TAB$ ratio. (From ref. 49.)

of both normal spherical micelles and cylindrical micelles in equilibrium. In the latter case, the possibility for any viscoelasticity due to the presence of a microemulsion is eliminated because there is no two-liquid region in this system.

C.   Mixed Micelles of Two Different Surfactants

1.   Micelle Composition as Functions of Concentration and Mixing Ratio

When two different surfactants are present in solution, mixed micelles composed of the two species are formed; their composition depends on concentration and mixing ratio, but simple micelles composed of a single species are not formed (51–53). The nMr spectral behavior of mixtures of two surfactants is essentially the same as the behavior of a single surfactant, described in the preceding section, to the extent that rapid exchange of surfactant molecules takes place between the aqueous and micellar phases. Thus, it will be possible to estimate the composition of mixed micelles of

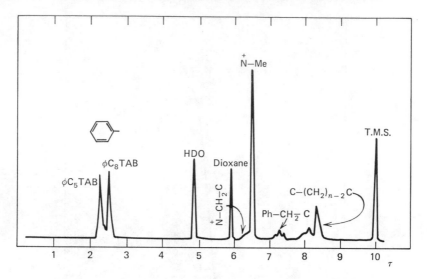

Fig. 13.   The nMr spectrum of a mixture of $\phi C_5 TAB$ and $\phi C_8 TAB$ in D_2O: total concentration, 15g/100ml; $\phi C_5 TAB / \phi C_8 TAB = 1/1$. (From ref. 54.)

two different surfactants as functions of concentration and mixing ratio from their chemical shifts.

Figure 13 shows a typical spectrum of an equimolar mixture of $\phi C_5 TAB$ and $\phi C_8 TAB$ obtained above the CMC of the mixture (54). The phenyl

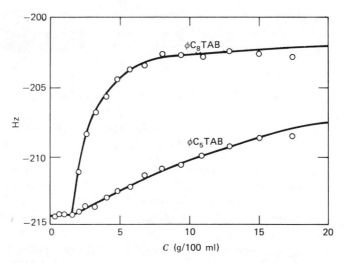

Fig. 14.   The shifts of phenyl proton signals as a function of the total concentration: $\phi C_5 TAB/\phi C_8 TAB = 1/1$. (From ref. 54.)

proton signal, which is a sharp singlet below the CMC, splits into two peaks in the concentration region above the CMC. The peaks at the higher and the lower fields are due to the phenyl protons of $\phi C_8 TAB$ and $\phi C_5 TAB$, respectively. The shifts of these phenyl proton signals with reference to the 1,4-dioxane peak (internal standard) are plotted in Fugure 14 as a function of the total concentration. Similar plots were obtained for the mixtures with mole ratios of 3:1 and 1:3. With increasing concentration both phenyl proton signals approach asymptotically definite but different positions for different mole ratios. The values of the two asymptotes may be regarded as the chemical shifts of phenyl protons of $\phi C_5 TAB$ and of $\phi C_8 TAB$, where both are in the mixed micelle whose composition is the same as the mixture.

In the mixed solutions of the two surfactants, the micelle composition changes with concentration and approaches the mole ratio of the mixture at sufficiently high concentrations. The concentrations and mole fractions of each surfactant in the bulk water and micellar phases at a given mixing ratio can be calculated as a function of the total concentration by successive approximation (54) using the observed chemical shifts ν, ν_m, and ν_M.

As an example, the results for the mixture with mole ratio 1:1 are shown in Figure 15, in which C is the concentration, X the mole fraction, and sub-

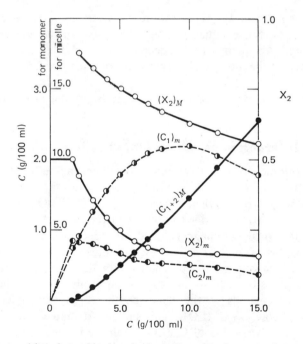

Fig. 15.   Composition changes in the mixed micelle and in the water phase as a function of total concentration: $\phi C_6 TAB/\phi C_8 TAB = 1/1$. (From ref. 54.)

scripts 1 and 2 refer to $\phi C_5 TAB$ and $\phi C_8 TAB$, respectively. Similar curves have been obtained for the mixtures with mole ratios 1:3 and 3:1, although they are not shown.

Since the surfactants, $\phi C_5 TAB$ and $\phi C_8 TAB$, are homologs and differ only in the length of the alkyl chain, we may describe the general features of surfactant mixtures evident from the curves in Figure 15 (and similar curves) as follows:

1. The concentration $C_{2,m}$ of the more lipophilic component increases with increasing total concentration C until the CMC is reached, and thereafter decreases monotonically.

2. The concentration $C_{1,m}$ of the more hydrophilic component continues to increase to a much higher concentration, then falls off gradually because of the volume effects of micelles and of the bound water captured by the micelles.

3. The concentration $C_{1+2,M}(= C_{1,M} + C_{2,M})$ is zero below the CMC and increases almost linearly with increasing C above the CMC.

4. The mole fraction $X_{2,M}$ of the more lipophilic component in a micelle decreases from its peak value at the CMC and approaches the mole fraction of the mixture.

5. The mole fraction $X_{2,m}$ of the more lipophilic component in the water phase must be, of course, equal to the mole fraction of the mixture until the CMC is reached. Above the CMC, it decreases to a definite value.

6. The CMC of the mixture lies somewhere between the CMC's of the two components.

7. The extent of the above features depends on the mole ratio of the mixture.

2. Interaction of Surfactants in Mixed Micelles

When an anionic surfactant, sodium p-octylbenzene sulfonate $C_8H_{17} - \phi - SO_3Na$ (abbreviated as $NaC_8\phi S$), is added to a solution of a nonionic surfactant, dodecyl polyoxyethylene ether $C_{12}H_{25}O(CH_2CH_2O)_pH$ (abbreviated as $C_{12}(EO)_p$), the nMr signal due to protons of the polyoxyethylene chain of $C_{12}(EO)_p$ shifts to higher magnetic field and becomes broader (55). The broadening in this case is not caused by an increase in viscosity but by the interaction of the polyoxyethylene chains with the $NaC_8\phi S$ molecules, because the sharpness of the other proton signals is unchanged. The extent of the upfield shift depends on the chain length of polyoxyethylene and the mixing ratio of $NaC_8\phi S/C_{12}(EO)_p$. This shift is attributed to an interaction of the polyoxyethylene chain of $C_{12}(EO)_p$ with the benzene ring of $NaC_8\phi S$, since the addition of sodium dodecyl sulfonate $C_{12}H_{25}SO_3Na$ to the $C_{12}(EO)_p$

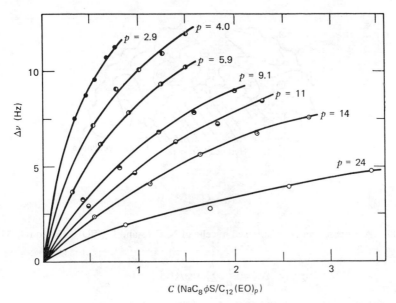

Fig. 16. The upfield shift $\Delta\nu$ of the polyoxyethylene proton signal plotted against C $(NaC_8\phi S/C_{12}(EO)_p)$ for $C_{12}(EO)_p$ with different p. (From ref. 55.)

solution causes no shift and no broadening of the polyoxyethylene signal. As further evidence for this interaction, one may cite a quite similar phenomenon observed when polyethylene glycol is mixed with benzene; the polyoxyethylene signal shifts to the higher magnetic field and the benzene signal to the lower field.

The curves of upfield shift $\Delta\nu$ versus concentration C for $C_{12}(EO)_p$ with different p are shown in Figure 16, where C is expressed as the mole ratio $NaC_8\phi S/C_{12}(EO)_p$ while the concentration of $C_{12}(EO)_p$ is kept constant at 5 % (by wt/vol). When compared at a constant C, $\Delta\nu$ decreases with increasing p. In order to discuss the results quantitatively, the mathematical treatment given below has been used for the data of Figure 16. The $C_{12}(EO)_p$ and $NaC_8\phi S$ molecules will form a mixed micelle, a schematic model of which is illustrated in Fugure 17. The benzene rings are probably located in a region between the hydrocarbon core and the polyoxyethylene shell. If this is the case, the result shown in Figure 16 suggests that in the mixed micelle, the influence of the benzene ring of $NaC_8\phi S$ cannot extend to the entire range of the polyoxyethylene chain when the latter is sufficiently long. If one assumes that the ith oxyethylene unit counted from the alkyl group is affected by a shift value of $\Delta\nu_i$, then $\Delta\nu$ is proportional to $(1/p)\sum_{i=1}^{p}\Delta\nu_i$ at a definite C. Under this and additional assumptions, equation 5 has been derived (55).

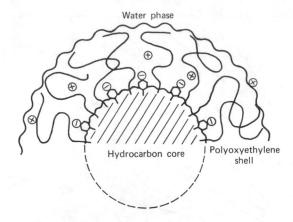

Fig. 17.   A proposed model of a mixed micelle of $NaC_8\phi S$ and $C_{12}(EO)_p$. (From ref. 55.)

$$\Delta\nu = \frac{1}{p} \sum_{i=1}^{p} \Delta\nu_i f(C) \tag{5}$$

$$f(C) = 1 - e^{-qc}$$

where $f(C)$ is a function related to C, and q is the number of $C_{12}(EO)_p$ molecules influenced by one $NaC_8\phi S$ molecule in the mixed micelle. Since the quantity $(1/p) \sum_{i=1}^{p} \Delta\nu_i$ expresses a saturation value, $\Delta\nu_{sat}$, of $\Delta\nu$ at sufficiently high C, that is,

$$\Delta\nu_{sat} = \frac{1}{p} \sum_{i=1}^{p} \Delta\nu_i \tag{6}$$

equation 5 is rewritten in the form

$$\Delta\nu = \Delta\nu_{sat}(1 - e^{-qc}) \tag{7}$$

From equation 7, one can easily derive

$$\Delta\nu_{at\,C+h} = e^{-qh}\Delta\nu_{at\,C} + (1 - e^{-qh})\Delta\nu_{sat} \tag{8}$$

which holds for any arbitrary constant h.

According to equation 8, the plot of $\Delta\nu_{at\,C+h}$ versus $\Delta\nu_{at\,C}$ should give a straight line, thus allowing one to evaluate $\Delta\nu_{sat}$ and q from its slope and intercept. Figure 18 shows the $\Delta\nu_{at\,C+h}$ versus $\Delta\nu_{at\,C}$ plots which have been obtained from the $\Delta\nu$ versus C curves given in Figure 16 by taking $h = 0.1$. The values of $\Delta\nu_{sat}$ and q calculated from the straight lines in Figure 18 are summarized in Table 8.

From equation 6, the shift value $\Delta\nu_p$ can be expressed as

$$\Delta\nu_p = p \cdot (\Delta\nu_{sat})_{for\,p} - (p - 1) \cdot (\Delta\nu_{sat})_{for\,p-1} \tag{9}$$

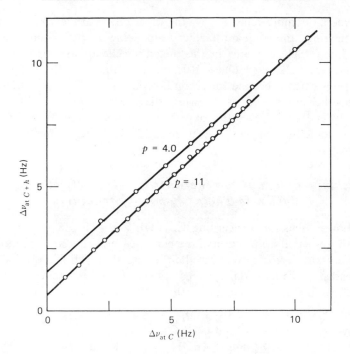

Fig. 18. Plots of $\Delta\nu_{at\,C+h}$ against $\Delta\nu_{at\,C}$ for $C_{12}(EO)_p$ with p of 4.0 and 11; $h = 0.1$. Similar straight lines were also obtained for other $C_{12}(EO)_p$. (From ref. 55.)

The values of $\Delta\nu_p$ calculated for the present system in terms of equation 9 (from the $\Delta\nu_{sat}$ versus. p curve for the system I, shown in Figure 20) are as follows: $(\Delta\nu_1 + \Delta\nu_2 + \Delta\nu_3)/3 = 15.8$, $\Delta\nu_4 = 13.1$, $\Delta\nu_5 = 11.7$, $\Delta\nu_6 = 10.4$, $\Delta\nu_7 = 9.0$, $\Delta\nu_8 = 7.7$, $\Delta\nu_9 = 6.3$, $\Delta\nu_{10} = 4.8$, $\Delta\nu_{11} = 3.6$ Hz; the values for $\Delta\nu_{12}$ to $\Delta\nu_{24}$ fluctuate in the range of $+ 2 \sim - 2$ Hz. These values show that the

TABLE 8

Values of $\Delta\nu_{sat}$ and q^a

Sample $C_{12}(EO)_p$	$\Delta\nu_{sat}$ (Hz)	q (molecules)
$p = 2.9$	15.9	1.6
4.0	14.3	1.2
5.9	13.5	1.1
9.1	12.0	0.62
11	10.9	0.58
14	8.5	0.54
24	4.6	0.55

^{a}From ref. 55.

oxyethylene units more remote from the alkyl group are less influenced by the benzene rings, the tenth and farther units being practically uninfluenced. If the polyoxyethylene chains in the mixed micelle are assumed to stretch straight into the bulk water phase, it is not probable that the effect of the benzene rings extends beyond the second or third unit. Therefore, the above result may be interpreted as showing that because of flexibility of the polyoxyethylene chains even the oxyethylene units remote from the alkyl group can sometimes approach near enough the benzene rings to interact with them.

3. Interaction of Surfactants in Mixed Micelles in Relation to Their Chemical Structures

The above discussion concerning $C_{12}(EO)_p$ and $NaC_8\phi S$ in their mixed micelles suggests that the position of the benzene ring in the $NaC_8\phi S$ molecule is an important factor governing their interaction. This study has been then extended to the following systems in order to better elucidate the interactions of two different surfactants in their mixed micelles (56).

System   I: $NaC_8\phi S$ and $C_{12}(EO)_p$
System   II: $NaC_4\phi C_4 S^*$ and $C_{12}(EO)_p$
System III: $Na\phi C_8 S^*$ and $C_{12}(EO)_p$
System   IV: $NaC_8\phi S$ and $NaC_{12}(EO)_p S^*$

a. Systems I, II, and III.

Addition of $NaC_4\phi C_4 S$ or $Na\phi C_8 S$ to $C_{12}(EO)_p$ solutions also causes an upfield shift of the polyoxyethylene signal, as in the case of $NaC_8\phi S$. Figure 19 shows the plots of $\Delta\nu$ versus C for systems I, II, and III, with the concentration of $C_{12}(EO)_6$ kept constant at 5%. The magnitudes of $\Delta\nu$ in system II and especially in system III are much smaller than the $\Delta\nu$ observed in system I, when compared at an equal value of C. This result indicates that in the mixed micelle, the benzene ring of $NaC_4\phi C_4 S$ or $Na\phi C_8 S$ which is located in the center or at the end of the hydrocarbon chain can also interact with the polyoxyethylene chain of $C_{12}(EO)_p$, although the degree of the interaction is not as strong as in the case of $NaC_8\phi S$. It also suggests that the hydrocarbon chains are mobile or flexible in the micelle. The interior of the micelle behaves as a liquid hydrocarbon because the interaction will not take place unless the benzene rings approach near the polyoxyethylene shell of the micelle. The

---

*$NaC_4\phi C_4 S$: $C_4H_9-\phi-(CH_2)_4SO_3Na$
$Na\phi C_8 S$: $\phi-(CH_2)_8SO_3Na$
$NaC_{12}(EO)_p S$: $C_{12}H_{25}O(C_2H_4O)_pSO_3Na$

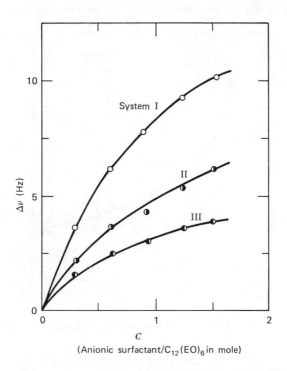

Fig. 19. Curves of $\Delta\nu$ vs. C for systems I, II, and III with $C_{12}(EO)_6$. (From ref. 56.)

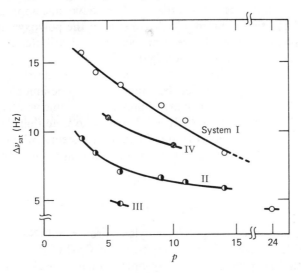

Fig. 20. Values of $\Delta\nu_{sat}$ plotted against p for systems I, II, III, and IV. (From ref. 56.)

101

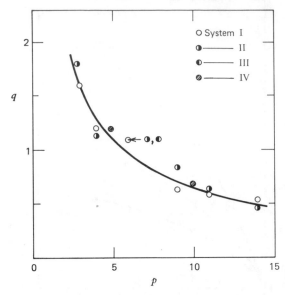

Fig. 21.   Values of q plotted against p for systems I, II, III, and IV. (From ref. 56.)

fact that $\Delta\nu$ at a given C decreases in the order of system I > II > III may be interpreted as showing that the number of benzene rings existing in the poly-oxyethylene shell decreases in this order. A thermodynamic approach, based on the assumption that a mixed micelle consists of the polyoxyethylene and hydrocarbon phases and that benzene molecules are distributed in these two phases, also supports this explanation (56).

Figures 20 and 21 show the saturation value of the upfield shift, $\Delta\nu_{sat}$, and the number of the nonionic molecules influenced by one anionic molecule, q, plotted as a function of p, which are obtained in the same way as described in Section II.C.2 (see equation 8). It is interesting to note that the value of $\Delta\nu_{sat}$ is much smaller in systems II and III than in system I when compared at constant p, whereas the value of q depends on p in a similar way in all three systems.

b. System IV.

The nMr spectrum of the polyoxyethylene protons of $NaC_{12}(EO)_pS$ in water is somewhat different from the spectrum of $C_{12}(EO)_p$. With $C_{12}(EO)_p$, the polyoxyethylene protons give only one resonance peak. With NaC_{12}-$(EO)_pS$, on the other hand, the signal of the first oxyethylene unit next to the ionic head $-OSO_3^-$ is distinguishable from the signal of the other oxyethylene units. On the addition of $NaC_8\phi S$ to $NaC_{12}(EO)_pS$ solutions, the signal of the first oxyethylene unit shows no appreciable change and remains at the same position, while the signals of other oxyethylene units shift to higher magnetic field. In this mixed micelle, the benzene ring of $NaC_8\phi S$ cannot exert an influence on the first oxyethylene unit of $NaC_{12}(EO)_pS$, probably because of the steric effect of the ionic head $-OSO_3^-$ or the repulsion between the ionic heads of the two surfactants.

In Figures 20 and 21 are given the values of $\Delta\nu_{sat}$ and q for system IV with p of 5 and 10; in this case, the effective number of oxyethylene units for the upfield shift is p-1. The dependence of q on p in system IV is similar to that in systems I, II, and III. On the other hand, the value of $\Delta\nu_{sat}$ in system IV is smaller than the value in system I when compared at an equal p, as seen in Figure 20. For the mixed micelle in system IV, the polyoxyethylene chain of $NaC_{12}(EO)_pS$ is considered to be more extended than the chain of C_{12}-$(EO)_p$ in system I, owing to the repulsion of the attached charges.

c. Solubilization Behavior of Systems I, II, and III.

The interaction of the polyoxyethylene chains with the benzene rings in the above-mentioned mixed micelles is expected to affect the solubilization behavior of the micelles (57, 58). Solubilizing capacities, S, for an oil-soluble dye, Yellow OB, were measured at 30°C in systems I, II, and III (with $p = 9$), and for comparison, in the mixture of $C_{12}(EO)_9$ and sodium decyl sulfonate $C_{10}H_{21}SO_3Na$ (abbreviated as $NaC_{10}S$), which have no benzene rings in the molecules. The synergistic effect in solubilization, ΔS, was then defined as the increase in solubilizing capacity caused by the interaction. The calculation of ΔS (see ref. 57 for details) for systems I, II, and III was based on the assumptions that (a) $NaC_{10}S$ does not interact with $C_{12}(EO)_9$, (b) the polyoxyethylene shell of the $C_{12}(EO)_9$ micelle is made less compact by mixed micelle formation with anionic surfactant, and (c) $\Delta S = 0$ for the system of $NaC_{10}S/C_{12}(EO)_9$.

In Figure 22, the values of ΔS for each system are plotted as a function of the concentration of anionic surfactant, C; the concentration of $C_{12}(EO)_9$ is kept constant at $5 \times 10^{-3}M$. A remarkable synergistic effect in solubilization is observed in system I, which may be attributed to the strong interaction of the benzene rings with the polyoxyethylene chains. On the othre hand, the ΔS values are smaller in systems II and III than in system I at all concentrations, and no appreciable difference is seen between these two systems. Probably this can be explained by the weaker interaction, although the interaction observed by nMr measurement is slightly stronger in system II than in system III.

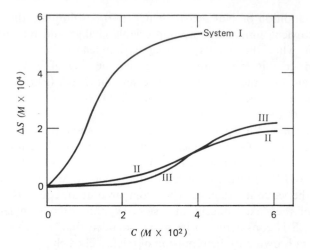

Fig. 22. Synergistic effect ΔS in solubilization as a function of anionic surfactant concentration, C; the concentration of $C_{12}(EO)_9$ is kept constant at $5 \times 10^{-3}M$. (From ref. 57.)

D. Solubilization of Organic Compounds

1. Solubilization of Benzene

Surfactant micellar solutions can solubilize organic compounds of relatively low molecular weights, which are insoluble or only slightly soluble in water. In this section the solubilization mechanism of benzene in micellar solutions on the basis of the data obtained by nMr measurements (23) will be discussed first.

When benzene molecules are solubilized in a surfactant solution above the CMC, some of the molecules are in the micelles and the others are in the bulk water; they are constantly exchanged between these two environments. In principle, the chemical shift of the benzene protons will be different in the two environments and, therefore, two sharp signals will be observed at different positions in the nMr spectrum if the exchange is slow. However, only one sharp signal is observed in the spectrum of the benzene saturated in the solution of sodium dodecyl sulfate (abbreviated as $NaC_{12}S$), indicating that the exchange of benzene molecules between the micellar and water phases is very rapid. The signal shifts to higher magnetic field with increasing concentration of $NaC_{12}S$ in the solution saturated with benzene, while the signal width changes little. Figure 23 shows the shift of the benzene signal, with reference to the dioxane peak (internal standard), as a function of $NaC_{12}S$ concentration in the D_2O solution saturated with benzene. The benzene signal does not shift until the $NaC_{12}S$ concentration reaches a value of about

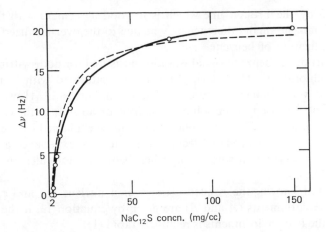

Fig. 23. The shift of the benzene signal as a function of $NaC_{12}S$ concentration in D_2O solutions of $NaC_{12}S$ saturated with benzene: ○ shift observed, --- shift calculated from equation 12. (From ref. 23.)

0.2%, corresponding to the CMC of $NaC_{12}S$, at which an abrupt shift is observed. The shift increases asymptotically up to about 20 Hz with further increase in concentration.

The shift–concentration curve shown in Figure 23 has been analyzed making the customary assumptions (23). The benzene content b (per unit volume of solution) in the surfactant solution is given by

$$b = b_w(1 - \bar{v}C) + b_M(C - C_0) \tag{10}$$

where b_w is the solubility of benzene in g/ml water, b_M the saturation amount of solubilized benzene in g/g micellar surfactant, C and C_0 the concentrations of the total and monomeric surfactant expressed in g/ml solution, respectively, and $\bar{v}$ the partial specific volume of the surfactant. If the exchange of benzene is sufficiently rapid, as in the present case, the signal above the CMC will appear at a position ν:

$$\nu = \frac{\nu_w b_w (1 - \bar{v}C) + \nu_M b_M (C - C_0)}{b_w (1 - \bar{v}C) + b_M (C - C_0)} \tag{11}$$

where ν_w and ν_M are the signal positions of benzene in water and in the micelles, respectively. Then, the shift x measured from ν_w is given by

$$x = \nu - \nu_w = \frac{b_M (C - C_0)}{b_w (1 - \bar{v}C) + b_M (C - C_0)} (\nu_M - \nu_w) \quad (C > \text{CMC})$$
$$= 0 \qquad\qquad\qquad\qquad\qquad\qquad\qquad\quad (C \leq \text{CMC}) \tag{12}$$

Inserting appropriate values into equation 12, one obtains the curve shown in

Figure 23 by the broken line which is in good agreement with the experimental data. The small difference may be due to the over-idealized model for the solubilization of benzene.

The shift of the benzene signal was also measured for dodecyltrimethylammonium chloride, $C_{12}H_{25}N(CH_3)_3Cl$, solutions saturated with benzene (23), which was very similar to the curve shown in Figure 23. However, when cyclohexane was used as the solubilizate, little or no shift of the cyclohexane signal was observed. Thus the upfield shift can be explained by the view that the protons of the benzene molecules in the micelle experience a large diamagnetic anisotropy arising from other benzene molecules in the same micelle (59, 60).

As is well known, the residence times of a nucleus (τ_A and τ_B) in two different environments (A and B) are given by equation 13, if the exchange between the two environments is relatively rapid (1):

$$\frac{1}{T_2'} = \frac{p_A}{T_{2A}} + \frac{p_B}{T_{2B}} + 4\pi^2 p_A^2 p_B^2 (\nu_A - \nu_B)^2 (\tau_A + \tau_B) \qquad (13)$$

where p_A and p_B, ν_A and ν_B, T_{2A} and T_{2B} represent, respectively, the fractional populations, the signal positions, and the transverse relaxation times for the states A and B, while T_2' is the effective transverse relaxation time for the observed signal. Equation 13 can be transformed into equation 14 by making use of the relations $p_A/p_B = \tau_A/\tau_B$, $p_A + p_B = 1$, $1/T_2' = \pi h$, $1/T_{2A} = \pi h_A$ and $1/T_{2B} = \pi h_B$, where h, h_A, and h_B represent the half-height-widths of the observed signal and of the signals for the states A and B.

$$\tau_A = \frac{h - (h_A p_A + h_B p_B)}{4\pi p_A p_B^2 (\nu_A - \nu_B)^2} \qquad (14)$$

The half-height-widths obtained by slow scanning measurements are listed in Table 9, along with the fractional populations in the micellar environment,

TABLE 9

Half-height widths of the benzene signal in the $NaC_{12}S$ solutions saturated with benzene[a]

$NaC_{12}S$ Concn. $\times 10^2$ (g/ml)	Shift of C_6H_6 signal (Hz)	Half-height width[b]	p in micellar environment
0	0	$0.54 = h_w$	0
0.45	4.0	0.56	0.2
0.70	6.5	0.56	0.325
1.29	9.5	0.60	0.475
15.0	20.0	$0.50 \approx h_M$	≈ 1.0

[a]From ref. 23.

[b]h_w and h_M: half-height widths for benzene in water and micellar phases, respectively.

which were calculated from $(\nu - \nu_w)/(\nu_M - \nu_w) = x/20$. Inserting these values into equation 14, one obtains a residence time of the order of 10^{-4} sec or less. This residence time seems to be reasonable when compared with the value estimated from the Einstein equation (23) or with reported values (61).

2. Solubilization of Aromatic Alcohols and Phenols

As described in the above sections, organic compounds having a benzene ring in the molecule are suitable for nMr spectral studies. In this section the mode of solubilization of aromatic alcohols and phenols in micellar solutions is discussed on the basis of the nMr spectra of these compounds solubilized by a surfactant $NaC_{12}S$; special attention is given to the chemical shifts of their phenyl groups (62). The alcohols and phenols used as solubilizates have the chemical structures:

$$\phi - (CH_2)_n OH \qquad (\phi C_n OH), \quad n = 1 \sim 3$$
$$CH_3 - \phi - (CH_2)_n OH \quad (C_1 \phi C_n OH), \; n = 1 \text{ or } 2$$
$$H(CH_2)_n - \phi - OH \qquad (C_n \phi OH), \quad n = 1 \sim 3$$

The typical nMr spectra of phenyl protons for $C_2\phi OH$, $C_1\phi OH$, $C_1\phi C_1 OH$, and $\phi C_2 OH$ saturated in water and in the surfactant solution are given in Table 10. In the surfactant solution, the signals of their phenyl protons shift to higher magnetic fields owing to the solubilization of these materials in the micellar medium. The pattern of phenyl proton signals and the extent of their shifts to higher fields depend on the chemical structure of solubilizates and the number n, that is, the number of methylene units in the molecule. According to their signal patterns, all the solubilizates examined can be classified into four types, as shown in Table 10.

The signals of the H_A and H_B protons of $C_2\phi OH$ (type I), for example, remain at the same positions as those in water until the surfactant concentration reaches the CMC. Above the CMC, the shifts of the H_A and H_B signals to higher fields are promoted by the increase in concentration of $C_2\phi OH$. The spectral behavior in solubilization of other solubilizates is similar in tendency to that observed for $C_2\phi OH$, although the chemical shifts of their phenyl protons are different. $C_1\phi OH$ (type II) shows the H_A and H_B proton signals separately; each signal splits into two peaks in water and overlaps to a single peak in the surfactant solution. $C_1\phi C_1 OH$ (type III), on the other hand, shows only a single peak of the phenyl protons in water, but it splits into four peaks in the surfactant solution. The phenyl proton signal for $\phi C_2 OH$ (type IV) is a singlet in both water and the surfactant solution. Table 11 summarizes the chemical shifts of phenyl protons for all the solubilizates in water and in a 5.0% $NaC_{12}S$ solution. (Incidentally, it has been

TABLE 10

nMr spectra of phenyl groups of $C_2\phi OH$, $C_1\phi OH$, $C_1\phi C_1OH$, and ϕC_2OH saturated in water and in 5.0% $NaC_{12}S$ solution*

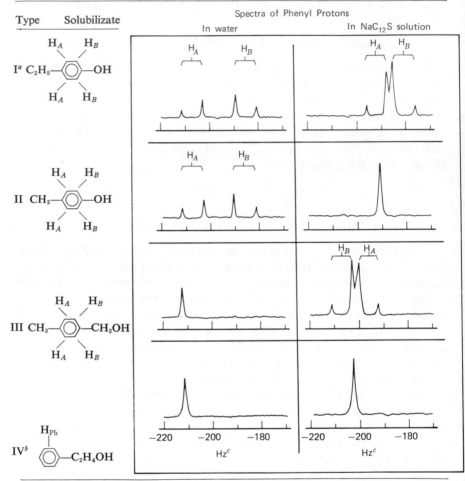

*From ref. 62.

[a] $C_3\phi OH$ shows a spectrum similar to the type I.

[b] ϕC_1OH, ϕC_3OH, and $C_1\phi C_2OH$ similar to the type IV.

[c] Hz from the dioxane peak.

recently reported that cationic surfactants of the type $Me(CH_2)_nN^+Me_2CH_2$-$\phi Me \cdot Cl^-$ ($n = 7$ or 9) show a pattern change of the type III when they aggregate into micelles (63).)

A large ring-current arising from other phenyl groups in the same micelle

Table 11

Chemical shifts of phenyl protons in water and in 5.0% $NaC_{12}S$ solution[a]

Solubilizate	n	Type of Spectra	Chemical Shifts[b] (Hz)	
			In Water	In $NaC_{12}S$ soln.
H_A H_B / C_n—phenyl—OH / H_A H_B	1	II	$H_A = -206$ $H_B = -187$	$H_{A,B} = -191$
	2	I	$H_A = -206$ $H_B = -186$	$H_A = -189$ $H_B = -183$
	3	I	$H_A = -201$ $H_B = -183$	$H_A = -192$ $H_B = -182$
H_A H_B / C_1—phenyl—C_nOH / H_A H_B	1	III	$H_{A,B} = -212$	$H_A = -198$ $H_B = -205$
	2	IV	$H_{A,B} = -211$	$H_{A,B} = -194$
H_{Ph} | phenyl—C_nOH	1	IV	$H_{Ph} = -216$	$H_{Ph} = -209$
	2	IV	$H_{Ph} = -212$	$H_{Ph} = -202$
	3	IV	$H_{Ph} = -214$	$H_{Ph} = -198$

[a]From ref. 62.
[b]Chemical shifts were measured with reference to the dioxane peak.

is considered to make a direct contribution to the upfield shifts of phenyl proton signals (59, 60). A difference in magnetic susceptibility between the water and micellar phases is also considered to have some effect on the upfield shifts. In addition to these two factors, electrical forces, for example, the electric field of the double layer surrounding the micelle and the polarization of π-electrons in the phenyl group at the charged surface of the micelle, seem to be making some contribution.

For solubilizates of the type $C_n\phi OH$, the upfield shifts, $\Delta\nu$, of phenyl protons in the $NaC_{12}S$ solution from the peaks in water are plotted in Figure 24 as a function of solubilizate concentration. With $C_1\phi OH$ and $C_2\phi OH$, both $\Delta\nu_{H_A}$ and $\Delta\nu_{H_B}$ increase with increasing concentration and reach a maximum, then decrease until these phenols are saturated in the solution. The extent of the initial increase in $\Delta\nu_{H_A}$ or $\Delta\nu_{H_B}$ is about the same. This may suggest that the manner of solubilization of these two phenols is similar, at least at the earlier stages of solubilization. The maximum observed in the $\Delta\nu$–concentration curve of $C_1\phi OH$ or $C_2\phi OH$ suggests that the mode of solubilization is different before and after the maximum. With $C_3\phi OH$, on the other hand, the spectral behavior is different from that of the above two phenols: No maximum is observed and $\Delta\nu_{H_B}$ is nearly zero, independent of the concentration. Probably, owing to its longer hydrocarbon chain, this phenol is solubilized in the micelle, with the hydrocarbon part inside and the

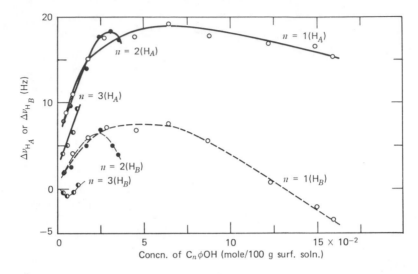

Fig. 24. Upfield shifts $\Delta\nu_{H_A}$ and $\Delta\nu_{H_B}$ of $C_n\phi OH$ in 5.0% $NaC_{12}S$ solution, from the peaks in water, as a function of concentration. (From ref. 62.)

polar group in the water phase, that is, in the form of the so-called palisade type micelle.

As illustrated in Table 11, $C_1\phi C_1OH$ shows the type III pattern and $C_1\phi C_2 OH$ shows that of type IV. The split of the peak of $C_1\phi C_1OH$ in the surfactant solution could be explained by the polarization of π-electrons in the phenyl group by the surface charge of the micelle. The electron density around the H_A protons would probably be larger than around the H_B protons and, therefore, the shielding effect of the electrons on the H_A protons would be greater. This effect will make H_A and H_B protons distinguishable by nMr. With $C_1\phi C_2OH$, the locus of the phenyl group in the micelle is probably not near to the charged surface because of the orientation of the tolyl group with access to the hydrocarbon part of the micelle. If this is the case, the surface charge will not exert any influence on the phenyl group and, therefore, the split of the phenyl proton peak will not take place in the micellar phase.

Figure 25 shows the $\Delta\nu$–concentration curves for solubilizates ϕC_nOH. The magnitude of $\Delta\nu$ increases with increasing n when compared at a definite concentration. This would indicate that the alcohol with larger n is solubilized with the phenyl group buried deeper in the micelle.

It is interesting to compare the three solubilizates, $C_2\phi OH$, $C_1\phi C_1OH$, and ϕC_2OH, which are position-isomers with respect to the phenyl group. On going from a water environment to a micellar environment, the phenyl proton signals of these three solubilizates shift to higher fields to varying extents and

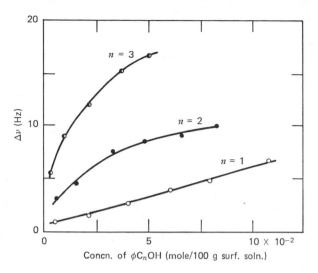

Fig. 25.   Upfield shifts $\Delta\nu$ of $\phi C_n OH$ in 5.0% $NaC_{12}S$ solution, from the peaks in water, as a function of concentration. (From ref. 62.)

the signal patterns change in different manners. The results suggest different modes of solubilization. $C_2\phi OH$ is probably incorporated in the micelle surface, whereas $\phi C_2 OH$ may penetrate into the hydrocarbon part of the micelle, and $C_1\phi C_1 OH$ may be located between the surface and the hydrocarbon part of the micelle.

3. Loci of Solubilization in a Micelle

Eriksson and Gillberg (64) have discussed the solubilization of aromatic solubilizates (benzene, N,N-dimethylaniline, nitrobenzene, and isopropylbenzene) and cyclohexane in a micellar solution of cetyltrimethylammonium bromide $C_{16}H_{33}N^+(CH_3)_3Br^-$ (abbreviated as $C_{16}TAB$). This constitutes an extension of Eriksson's previous work (65) on solubilization of benzene and bromobenzene in cetylpyridinium chloride solution.

The addition of an aromatic solubilizate to the $C_{16}TAB$ solution causes, in general, the signals of $C_{16}TAB$ and of the solubilizate itself to shift towards higher magnetic fields. For example, Figure 26 shows the signal shifts due to solubilization of N,N-dimethylaniline in a 0.17 M $C_{16}TAB$ solution as a function of the solubilizate concentration. With the solubilization of benzene, nitrobenzene, or isopropylbenzene, upfield shift–solubilizate concentration curves similar to Fugure 26 were also obtained. On the other hand, when a saturated hydrocarbon like cyclohexane is solubilized, only minor shifts of the resonance signals were produced.

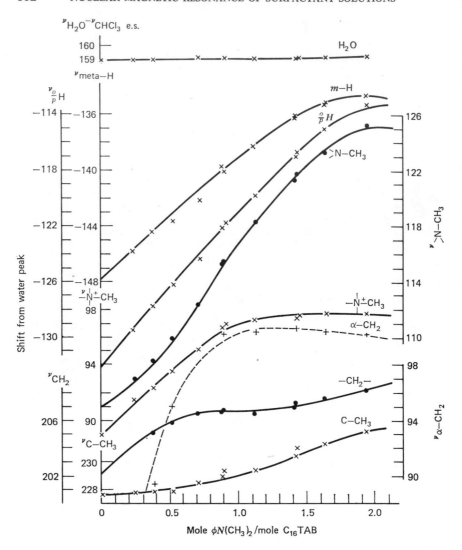

Fig. 26. Resonance signal shifts due to solubilization of *N,N*-dimethylaniline in 0.17 *M* $C_{16}TAB$ solution. (From ref. 64.)

As decribed in the preceding section, a benzene ring in aromatic substances usually gives a negative contribution to the local magnetic field, leading to upfield shifts for the protons at neighboring molecules. The main features of the curves of upfield shift–solubilizate concentration can be explained by this "aromatic" shift. Another process, which also may cause considerable upfield shifts, is a change of medium for the examined molecule from a polarizing to a more inert environment.

Useful information on the loci of solubilization may be derived from the signal positions at zero solubilizate concentration in the curves of Figure 26 for dimethyaniline and in similar curves for other solubilizates. If allowance is made for the difference in bulk susceptibility, the influence of the micelle medium can be compared with those of water and hydrocarbon media (Table 12). For all the aromatic solubilizates the signal positions extrapolated to zero concentration are lower in the $C_{16}TAB$ solution than in the cyclohexane solution. For benzene the figure obtained in the surfactant solution is very near to the resonance position of benzene in water. This indicates that at least at low solubilizate contents benzene is solubilized by adsorption at the micelle–water interface. The figures obtained for isopropylbenzene suggest that the first added molecules are adsorbed at the micelle–water interface and then orientated in such a manner that the benzene ring is preferentially directed to the surrounding water and the isopropyl group is positioned mainly in the hydrocarbon environment. For nitrobenzene and dimethylaniline the data are also consistent with the assumption of adsorption of the first solubilizate molecules at the micelle–water interface. The figures for cyclohexane show that solubilization by interfacial adsorption does not take place.

From the shifts of solubilizate proton signals and of the N–CH$_3$, –CH$_2$–, and α-CH$_2$ signals of $C_{16}TAB$ as a function of solubilizate concentration, the following solubilization mechanism has been proposed for each solubilizate. For benzene and dimethylaniline (Fugure 26), adsorption at sites close to the α-CH$_2$ groups is the predominant solubilization and the aromatic substances

TABLE 12

Signal positions at zero solubilizate concentration relative to the water signal[a]

Resonance Signal		In C_6H_{12} (corrected[b])	In 0.17 M $C_{16}TAB$ Solution	In water (corrected[b])
C_6H_6		−147.6	−162.2	−164.7
$\phi CH(CH_3)_2$	CH$_3$	210.5	208.0	
	arom. H	−142.7	−151.0	
$\phi N(CH_3)_2$	CH$_3$	113.5	105.0	
	ortho-H	−116.6	−132.2	
meta-,	para-H	−133.6	−147.9	
ϕNO_2	ortho-H	−201.1	−214.2	−217±2
meta-,	para-H	−165.7	−187.8	−187±2
C_6H_{12}		197.5	196.5	

[a]From ref. 64.

[b]Correction was made for the bulk susceptibility.

will force out the water molecules present initially at these adsorption sites. At a benzene content approximately equal to 1.0 mole C_6H_6/mole $C_{16}TAB$, the adsorption of benzene is completed and a transition of solubilization occurs through dissolution of benzene in the interior of the micelle. With dimethylaniline this transition occurs at about 0.7 mole ϕ-$N(CH_3)_2$/mole $C_{16}TAB$. For nitrobenzene, the molecules are adsorbed at the micelle–water interface primarily because of interactions between the nitro groups and the polar surfactant molecules and/or the water molecules present; then the benzene rings will be situated at sites close to the α-CH_2 groups. At a nitrobenzene content of 0.8 to 0.9 mole ϕ-NO_2/mole $C_{16}TAB$, the solubilization mechanism is changed to one of dissolution in the interior of the micelle. For isopropylbenzene, the molecules preferentially gather in the hydrocarbon core of the micelle except at very low solubilizate contents; in this case, however, some interfacial adsorption may also be possible. These solubilization mechanisms have been supported by the features of changes in signal width of the resonance peaks and of changes in viscosity of the solutions as a function of solubilizate concentration (64).

Solubilization sites of benzene and nitrobenzene have also been studied by Fendler et al. (66) in the micelle of a zwitterionic surfactant, 3-(dimethyl-dodecylammonio)propane-1-sulfonate. According to these authors the solubilizates interact primarily with the charged surface of the micelle but the dynamic solubilization site of nitrobenzene is somewhat closer to the micelle interior than that of benzene.

III. OTHER PROPERTIES OF SURFACTANT SOLUTIONS

A. Emulsions

Suppose that water is progressively added to an oil solution of a suitable surfactant. The water will be dissolved in the clear solution until the solubilization end point is reached, when emulsion droplets first appear. A question may be raised at this stage: Do two distinct water species exist, each having its own set of properties due to environmental differences, or are all the water molecules equivalent and therefore indistinguishable? This problem has been examined successfully by means of proton magnetic resonance (67). Figure 27 illustrates the spectral changes observed when increasing amounts of water were added to a cyclohexane solution of Duomeen T dioleate (tallow-1,3-propylenediamine dioleate, DTO).

The uppermost (partial) spectrum shows the signals assigned to the carboxyl protons at 7.22 ppm and to the olefinic hydrogens at 3.88 ppm. Upon addition of 0.15–0.60% H_2O, the former signal moves to higher magnetic field and increases in peak area. The existence of a single signal indicates

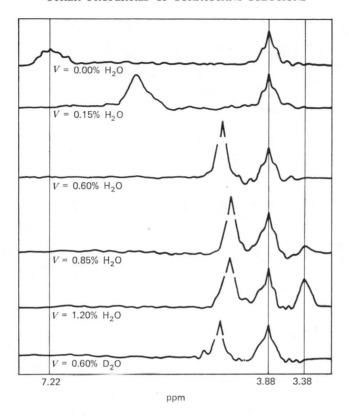

Fig. 27. Nuclear magnetic resonance spectra of mixtures containing 4.0 g of DTO in 100 ml of cyclohexane and varying quantities of water. Cyclohexane itself was used as an internal reference set at zero ppm. (From Fig. 1 of ref. 67.)

a very rapid exchange of protons between the solubilized water and the carboxyl groups. Once the solubilization end point (0.62% H_2O) is reached, the upfield movement and the increase in peak area cease. Additional water causes the sample to become turbid; the protons of separated water, actually emulsified water, give rise to a distinct signal at 3.38 ppm, as seen in the spectrum of the system containing 0.85% water. Increasing the water content to 1.20% increases the peak area of the emulsified species but does not affect that of the solubilized moiety. From these observations, it is concluded that two distinct water species coexist in these emulsions, and that the rate of proton exchange between the solubilized water and the emulsified species is very slow.

The bottom spectrum shows how heavy water, D_2O, affects the chemical shift of the carbonyl proton. Protons originally attached to the carbonyl

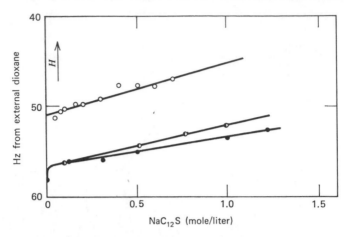

Fig. 28.   Proton chemical shift of H_2O vs. $NaC_{12}S$ concentration at 25°C: ○ emulsion, ● $NaC_{12}S$ solution, ◑ $NaC_{12}S$ solution saturated with benzene. (From p. 1072, Fig. 2 of ref. 68.)

groups can be exchanged with deuterons of D_2O, thus producing a mixture of -COOH, HDO, and H_2O, among which the protons move rapidly back and forth. Accordingly, the shift of signal does occur but without any increase in peak area since the proton content is invariable.

The following experiment has been carried out with the intention of shedding light on the structure of water at the immediate vicinity of oil droplets in O/W-type emulsions (68). A series of samples were prepared by shaking 1 : 1 mixtures of benzene and water that contained varying quantities of sodium dodecyl sulfate ($NaC_{12}S$) as an emulsifier, and the chemical shifts of water proton were measured in relation to an external reference, dioxane. The shifts are plotted in Figure 28, along with similar plots for solutions of $NaC_{12}S$ alone and for solutions of $NaC_{12}S$ just saturated with benzene. The last solution is considered to have a composition identical with that of the interdroplet medium in the corresponding emulsion having the same $NaC_{12}S$ concentration.

Irrespective of $NaC_{12}S$ concentration, a nearly constant difference is observed between the chemical shifts of the whole emulsion and its interdroplet medium. The difference has been attributed to surface effects of the emulsion droplets, such as the breaking of the hydrogen-bonded structure of water at the oil–water interface. This interpretation seems premature; the difference may be ascribed merely or mostly to the difference in bulk susceptibility, if we consider the high benzene content ($>50\%$ in volume) in the emulsions examined.

B. Phase Studies

1. One-Component Systems (Anhydrous Surfactant)

Using various techniques, such as optical microscopy, X-ray diffraction, calorimetry, and dilatometry, it was established that depending on temperature, anhydrous surfactants may exist in several mesomorphic phases, in addition to crystalline phases and an isotropic liquid phase. Broad-line nMr spectroscopy can detect phase transitions and also give some information about the molecular motions present in these states. Indeed, Dunell et al. have successfully made such studies on the alkali metal (Li, Na, K, Rb, Cs)

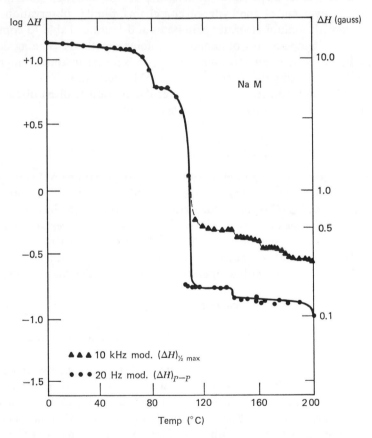

Fig. 29. Signal width of sodium myristate as a function of temperature. (From p. 4260, Fig. 3 of ref. 74.) ($\Delta H_{1/2max}$: width at the point of half-maximum intensity). ($\Delta H)_{p-p}$: separation of points of maximum and minimum slope.

salts of C_6–C_{18} fatty acids (69–73). Transitions between the crystalline phase and the first mesomorphic phase have been easily detected as an abrupt change in a singal width–temperature plot, and valuable information about molecular motions could be derived by the analysis of second moments. Transitions between two mesomorphic phases, however, have generally not been observed. This inability has been overcome by the application of the "first side band" method in a similar experiment of Lawson and Flautt on sodium salts of $C_{12} \sim C_{18}$ fatty acids (74). The latter investigation is outlined below to show the general procedure for analyzing broad-line nMr data.

Figure 29 illustrates the change of signal width of sodium myristate (74). The width decreases stepwise with increasing temperature, suggesting some promoted movement of the hydrocarbon chains. Similar observations have been made with sodium stearate, palmitate, and laurate. Table 13 compares the transition temperatures obtained by different methods. The agreement is satisfactory; besides, an additional transition is found in sodium myristate at 162°C, thus proving the usefulness of the nMr method.

The narrowing of nMr signals by molecular motion is often discussed in terms of equation 15, applicable to powder samples:

$$\tau_C = \frac{1}{\alpha\gamma\Delta H} \tan\left[\left(\frac{\pi}{2}\right)\frac{(\Delta H^2 - \Delta H''^2)}{(\Delta H'^2 - \Delta H''^2)}\right] \tag{15}$$

where τ_C is the correlation time, γ is the magnetogyric ratio, ΔH is the signal width expressed in gauss units, $\Delta H'$ is the width before motional narrowing (rigid lattice), and $\Delta H''$ is the width after motional narrowing has taken place. The factor α is of the order of unity and encompasses the uncertainties arising from our multivocal definition of the "width" with respect to signal shape (75). By assuming that the molecular motion can be described by a singgle correlation time that varies with temperature according to the Arrhenius relation

$$\tau_C = \tau_0 \exp\left(\frac{E_{act}}{RT}\right), \tag{16}$$

the activation energy E_{act} associated with the molecular motion can be estimated from the $\ln \tau_C$–$1/T$ plot.

The data for the four soaps in the temperature range below crystalline–subwaxy transitions did not fall on straight lines; however, from the slopes of the best straight lines drawn through points of low temperature ($<$ ca. 65°C) segments, values of E_{act} of about 3 kcal/mole were obtained for all four soaps. This result partially conflicts with that of an earlier experiment of Grant and Dunell (69) on sodium stearate. Their plot formed two distinct straight lines, thus giving two activation energies: 0.8 kcal/mole below 65°C and 19.2 kcal/mole above 65°C. The latter value seems to be of the right magnitude for a violent twisting or rotation of the paraffine chains (76), whereas 3 kcal/mole

TABLE 13

Phase transitions of anhydrous sodium soaps[a]

| | Transition Temperature (°C) | | | | | | | |
| | C_{18} | | C_{16} | | C_{14} | | C_{12} | |
	NMR	Other	NMR	Other	NMR	Other	NMR	Other
Crystalline–crystalline	85	86–96	82		80	79–80		
Crystalline–subwaxy	114	108–18	113	114–25	108	98–113	104	98–100
Subwaxy–waxy	131	129–34	143	134–40	141 (162)	133–42	148	130–42
Waxy–superwaxy	158	165–80	174	172–76	182	175–82	191	182–87
Superwaxy–subneat	180	188–210	>200	195–211	203	204–18	>200	200–20

[a]From ref. 74.

is approximately of the order of magnitude usually associated with the rotation and/or small-amplitude torsional oscillations of the ends of hydrocarbon chains (77).

For a polycrystalline sample without molecular motion, the expression for the second moment S of the proton magnetic resonance is given by

$$S_{\text{rigid}} = (6/5)\, I\, (I+1)\, g^2 \mu_0^2 N^{-1} \sum_{j>k} r_{jk}^{-6}$$
$$+ (4/15)\, \mu_0^2 N^{-1} \sum_{j,f} I_f(I_f + 1)\, g_f^2 r_{jf}^{-6} \tag{17}$$

where N is the number of protons whose interactions are considered, r_{jk} is the distance between any two protons j and k, r_{jf} is the distance between any proton j and other nucleus f, and μ_0 is the nuclear magneton. Notations I and g stand for spin number and nuclear g factor with the subscript f for a nucleus other than a proton. If the details of molecular arrangements are known, one can calculate the rigid-lattice moment S_{rigid}, taking into account both the intramolecular and intermolecular contributions (78).

The existence of molecular motions reduces the second moment. For example, the intramolecular second moment has been derived for free rotational motion:

$$S = (1/4)\, [(6/5)\, I\, (I+1)\, g^2 \mu_0^2 N^{-1} \sum_{j>k} r_{jk}^{-6}\, (3\cos^2\phi_{jk} - 1)^2$$
$$+ (4/15)\, \mu_0^2 N^{-1} \sum_{j,f} I_f\, (I_f + 1)\, g_f^2 r_{jf}^{-6}\, (3\cos^2\phi_{jf} - 1)^2] \tag{18}$$

where ϕ_{ab} is the angle between the axis of motion and the radius vector connecting the nuclei a and b (75). The intermolecular contribution is difficult to evaluate, but according to Andrew, the intra- and intermolecular contributions are reduced by about the same factor (79). Thus the total second moment can be estimated.

Another important motion may be a torsional oscillation, for which the reduction is given by

$$\rho = \frac{S}{S_{\text{rigid}}} = 1 - (3/4)\, [\{1 - J_0^2\, (\alpha)\}\, \sin^2 2\gamma$$
$$+ \{1 - J_0^2\, (2\alpha)\}\, \sin^4 \gamma] \tag{19}$$

where γ is the angular amplitude of the motion, α is the angle between the axis of reorientation and the internuclear vector, and J_0 is a Bessel function (79).

The data for sodium stearate shown in Table 14 suggest that some rotation of the end methyl groups is occurring even at 90°K. It appears that the hydrocarbon chains are rotating freely about their long axes just after the crystalline–crystalline transition. However, it is impossible to decide which kind of motion, rotation or torsional oscillation, is responsible for the reduction in

TABLE 14

Experimental and theoretical second moments for anhydrous sodium soaps[a]

Condition	Stearate	Palmitate	Myristate	Laurate
		Moments (gauss^2)		
		Experimental		
90°K	25.8 ± 1.5	27.9 ± 1.5	27.6 ± 1.6	19.7 ± 3.0
After the crystalline–crystalline transition	9.6 ± 0.2 (88.0°)	11.2 ± 0.2 (86.7°)	13.8 ± 0.2 (89.7°)	—[b]
Prior to the crystalline–subwaxy transition	7.8 ± 0.2 (95.4°)	11.6 ± 0.2 (105.5°)	12.7 ± 0.2 (95.2°)	—[c]
		Theoretical		
Rigid lattice	28.2 ± 1.0	27.9 ± 1.0	27.6 ± 1.0	27.4 ± 1.0
Rotation of the methyl groups	24.3	23.9	23.1	22.2
Rotation of the hydrocarbon chains as a unit about the long axes	9.6	9.4	9.3	9.1

[a]From ref. 74.
[b]This transition is not present in laurate.
[c]The line width change is not sufficiently sharp to allow an exact point to be chosen.

the second moment. A calculation by means of equation 19 indicates that torsional oscillations with amplitudes of about 150° alone can account for the observed moment of 9.6 gauss². Quite likely, both rotations and oscillations are present. The moment prior to the crystalline–subwaxy transition, 7.8 gauss², suggests some longitudinal flexing of the hydrocarbon chain at this temperature.

In the palmitate and myristate, the lattice appears to be rigid at 90°K. The values of moments at higher temperatures can be accounted for either by torsional oscillations or by rotations about the long axes of the hydrocarbon chains. The amount of motion decreases with a decrease in the chain length.

The behavior of the laurate is significantly different from that of other soaps. Under the assumption that the laurate has the same crystal structure as the longer chain soaps, the calculation postulates (a) fairly large molecular motions at 90°K, (b) violent oscillations, or almost free rotation, around the long axes of the hydrocarbon chains at about 80°C, the moment being 12 gauss², and (c) the presence of translational motions at 100°C, with 5.0 gauss².

Table 15 summarizes the second moments of the four mesomorphic phases encountered. In these phases, the molecular motion is more extensive than just rotational and/or oscillatory (expected to be several gauss²) motion, but less extensive than in an isotropic liquid (expected to be 10^{-4}–10^{-5} gauss²).

The spin–lattice relaxation behavior of soaps in the temperature range from -150 to $-50°C$ is mainly determined by the reorientation of the terminal methyl groups, which persists at low temperatures. Analyzing the spin–lattice

TABLE 15

Experimental second moments of the mesomorphic phases in anhydrous sodium soaps[a]

Phase	Moment (gauss²)	Temp. (°C)	Moment (gauss²)	Temp. (°C)
	Stearate		*Palmitate*	
Subwaxy	0.84	126.4	0.29	133.6
Waxy	0.36	139.7	0.25	157.0
Superwaxy	0.24	169.7	0.13	185.0
Subneat	0.11	196.5	—	—
	Myristate		*Laurate*	
Subwaxy	0.92	132.6	0.37	130.3
Waxy	0.60[b]	149.8	0.21	155.0
	0.39[b]	169.6		
Superwaxy	0.24	190.2	0.18	195.0
Subneat	0.16	200.0	—	—

[a] From ref. 74.

[b] Moment values obtained below and above the transition (162°C, cf. Table 13) discovered in this work.

relaxation data of protons of sodium and lithium soaps with an odd and even number of carbon atoms, van Putte has shown that the energy barrier to the reorientation is higher in the odd-membered than in the even-membered soaps (80).

2. Two-Component Systems (Surfactant + Water)

Many mesomorphic systems comprising two or more components are known. The most common is a system composed of water and an amphiphilic compound like a surfactant. One or two phases appear, depending on the ratio of the two components and on the temperature of the system. For example, progressive addition of water to anhydrous dimethyldodecylamine oxide ($C_{12}DAO$) at 30° causes a series of phase transitions: crystalline (C) → neat (N) → viscous isotropic (VI) → middle (M) → fluid isotropic (FI). Two-phase regions are found at the boundaries between N and VI, VI and M, and M and FI (36).

The FI phase is the ordinary micellar solution, which was the main subject of Section II. The N and M phases are considered to have the structures (a) and (b) as illustrated in Figure 30. The structure of the VI phase, which is encountered frequently in surfactant–water systems, is not fully understood, although in some cases x-ray studies have indicated that it consists of spherical units packed in a face-centered arrangement, with water molecules filling the voids among the spherical units.

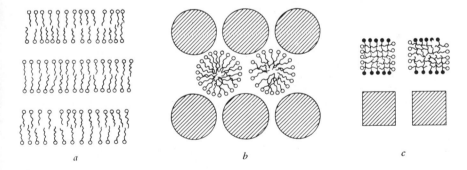

a   *b*   *c*

Fig. 30.   Schematic illustration of structures proposed for liquid crystalline (mesomorphic) phases. (*a*) Lamellar (neat) phase, alternating layers of water and amphiphile molecules (phases B and D in Fig. 34). (*b*) Hexagonal (middle) phase, amphiphile molecules lie in cylinders with circular cross-section, arranged in hexagonal array and water molecules lie in the spaces between the cylinders (phase E in Fig. 34). (*c*) Two-dimensional tetragonal structure, amphiphile molecules lie in rods of square cross-section arranged in tetragonal array with water molecules in the spaces between the rods (phase C in Fig. 34). (After Ekwall et al. (91), and p. 370, Fig. 1 of ref. 48.)

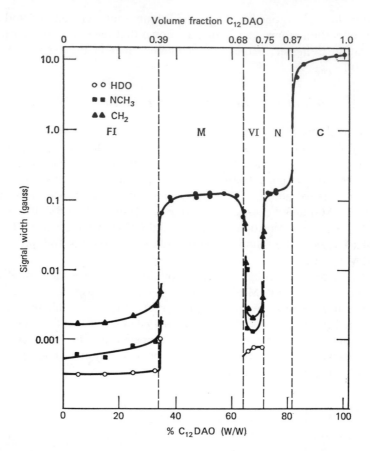

Fig. 31. The proton signal widths (gauss) of the $C_{12}DAO$–D_2O system at room temperature. (From p. 2067, Fig. 1 of ref. 36.)

Figure 31 shows the widths of signals from the surfactant protons as a function of the concentration of $C_{12}DAO$ in D_2O. Both the FI and VI phases produce high-resolution spectra on which N-methyl plus N-methylene, intermediate nine-methylene, terminal C-methyl, remaining one-methylene, and HDO signals are observed. The widths of the first two signals are shown with solid squares and triangles while unfilled circles illustrate the signal widths of HDO present as an isotopic impurity in D_2O used and moisture in the surfactant. The solid circles in C, N, and M regions have been obtained from broad-line nMr. In the two-phase regions, both the well-resolved high-resolution spectra and the broad-line spectra are observable. Although the high-resolution spectra of surfactant protons cannot be obtained in the N and M phases, the signals from the residual HDO remain relatively sharp: 15–30 mG in N and 8–12 mG in M phase.

The broad-line spectrum of the C phase consists of a broad structureless signal; at its center is a weak, narrow component, probably due to the chain methyl groups possessing some rotational mobility. At the transition from C to N phase, the width decreases to about one-hundredth. The signals of the N and M phases have a shape designated as "super-Lorentzian," that is, wider in the wings when compared with a Lorentzian signal with the same half-height width. The ratio of the width at one-eighth height to the width at one-half height, $R(8/2)$, is 6.6 ± 0.2 in the N phase and 5.2 ± 0.2 in the M phase while that of a pure Lorentzian signal is 2.64. This super-Lorentzian shape has been attributed to a distribution of correlation times in the hydrocarbon chains of the surfactant molecules. Not only the hydrophilic groups of the surfactant molecules but probably two or three of the methylene groups in the hydrocarbon chains make up the "boundary" of the structural unit. These groups are probably less mobile than the other methylene groups in the chain and are thus responsible for the wings of the signal. The larger $R(8/2)$ of the N phase implies that the boundary of structural units of N phase is more rigid than that of M phase.

Additional information has been obtained from the spectra of D_2O in the same system, $C_{12}DAO$–D_2O. In the FI and VI phases, the spectra consist of a single sharp line. In the neat and middle phases, the spectra are of the "power-type" expected from randomly oriented crystallites of nuclei having a spin value of unity. The agreement of the experimental spectra with the theoretical spectra calculated by assuming suitable parameters was excellent. In the two-phase regions, the spectra are a superposition of the single sharp signal and the power-type spectrum, thus indicating that the exchange of D_2O between the two phases is slow.

It is well known (81) that mixtures of soap and water with a suitable ratio can afford a lamellar mesophase, where water layers and soap bilayers are alternatively stacked in a one-dimensional periodic lattice (cf. Figure 30a). Such a sample consists ordinarily of randomly arranged crystallites and, therefore, the lamellae take all orientations with equal probabilities. When a thin film of this unoriented mixture is pressed between two microscope cover glasses with lateral sliding motions, a sandwich is obtained, in which all the lamellae are parallel (82).

Using a pile of about 30 such sandwiches of $C_{11}D_{23}COOK$–H_2O mixture, Charvolin et al. have recorded deuteron nMr spectra at various angles, ϕ, between the external magnetic field and the normal to the glass plates (83). The spectrum obtained at $\phi = 90°$ is reproduced in Figure 32a. It exhibits six well-resolved doublets split by quadrupolar coupling, with the following

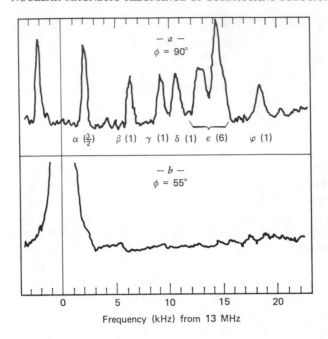

Fig. 32.   Thirteen-megahertz dMr spectra obtained with an oriented lamellar sample for two orientations (*a*) $\phi = 90°$, (*b*) $\phi = 55°$ (from p. 346, Fig. 1 of ref. 83). The sample is a mixture of $C_{11}D_{23}COOK$ 79% and H_2O 21% at 82°C. Only one-half the spectrum, which is symmetrical about frequency zero, is shown. Relative intensities are given in parentheses.

assignments: terminal CD_3 to α, one CD_2 each to β, γ, δ, and φ, and the remaining six CD_2 to ε. The doublet spacings, $\Delta\nu$, among which only that of the terminal methyl is seen in Figure 32*a*, varied with ϕ as shown in Figure 33, in accordance with the theoretical equation 20 derived by assuming that the axis of symmetry is perpendicular to the lamellae:

$$\Delta\nu = \left(\frac{3}{4}\right)\left(\frac{e^2qQ}{h}\right) S \,(3\cos^2\phi - 1) \tag{20}$$

where e^2qQ/h is the quadrupolar coupling constant and S is the "order parameter." At 55°, near the "magic angle" 54° 44', the spacings converged to zero, as illustrated in Figure 32*b*. These observations led the authors to conclude that a molecule, or a segment of it, cannot stay oriented in any direction other than the normal to the lamellae for a time longer than a few 10^{-6} sec. This and similar kinds of studies on oriented bilayers (82, 84) may contribute to the elucidation of cell membranes, which have similar bi-layer structures (81).

The chain motions of surfactant in lamellar mesophases would be under-

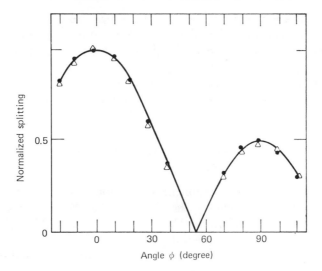

Fig. 33. Variations of normalized splittings as functions of the sample orientation ϕ in the magnetic field (from p. 346, Fig. 2 of ref. 83). For clarity, only splittings of signals α and φ are represented; the others are similar. The continuous line shows the function $|(3\cos^2\phi - 1)/2|$.

stood in more detail, if the mobility could be measured individually at different parts of the same molecule. Measurements of spin–lattice relaxation time are useful in estimating the chain rotational mobility, but because of spin diffusion only the average rotation of the chain is determined. If, however, a surfactant containing a mixed fluorocarbon and hydrocarbon alkyl chain is employed, T_1 measurements of ^{19}F and ^{1}H nuclei will bring to light the distribution of rotational motions along the chain; fluorine and protons have very different nMr frequencies and, therefore, spin diffusion does not occur between them.

A lamellar phase sample composed of sodium, 12,12,13,13,14,14,14-hepta-fluoromyristate, $CF_3(CF_2)_2(CH_2)_{10}CO_2Na$, and D_2O has been examined by Tiddy (85). The spin–lattice relaxation rates obtained were considerably different for ^{19}F and for ^{1}H. However, in discussing the mobility, the direct comparison is meaningless. The spin–lattice relaxation rates must be compared after considering the difference of the nuclear magnetic moments (μ_F and μ_H), the difference of the internuclear distances (r_{F-F} and r_{H-H}), and the effect of the CF_3 group, as opposed to a CF_2 group. A calculation taking these factors into account indicated that there was little or no distribution of CH_2 or CF_2 rotational motion as distance from the head group increased in the surfactant molecule. A similar comparison has been made between the spin–spin relaxation times of ^{19}F and ^{1}H, which are considered to reflect motions of the chains with lower frequencies about the short molecular axis. The result demonstrated comparable rates of motion in the fluorocarbon and hydrocarbon portions. As noticed by the author, the nonexistence of a distribution of chain motions in this particular system should not be taken as direct evidence that a distribution of CH_2 motions does not exist in lamellar phases formed from hydrocarbon amphiphiles.

3. Three-Component Systems (Surfactant + Water + Organic Compound)

Coexistence of a third component with amphiphilic but predominantly lipophilic properties promotes the formation of aqueous mesomorphic phases. A number of papers (4, 48, 86–89) have been published on three-component systems, among which the earliest is that of McDonald (4). This is also the first paper dealing with the nMr of surfactants. High-resolution spectra were recorded for sodium dodecyl sulfate–water–octanol or caprylic acid mixtures, and octylamine–water (two-component) mixture. In all these systems, alkyl proton signals observable in isotropic liquid phases disappeared in liquid–crystalline mesophases. Because of the strong restriction of alkyl chain motions, the signals have been broadened out over the spectral range.

The phase diagram of the sodium caprylate–decanol–water system, a typical example of three-component systems, is presented in Figure 34. This diagram shows the existence of three different lamellar phases (B, C, and D) in addition to normal and reversed haxagonal phases (E and F). Proposed structures for the phases are illustrated in Figure 30. The spin–lattice (T_1) and spin–spin (T_2) relaxation times were measured by Tiddy for a number of samples whose compositions are shown with dots in the B, C, and D phase regions (48).

The $90°$-τ-$90°$ pulse method gave a single T_1 relaxation curve both in samples containing H_2O and D_2O, respectively. In the latter case, spin diffusion through alkyl chain protons leads to a single T_1 for all the alkyl chain protons, which is designated as T_1(alkyl). In the former samples, spin diffusion can take place between water and alkyl chain protons by virtue of the rapid exchange of hydroxyl protons between water and decanol. The observed T_1(alkyl + water) is thus a weighted average of T_1(alkyl) and T_1(water) due to water protons.

$$\frac{1}{T_1 \text{ (alkyl + water)}} = \frac{P_A}{T_1 \text{ (alkyl)}} + \frac{P_B}{T_1 \text{ (water)}} \tag{21}$$

where P_A = fraction of protons in alkyl chains and P_B = fraction of protons

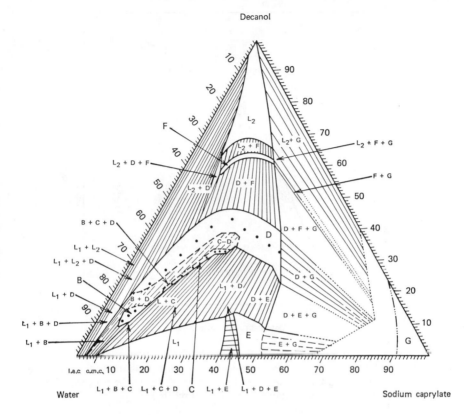

Fig. 34. Phase diagram of the three-component system, sodium caprylate + water + de-canol. Points indicate sample compositions. (From p. 370, Fig. 2 of ref. 48.)

in water. Equation 21 enables one to calculate T_1(water) from the relaxation times obtained experimentally.

The Gill–Meiboom technique (90) gave two separate T_2 curves for the H_2O samples, having T_2 values of the order of 100 μsec and of milliseconds. In the samples containing D_2O, only the microsecond decay was observed, showing its origin in the alkyl chain protons. T_2(alkyl) and T_2(water) can thus be distinguished. Figure 35 and 36 illustrate the dependences of T_1^{-1} (water) and T_2^{-1}(water) on the molar ratio of sodium caprylate/water in the D-phase region, when the decanol/caprylate ratio was kept constant.

The dependence of T_1^{-1}(water) has been interpreted as follows: There are several species of water molecules in the system. Some are attached by hy-drogen bonds to the caprylate ions, some are in the solvation shell of the sodium ions, some others are bound to the decanol hydroxyl groups (but may exert little effect) while the rest is in a "free" state. The molecules bound to

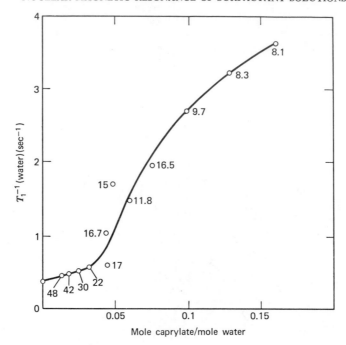

Fig. 35.   The dependence of T_1^{-1} (water) on the molecular ratio of sodium caprylate/water for D phase. The numbers adjacent to the points refer to the water spacings (Å) calculated by using x-ray data. (From p. 373, Fig. 4 of ref. 48.)

the sodium caprylate have larger T_1^{-1} values than that of the free water, owing mainly to the restricted motion of the caprylate ions in the lamellar bi-layer. In the range of lower caprylate content, where the water spacing is larger than 20 Å, an increase in caprylate content causes a gradual increase in T_1^{-1}(water). As the caprylate content exceeds a limit giving the water spacing of about 20 Å, an additional restriction begins to operate on the motion of hydrated sodium ions, and T_1^{-1}(water) increases more rapidly. The dependence of T_2^{-1}(water) shown in Figure 36 also supports this view.

Comparing the T_1(water) and T_2(water) values obtained for B, C, and D phases, the author inferred that the molecular motion of water in C phase is not so strongly restricted as in D phase, and that the decanol head group is less restricted in B phase than in C and D phases.

As described previously, T_1(alkyl) values are determined by rotations of a part of the alkyl chain or of the complete alkyl chain about the long axis of the molecule, whereas T_2(alkyl) values arise from the slower motions of the chains about the short molecular axes (46, 47). The fact that all T_1(alkyl) values obtained for B, C, and D phases were identical within experimental error reveals that the chain motions, though more strongly restricted than in

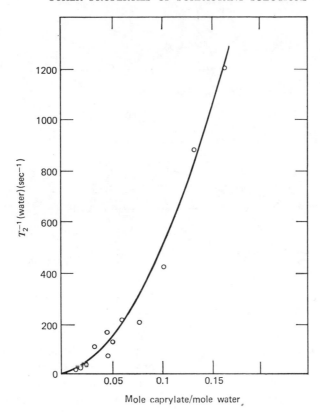

Fig. 36. The dependence of T_2^{-1} (water) on the molecular ratio of sodium caprylate/water for D phase. (From p, 375, Fig. 5 of ref. 48.)

the liquid state, are similar over a wide range of composition and in different phases. The curve of T_2^{-1}(alkyl) against the ratio of sodium caprylate / D$_2$O, although not reproduced here, indicates that the average vibrational and rotational motions of the chains about the short molecular axis become slower with decrease of water content. Finally, it was pointed out that the structure of C phase may be either of the lamellar type or an emulsion of D phase in micellar solution; the evidence for the structure proposed (91) previously, Figure 30c, is inconclusive.

4. Four-Component Systems (Surfactant + Water + Organic Compounds)

Microemulsions, first introduced by Schulman (92), are in general optically clear and thermodynamically stable. The droplet diameter varies from less than 100 to 600 Å, as distinct from ordinary emulsions, or macroemulsions,

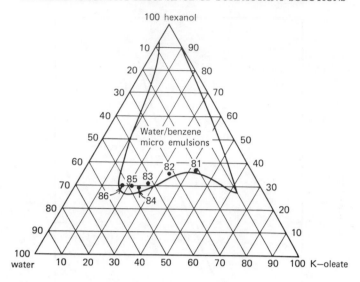

Fig. 37.   Phase diagram for the system: 1 g of potassium oleate, 7.5 ml of benzene, water, and 1-hexanol. The axes are labeled in weight percent of each component. The circles show the location of the samples HR 81–86. (From p. 257, Fig. 1 of ref. 95.)

in which the lower limit of particle size is probably 10^3 Å (93). Nuclear magnetic resonance behaviors have been examined by Gillberg et al. (94) and later by Hansen (95) with water-in-oil type microemulsions prepared by mixing potassium oleate, hexanol, benzene, and water in various ratios. The following summarizes Hansen's work:

Deuteron relaxation times were measured with six samples, each containing 7.5 ml of benzene, 1 g of potassium oleate, and various amounts of 1-hexanol and D_2O with the compositions shown in Figure 37. The results are summarized in Table 16, along with molar ratios of water to surfactant n_w/n_s and radii of water droplets r_w calculated by the method of Bowcott and Schulman (96). The increase in water relaxation times with increasing droplet size has been interpreted with the microemulsion model illustrated in Figure 38. In the droplet, there are two species of water molecules, which interchange rapidly. Water molecules inside the droplet core behave similarly to those in liquid water, and have a relaxation time $(T_1)_{\text{free water}} = 0.50$ sec. Water molecules near the interfacial film would interact strongly with the surfactant polar groups, resulting in a decreased reorientation rate. The relaxation time of these "bound" molecules, $(T_1)_{\text{bound water}}$, may be equated to 0.033 sec (cf. Table 16). The observed relaxation time is given by

$$\frac{1}{T_1} = \frac{f_B}{(T_1)_{\text{bound water}}} + \frac{f_F}{(T_1)_{\text{free water}}} \tag{22}$$

TABLE 16

Water to surfactant ratio, radius of droplet, relaxation times, and
fraction of bound water for microemulsion samples HR81–86[a]

| Samples | n_w/n_s | $r_w(\text{Å})$ | $T_1(\text{sec})$ | $T_2(\text{sec})$ | f_B % calcd. by eq. | | |
| | | | | | (22) | (23) | |
						$d = 1.0$	1.5Å
HR81	8.0	13	0.12	0.10	20.7	20.9	
HR82	16.0	24	0.17	0.15	12.3	12.1	17.7
HR83	24.0	36	0.19	0.20	10.3	8.1	11.9
HR84	30.4	45	0.23	0.22	7.3	6.5	9.6
HR85	36.8	54	0.24	0.23	6.8	5.5	8.1
HR86	48.0	65	0.25	0.25	6.4	4.5	6.8
Extraporated to		0	0.033				
Liquid D_2O			0.5	0.5			

[a]Extracted from Table I, Figure 4, and Figure 5 in the original paper (ref. 95).

where f_B and f_F are the mole fractions of bound and free water. A droplet
with a larger size has a larger f_F, and therefore, a longer T_1.

Using equation 22, one can calculate f_B for each sample. From purely
geometrical considerations, f_B can also be calculated as a function of water
droplet size, if a thickness d of the bound-water layer is assumed.

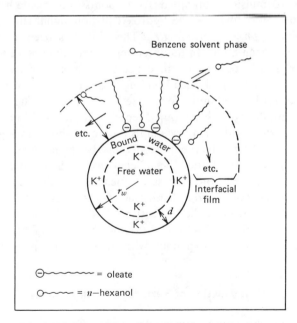

Fig. 38.   Schematic of a microemulsion model (from p. 260, Fig. 6 of ref. 95). $r_w = 13$–65
Å, $d = $ ca. 1.0 Å, $c = 25$ Å.

TABLE 17

Proposed structures for potassium oleate–hexanol–water–hexadecane system[a]

Volume Ratio water/oil	Optical	Structure
0–0.1	Clear	K oleate–hexanol–hexadecane mixture with molecularly dissolved water
0.1–0.6	Clear	W/O microemulsion, spherical[b]
0.7–1.0	Turbid, biref.	W/O (micro) emulsion, cylindrical[b]
1.0–1.3	Turbid, biref.	Water and oil lamellae
1.3–1.4	Clear	O/W microemulsion
1.4	Opaque, milky	O/W (micro) emulsion

[a]Extracted from ref. 97.
[b]Shape of water droplet.

$$f_B = \left(1 - \frac{d}{r_w}\right)^3 \tag{23}$$

The f_B values calculated by equations 22 and 23 show fairly good agreement for each sample (cf. Table 16) when d is assumed to be 1.0 to 1.5 Å, a value corresponding to a "monolayer" of water (molecular diameter 1.2 Å).

For this series of samples of potassium oleate/hexanol/D$_2$O/benzene, no variation was observed in the chemical shift of the hydroxyl proton. For series of microemulsions obtained from potassium oleate/hexanol/H$_2$O/ benzene or perchloroethylene, the hydroxyl proton signal shifted downfield with increasing n_w/n_s or increasing r_w. This shift has been ascribed to the breaking of hydrogen bonds in the interfacial water in contact with the surfactant polar groups. Because of the expected differences between H$_2$O and D$_2$O with respect to the hydrogen bonding with the surfactant polar groups, no shift could be observed in the former series of samples. The chemical shift and signal width data of other groups, reported in this and in an earlier (94) paper, are consistent with the microemulsion model shown in Figure 38.

Such water-in-oil type microemulsions may be, sometimes, inverted into oil-in-water type by the addition of sufficient water. Shah and Hamlin (97) have examined the inversion process by means of optical clarity, birefringence, electrical resistance, and nMr measurements. These measurements were made on samples formed by the successive addition of water to a mixture of 0.2 g of potassium oleate, 0.4 ml of hexanol, and 1 ml of hexadecane. The results could be reasonably explained with the structures proposed in Table 17.

C. Interaction of Surfactants with Proteins

The interaction between croteins and anionic suifactants, especially sodium dodecyl sulfate (NaC$_{12}$S), has been the subject of a large number of investigations. Early studies have suggested the following model (98): The surfactants are bound to proteins through ion-pair formation with cationic sites on the protein, with the hydrocarbon portion of the surfactant free. At higher concentrations, a second layer of surfactant molecules is formed through hydrophobic interaction (see ref. 99 for detailed discussion of the hydrophobic interaction) with the hydrocarbon chains of the first-layer molecules. More recent studies by a variety of methods have indicated that in fact the hydrocarbon chains of anionic surfactants are involved in the binding to protein, and that the binding of anionic surfactants produces conformational changes in the protein molecule (100). Additional information has been obtained by Rosenberg et al., who investigated by nMr the interaction between $NaC_{12}S$ and C-phycocyanin, a photosynthetic protein pigment isolated from *Phormidium luridum* (101).

The protein used was fully deuterated C-phycocyanin extracted from the alga grown in 99.8% D_2O nutrient medium. Thus, the proton magnetic resonance of $NaC_{12}S$ could be observed without interference from signals arising from the protein. The spectrum has three well-defined main peaks due to the C-1 methylene, the intermediate nine-methylene, and the terminal methyl groups, and, besides these three peaks, a fourth broad, minor peak due to the remaining one-methylene group. On the addition of phycocyanin to 0.01 M $NaC_{12}S$, all peaks were shifted upfield and became broader, as shown by Figure 39. These observations have been interpreted as quoted below (102):

That all the main peaks are shifted upfield provides evidence that the entire hydrocarbon chain of the surfactant interacts with the protein; that the peaks are shifted upfield indicates that the surfactant is in a more hydrophobic environment when bound to the protein than when free in aqueous solution; and that the peak assigned to the C–1 methylene becomes broadened at lower protein concentration than the other sharp peaks indicates that the adjacent ionic group of the surfactant is also involved in the interaction.

When $NaC_{12}S$ concentration is lowered to $1 \times 10^{-3}M$, the broadening of signals caused by phycocyanin becomes more prominent and the proton peaks are observable only at comparatively low concentrations of phycocyanin (O to ca. $1.55 \times 10^{-4}M$). At a $NaC_{12}S$ concentration of $1 \times 10^{-4}M$, the peaks are broadened to the extent that they are no longer observable except at extremely low protein concentrations. It was possible, however, to observe a very broad resonance peak of bound $NaC_{12}S$ at very high protein concentrations (above $1.6 \times 10^{-3}M$). These results, in conjunction with available physicochemical data from various sources, seem to imply that the interaction of small amounts of $NaC_{12}S$ tightens the structure of the protein while high concentrations of $NaC_{12}S$ loosen the protein structure (cf. ref. 102 and references cited therein). As pointed out by the original authors,

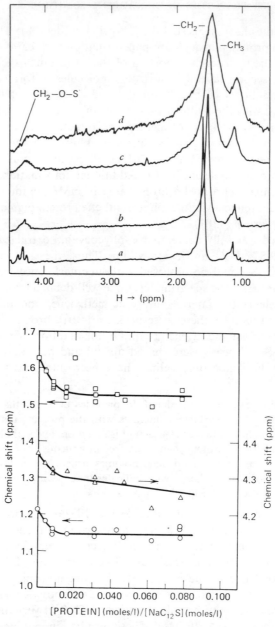

Fig. 39. Nuclear magnetic resonance spectra of 0.01 M NaC$_{12}$S solutions in 0.01 M phosphate, pH 7.3, with (a) no, (b) 1.55 × 10^{-4} M, (c) 4.65 × 10^{-4} M, (d) 6.20 × 10^{-4} M phycocyanin. The scale is referred to external hexamethyldisiloxane. Probe temperature, 32°C. (From p. 34, Figs. 1 and 2 of ref. 101.)

complications may arise from the fact that phycocyanin forms an associating system of monomer, trimer, and hexamer.

Fluorine nMr measurements on surfactants labeled with fluorine atoms are useful means not only for investigating the properties of micelles but also for studying the protein–surfactant interactions, because of the pronounced solvent dependence of ^{19}F chemical shifts.

The fluorine chemical shift of sodium p-(8,8,8-trifluoro)octylbenzene sulfonate $(NaF_3C_8\phi S)$ has been measured as a function of concentration C in water and also in buffer solutions like those used to prepare the BSA (bovine serum albumin)-containing samples (103). Since only one signal is found in the spectra, the exchange of surfactant molecules between water and micelles is considered to be very rapid. The observed shift expressed in ppm unit, δ, is then expected to be (cf. equations 1a, 1b, and 2).

$$\delta = \delta_m \qquad \text{for } C \leq \text{CMC}$$

$$\delta = \frac{\delta_m \cdot \text{CMC} + \delta_M \cdot (C - \text{CMC})}{C} \qquad (24)$$

$$= \delta_M + (\delta_m - \delta_M) \cdot \frac{\text{CMC}}{C} \qquad \text{for } C > \text{CMC}$$

where δ_m and δ_M are the chemical shifts of monomeric and micellar species, respectively. The plot of δ against $1/C$ should then consist of two straight-line segments intersecting at $1/C = 1/\text{CMC}$. Extrapolation to the intercept at $1/C = 0$ yields the value of δ_M, while δ_m is obtained directly from the measurements at concentrations lower than CMC. The CMC thus estimated decreased from 0.0218 M in water to 0.0082 M in the buffer solution, whereas no significant effect was exerted by the buffer on the chemical shifts $\delta_m = 3.65$ and $\delta_M = 4.63$ ppm from the external reference 1,1,2–trichlorotrifluoro-1-propene.

For the BSA-containing samples, the free-surfactant concentration, D_f, and the average number of surfactant ions bound per BSA molecule, $\bar{n}$, were determined by equilibrium dialysis experiments. In all the solutions examined, D_f was lower than the CMC 0.0082 M; therefore, ordinary micelles are considered to be absent or negligible if present. A plot of $\bar{n}/D_f$ against $\bar{n}$ (Scatchard's plot) in the region of low $\bar{n}$ suggests that BSA has about 11 equivalent and noninteracting binding sites for $F_3C_8\phi S^-$. With these results in mind and assuming a rapid exchange of free and bound surfactant molecules, one may write the relation

$$\delta = \frac{(\delta_m D_f + \delta_b D_b)}{C}$$

or

$$\delta_b = \delta + \left(\frac{D_f}{D_b}\right)(\delta - \delta_m) \qquad (25)$$

where $D_b = C - D_f$ and δ_b are, respectively, the concentration and chemical shift of bound $F_3C_8\phi S^-$. The values thus obtained are presented in Figure 40, where δ_b is plotted as a function of $1/\bar{n}$. There is no a priori reason to plot the data in this way, but it is noteworthy that when this is done the points for $\bar{n} > 17$ fall on or near a straight line

$$\delta_b = 4.83 - \frac{23.0}{\bar{n}} \tag{26}$$

while for $\bar{n} < 17$ the shift is approximately constant.

$$\delta_b = 3.475 \tag{27}$$

The following model is probably the simplest that can account for equations 26 and 27.

The BSA molecule has 17 sites at which $F_3C_8\phi S^-$ can be tightly bound, which are not necessarily equivalent but which all cause the surfactant to resonate at or near 3.475 ppm. When these sites are filled, the excess surfactant is taken up into a micelle-like environment with a chemical shift of 4.83 ppm. Thus, in the solution coexist a series of protein–surfactant complexes, each being denoted by the symbol AD_r to represent one molecule of

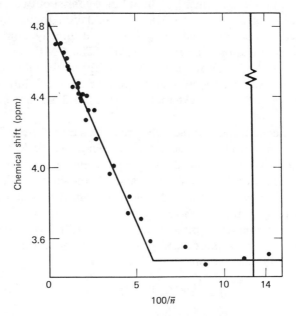

Fig. 40. ^{19}F chemical shift of bound $F_3C_8\phi S^-$ as a function of $100/\bar{n}$. The shift value is expressed in ppm to higher field from the external reference, 1,1,2-trichlorotrifluoro-1-propene. (From p. 1945, Fig. 2 of ref. 103.)

BSA bound with r surfactant molecules. Let δ_r be the chemical shift for this species; then one gets

$$\delta_b = \frac{\sum\limits_r r\delta_r\,[AD_r]}{\sum\limits_r r\,[AD_r]} \tag{28}$$

$$\bar{n} = \frac{\sum\limits_r r\,[AD_r]}{\sum\limits_r [AD_r]} \tag{29}$$

In the range of low surfactant concentration where $\bar{n} < 17$, only species with $r < 17$ are present in significant amounts and, therefore, equation 27 holds. In the high concentration range where $\bar{n} > 17$, only species with $r > 17$ are present, and such species would have

$$\delta_r = \frac{17\,(3.475) + (r - 17)(4.83)}{r} = 4.83 - \frac{23.0}{r} \tag{30}$$

with rapid exchange of surfactant ions among all binding sites. Combining equations 28–30, one can easily derive equation 26.

The two estimates for the number of initial binding sites, 17 obtained from nMr data, and 11 from the Scatchard's plot, are not necessarily in conflict. Scatchard procedure yields the number of equivalent noninteracting sites, whereas the chemical shifts suggest only that 17 sites, which may not be equivalent or noninteracting, can bind surfactant with the chemical shift of 3.475 ppm, and that most of these sites are filled before others are used. There appear to be no compelling reasons to reject the above-mentioned simple model, although other interpretations may be possible. It is striking that the chemical shift at low n (3.475) is smaller than δ_m, whereas the micelle shift is considerably larger than δ_m. This has been interpreted as showing that the CF$_3$ groups in complexes with $r < 17$ are in a highly aqueous environment. Another spectral feature, that the maximum broadening of the signal is observed at low $\bar{n}$, has been attributed to the restricted motional freedom of the CF$_3$ groups in the tightly bound surfactant molecules. In this connection, a trivial mistake is found in the original paper (103). In equation 11 of the paper, $1/\bar{n}$ must be replaced by the weighted mean of $1/r$, that is, by $\sum_r (1/r)[AD_r]/\sum_r [AD_r]$.

Similar experiments (104) have recently been carried out with sodium 12,12,12-trifluorododecyl sulfate (NaF$_3$C$_{12}$S) and 13,13,13-trifluorotridecyl sulfate (NaF$_3$C$_{13}$S). For NaF$_3$C$_{12}$S, the results were essentially the same as those for NaF$_3$C$_8\phi$S, although in this case the number of initial binding sites was 15. The chemical shift of bound F$_3$C$_{12}$S$^-$ can be well expressed by $\delta_b = 3.42$ for $\bar{n} < 15$, and $\delta_b = 5.11 - 25.4/\bar{n}$ for $\bar{n} > 15$. Again in the case of NaF$_3$C$_{13}$S, similar relations hold in regions of low $\bar{n}$; $\delta_b = 3.40$ for $\bar{n} < 14$

and $\delta_b = 5.15 - 24.5/\bar{n}$ for $14 < \bar{n} < 84$. The situation is, however, strikingly different for binding numbers higher than 84, where δ_b reaches a plateau at 4.86 ppm. This behavior has been interpreted as showing that $F_3C_{13}S^-$ ions are bound at 14 magnetically equivalent initial sites but these sites are destroyed by a conformational transition that is essentially complete when $\bar{n} = 84$.

Unlike the proton and ^{19}F (spin value $\frac{1}{2}$), the ^{35}Cl (spin value $\frac{3}{2}$) nucleus has an electric quadrupole moment. The spin–spin relaxation time and accordingly the signal width are affected principally by the electric field gradient around the nucleus and its rotational correlation time (cf. Section IV.A.6). When compared for ^{35}Cl ions free in a solution and bound to a protein molecule, both the field gradient and the correlation time must be different for the two species. The difference in the field gradient is a measure of the type of bond formed and that in the correlation time depends on the molecular motion of the protein–surfactant complex. In effect, bound ions show a much broader signal. This fact has been utilized to investigate the competitive binding of chloride and dodecyl sulfate ions onto BSA molecules (105).

Consider the H_2O–BSA–$NaC_{12}S$–NaCl system. Under a condition of rapid exchange, the observed signal width $\Delta\nu$ is given by

$$\Delta\nu = P_{free} \Delta\nu_{free} + \sum_i P_{bi} \Delta\nu_{bi} \tag{31}$$

where (P_{free}, P_b) and $(\Delta\nu_{free}, \Delta\nu_b)$ represent the mole fractions and the signal widths of free and bound chlorine ions. The summation over i indicates the possibilities of many different sites. In most cases, P_{free} is close to 1.0 so that the contribution from the bound species can be determined by simply subtracting the signal width of the chlorine standard from the observed width. As seen in Figure 41, the observed width decreases rapidly and almost linearly with an increase in $NaC_{12}S$/BSA ratio when the ratio is less than 10. At greater ratios, the width decreases slowly until it reaches that of the standard. Similar plots have been obtained by using different fractions of BSA and chemically modified BSA's. Hexyl, octyl, and decyl sulfates are less effective than $NaC_{12}S$ in reducing chloride binding.

IV. MISCELLANEOUS APPLICATIONS AND BIBLIOGRAPHIES

A. Miscellaneous Applications

Brief surveys are given in this section about the application of nMr techniques to miscellaneous problems or topics that we have not dealt with in previous sections.

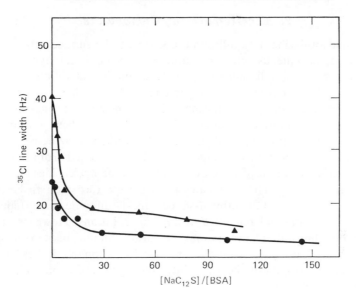

Fig. 41. ^{35}Cl signal width vs. $NaC_{12}S/BSA$. Titration with $NaC_{12}S$ of 2 mg/ml of BSA (●) and 4 mg/ml of BSA (▲), in 1 M NaCl. (From p. 5463, Fig. 2 of ref. 105.)

1. Characterization of Surfactants

The most fruitful application of nMr in chemistry is probably the structure determination and analysis of organic compounds. In the field of surfactants, this kind of application was first made by Walz and Kirschnek (106), who determined the structures of nonionic surfactants containing polyoxyethylene and/or polyoxypropylene residues. Succeeding studies (107–112) have led to a simple technique that allows complete characterization of alkyl ethoxylates in terms of average chain lengths of the alkyl and polyoxyethylene portions (112). Hydrophilic lipophilic balance (HLB) values of various nonionic surfactants could be determined by using only small samples (113). Lanthanide shift reagent was useful for the identification of *o,p*-isomers in ethoxylated alkylphenol nonionic surfactants (114). Characterization of ionic and ampholytic surfactants by nMr has also been reported (115–118). The publications of Koenig contain many spectral data and charts, and are helpful for the identification and signal assignment of surfactants (117, 118). The formation of molecular complexes between fatty acids and alkali soaps (119), and between fatty acids and polyoxyethylene alkyl ethers (120) has been investigated by nMr in combination with infrared spectroscopy.

2. Micelle-forming Compounds

Many phenothiazine tranquillizers are known to be surface active and to form micelles in aqueous solutions, although the biological significance of such properties is not well understood (121). From the data of chemical shift as a function of solute concentration, Florence and Parfitt have confirmed the micelle formation of promethazine, chlorpromazine, promazine, and thioridazine hydrochlorides in D_2O or H_2O (122, 123). These authors have proposed a model with a vertical stacking of the phenothiazine rings in the micelle interior. Important information has been obtained from nMr data about the self-association of purines (59), pyrimidine nucleosides (124), and acridine orange (125). The aggregates formed were too small to be called "micelles." Sodium 4,4-dimethyl-4-silapentane-1-sulfonate has frequently been used as an internal reference in aqueous solutions. However, in this case, the possibility that micelles may be formed should be kept in mind, particularly in the presence of electrolytes. The formation of micelles has been demonstrated by studies of chemical shifts and signal widths in the presence of paramagnetic species (126).

3. Residence Time of a Surfactant or Solubilizate Molecule in a Micelle

Attempts to determine the exchange rate of surfactant or solubilizate molecules between micelles and intermicellar solution by analyzing the nMr signal shape have achieved only partial success (cf. Sections II.A.1.c and II. D.1). The exchange is rapid on the nMr time scale and only the higher limit can be estimated for the mean residence time of the molecule in a micelle (22, 23). If the surfactant or solubilizate is a stable radical, the residence time can be determined by analyzing the ESR (electron spin resonance) signal shape. Table 18 lists the results thus obtained (61).

TABLE 18

Residence time, τ_B, in micelle[a]

Surfactant or Solubilizate[b]		Temp. (°C)	τ_B (sec)
$C_{12}H_{25}$-N$^+$(CH$_3$)$_2$—⟨ ⟩—$\dot{N}$O · Br$^-$		24	6×10^{-6}
H(CH$_2$)$_4$C(CH$_3$)$_2$-$\dot{N}$O–C(CH$_3$)$_3$ in	NaC$_8$S	23–5	1.3×10^{-6}
	NaC$_{10}$S	23–5	1.7×10^{-6}
	NaC$_{12}$S	23–5	2.7×10^{-6}
	NaC$_{14}$S	35	4.5×10^{-6}

[a]From ref. 61.
[b]NaC$_8$S, NaC$_{10}$S, NaC$_{12}$S, NaC$_{14}$S = sodium octyl-, decyl-, dodecyl-, tetradecyl sulfate.

4. Water Penetration into Micelle Core

Muller and collaborators have published a series of papers that reported fluorine nMr results obtained from various ^{19}F-labeled surfactants, such as $CF_3(CH_2)_n$ COONa (127–132). By plotting the observed chemical shift against reciprocal concentration, CMC, the chemical shifts of monomeric species δ_m and of micellar species δ_M were determined (cf. Section III.C, equation 24). These authors noticed the remarkable fact that δ_M is situated about midway between δ_m and the chemical shift of $CF_3(CH_2)_8CF_3$ in ethanol. The value of the latter, 3.8 ppm, was considered, based on much evidence, to be the shift value that would be displayed by the soap molecules in a medium consisting of either hydrocarbon chains or fluoroalkane chains. In other words, it would be the shift displayed by CF_3 groups in the interior of an idealized micelle, from which region water is rigorously excluded. The above-mentioned finding then suggests that in fact the CF_3 group "inside" the micelle remains exposed to water to quite a significant extent. The extent may be expressed numerically by the empirical parameter

$$ Z = \frac{(\delta_M - \delta_m)}{(3.8 - \delta_m)} \tag{32} $$

which measures the "degree of hydrocarbon-like character." The Z values of $CF_3(CH_2)_n$COONa ($n = 8$, 10, and 11) were all within 0.53 ± 0.03. Similar observations and calculations on solubilized benzotrifluoride, $C_6H_5CF_3$, gave Z values of 0.5 in sodium or potassium caprate, 0.5 to 0.6 in sodium laurate, and 0.6 to 0.7 in the fluorinated soaps. The Z value of sodium perfluorooctanoate increased from 0.37 at C2 through 0.84 at C8, thus indicating that the average environment becomes progressively more aqueous as one moves along the chain toward the ionic end.

Gordon et al. (133) have extended the above ideas to the proton chemical shift of solubilized benzene, and have obtained Z values of 0.06 to 0.16 in polyoxyethylene—(23)—lauryl ether, 0.18 in hexadecyltrimethylammonium bromide, 0.20 to 0.22 in sodium dodecyl sulfate, 0.52 in sodium dodecanoate, and 0.29 to 0.33 in sodium decanoate. Direct comparison between the two sets of Z values (Muller and Gordon) may be inadequate, yet the difference seems too large to be ascribed solely to the differences of the systems examined. On the other hand, preliminary [13]C nMr experiments (8) on heavy water solutions of sodium octanoate suggested that the water penetration into the micelles does not vary appreciably along the alkyl chain, in contrast to the findings of Muller et al.

In all the experiments mentioned above, the chemical shift was used as a measure of the environment. Spin-lattice relaxation may serve as another measure, because T_1 varies with the magnetic moments of the nuclear species surrounding the nucleus under investigation. Early measurements on methylene protons of sodium alkyl sulfates in D_2O solutions supported the assertion that some of the alkyl chains in the micelles were exposed to water; these measurements were compatible with a micelle model in which part of the hydrocarbon chains protruded into the water (33). On the other hand, the alkyl protons of n-decyl pentaoxyethylene glycol monoether showed the same T_1 in H_2O and D_2O at all temperatures (37). This suggests that the micellar core is not penetrated to any great extent by water molecules; otherwise the lower magnetogyric ratio of deuterium compared with hydrogen would increase T_1 in D_2O solution by an easily detectable amount. Little or no penetration of water can also be inferred from recent elaborate experiments on p-tert-octylphenoxy(polyethoxy)ethanols, $(CH_3)_3CCH_2C(CH_3)_2$-ϕ-$(OCH_2CH_2)_nOH$ ($n = 9$–10, 12–13, 30), as judged both from chemical shifts and from spin–lattice relaxation times of alkyl protons (34).

The ^{19}F spin–lattice relaxation time of heptafluorobutylic acid C_3F_7COOH dissolved in H_2O and D_2O above its CMC fitted well to equation 3 described in Section II.B.1.

$$\left(\frac{1}{T_1}\right)_C = \frac{(CMC)/C}{T_{1,m}} + \frac{[C - (CMC)]/C}{T_{1,M}} \tag{3}$$

The $T_{1,M}$ calculated by assuming $T_{1,m} = T_1$ at CMC was larger in D_2O than in H_2O. This has been attributed to the exposure of fluorocarbon chains in the micelles to water; because of the smaller magnetic moment of deuterons, D_2O is far less effective in relaxing ^{19}F nuclei. The stronger hydrogen-bonding ability or higher viscosity of D_2O compared with H_2O cannot explain the observation, since these would cause more rapid relaxation in D_2O. It should be noticed that the considerable exposure of the fluorocarbon chains to water was deduced from measurements on a short-chain surfactant, and care should be used in generalizing the conclusion. In a similar experiment on sodium pentadecafluorooctanoate, no significant difference of T_1 could be detected between H_2O and D_2O (45).

To discuss the value of Z, two factors must be taken into account. Each nucleus in the micelle is under the influence of water not only when it is in the micelle core and in contact with a water molecule or molecules penetrated therein, but also when it is at the micelle surface and in contact with the intermicellar water. The contribution of the former, in which we are interested, cannot be estimated separately. In considering these complications and (at least partially) conflicting observations, it seems still premature to draw a conclusion with respect to the water content in the micelle core (cf. also Section II.B.3).

5. Carbon-13 Fourier Transform nMr

As noted in Section I, ^{13}C spectroscopy is a promising technique for elucidating physicochemical properties of surfactants, although it has not yet been used extensively. Table 19 lists the ^{13}C spin–lattice relaxation times of individual carbon resonances observed with n-alkyltrimethylammonium bromides (6). The results are expressed in terms of NT_1 (not in T_1 itself), because ^{13}C relaxation of a protonated carbon in a large molecule is overwhelmingly dominated by ^{13}C–^1H dipole–dipole interaction, and the effective correlation time for rotational reorientation of the pertinent C–H vector can be expressed by

$$\tau_{eff} = r_{CH}{}^6/\{(\hbar\gamma_C\gamma_H)^2 \times (NT_1)\} \tag{33}$$

where N is the number of directly attached hydrogens, γ_C and γ_H are the magnetogyric ratios of ^{13}C and ^1H respectively, and r_{CH} is the C–H bond length. It is seen that the segmental motion of the carbon chain in monomolecularly dispersed surfactant increases slightly as one moves away from the polar head. Micellization restricts the mobility of all carbon atoms, but in different degrees; the maximum restriction is imposed on the polar end of the molecule. Similar results have also been reported for potassium caprate (10).

Table 20 shows the effect of added sodium chloride on the T_1's of sodium dodecyl sulfate $NaC_{12}S$ whose concentration is far higher than the CMC (9). It is apparent that the methylene group next to the polar head gains more freedom of motion as the electrolyte concentration is increased, whereas the mobilities of the other groups remain almost unchanged. Two factors may be considered to explain this finding: the change of electrical double layer around the micelle and the change of surface area per polar group accompanying the micelle growth. The original authors neglected the latter factor on the assumption that the aggregation number of the micelle in a 0.86 M solution of $NaC_{12}S$ alone is nearly equal to that in a 0.43 M solution of $NaC_{12}S$ containing 0.3 M NaCl (9). However, this assumption does not seem reasonable in the present authors' interpretation of the literature.

It is now well known that molecules of desoxycholic acid associate with one another to form channels, which can include a variety of guest molecules. Depending upon the molecular sizes of the guest substances, the combinations may crystallize in constant stoichiometric ratios. The association clearly persists in solution and, therefore, aqueous solutions of sodium desoxycholate can take up a variety of water-insoluble substances. Both p-xylene and 2-methylnaphthalene are taken up by sodium desoxycholate in about 2 : 1 molar ratio, and the signal widths of ^{13}C spectra indicate that there is

TABLE 19

^{13}C Spin–lattice relaxation times of $[RN(CH_3)_3]^+Br^-$ in aqueous solution at 34°C[a]

R		Concn. (M)	Type of Solution	NT_1 (sec)[b] CH_3–CH_2–CH_2–$(CH_2)_k$–CH_2–CH_2–CH_2–$N(CH_3)_3$							
n-Hexyl,	$k = 0$	1.0	Molecular	14.3	8.6	6.3		5.2	5.0	4.4	6.0
n-Octyl,	$k = 2$	0.2	Molecular[c]	12.9	7.8		4.7[d]	4.7	4.7		6.3
n-Octyl,	$k = 2$	2.0[e]	Spherical micelles	10.3	2.9	2.4	1.6[d]	1.0	1.0	.90	2.6
n-Hexadecyl[f]	$k = 10$	0.4[g]	Rod-shaped micelles	8.4	1.2				.68	.54	1.8

[a] From ref. 6.

[b] N is the number of directly attached hydrogens. T_1 values are accurate to $\pm 10\%$. Unless otherwise indicated, all NT_1 values are those of totally resolved carbon resonances.

[c] Predominantly molecular. Several values of CMC in the range 0.1–0.3 M have been reported: P. Mukerjee and K. J. Mysels, *Nat. Stand. Ref. Data Ser., Nat. Bur. Stand.*, No. 36, 103 (1971).

[d] Two-carbon resonance.

[e] CMC is about 0.1–0.3 M (see footnote c).

[f] At 41°.

[g] CMC is about 10^{-3} M (see reference in footnote c).

TABLE 20
^{13}C Spin-lattice relaxation time of sodium dodecyl sulfate[a]

	NT_1(sec)							
	$CH_3-CH_2-CH_2-CH_2-C_2H_4-C_4H_8-CH_2-CH_2-SO_4$							
σ (ppm from TMS)	14.44	23.50	32.63	29.67	30.15	30.52	26.16	69.86
NaC$_{12}$S 0.86 M	6.60	2.40	1.10	0.70	0.86	0.76	0.70	0.70
NaC$_{12}$S 0.43	6.30	1.50	1.32	0.82	0.82	0.78	0.74	0.70
NaC$_{12}$S 0.43, NaCl 0.15 M	6.00	1.80	1.20	0.96	0.90	0.70	0.74	1.06
NaC$_{12}$S 0.43, NaCl 0.30	6.60	1.70	1.00	0.96	0.80	0.80	0.80	1.40

[a]From ref. 9.

considerable restriction of the molecular motion of 2-methylnaphthalene but relatively little of *p*-xylene (7).

6. Binding of Counter-Ions

Nuclear magnetic resonance of counter-ions can afford valuable information about the counter-ion binding onto micelles in micellar solutions, and onto structural units in mesomorphic phases. If the ion has a spin quantum number I larger than or equal to unity, the predominant cause of relaxation is the interaction between the nuclear quadrupole moment and time-dependent electric field gradients at the nucleus. Under conditions of extreme narrowing, the relaxation times (and hence the signal width) are determined by

$$\frac{1}{T_1} = \frac{1}{T_2} = \frac{3}{40} \frac{2I + 3}{I^2 (2I - 1)} \left(1 + \frac{\eta^2}{3}\right)\left[\frac{eQ}{\hbar} \cdot q\right]^2 \tau_c \qquad (34)$$

where eQ is the electric quadrupole moment, q the largest of the components of the field gradient tensor along the principal axes, τ_c the correlation time describing the random motions that give rise to the electric field gradient fluctuation, and η the asymmetry parameter (which can usually be omitted).

Eriksson et al. (134) were the first to study the counter-ion binding by measuring the quadrupole relaxation. The ^{81}Br signal width has been measured as a function of octylammonium bromide concentration in comparison with sodium bromide. Below the CMC, the width is approximately constant, though somewhat larger than the width observed for a dilute NaBr solution. At the CMC the width starts to increase rapidly and at higher concentrations becomes much broader than below the CMC. Upon micelle formation, a part of the counter-ions will be adsorbed in the Stern layer of the micelle, accompanied by some variation in both τ_c and q. These authors considered that the variation of τ_c is mainly responsible for the observed broadening.

Lindman et al. have compared the signal widths of ^{79}Br and ^{81}Br among mono-, di-, and trialkylammonium bromides with different carbon numbers

(135). In all cases, the ^{79}Br signal widths are significantly greater than those observed in solutions of the same concentration of alkali bromides. Probably the alkyl groups in the cations tend to stabilize the water lattice, and close to the cations a strengthened coupling is operative between the anions and the water molecules. Micelle-forming species display, in addition, a prominent increase of signal width above the CMC, thus suggesting the counter-ion binding. The comparison of ^{79}Br and ^{81}Br data indicates that there is a fast exchange of bromide ions between the surface of micelles and the bulk of the solution. Therefore, the relation formerly expressed as equation 31 holds:

$$\Delta H = P_{\text{free}} \, \Delta H_{\text{free}} + \sum_i P_{bi} \, \Delta H_{bi} \tag{35}$$

Here, H is used in place of ν to convert the width value from hertz to gauss units. If there is only one kind of micelle, or two or more kinds of micelles having equal ΔH_b in the solution, equation 35 can be transformed into

$$\Delta H = P_{\text{free}} \, \Delta H_{\text{free}} + P_{\text{bound}} \, \Delta H_{\text{bound}} \tag{36}$$

where ΔH_{bound} and $P_{\text{bound}} = 1 - P_{\text{free}}$ are the signal width and the mole fraction of the counter-ions bound to the micelle surface. By estimating P_{bound} from other data (136) and taking $\Delta H_{\text{free}} = $ signal width just below CMC $= 0.7G$, the values of ΔH_{bound} are calculated to be $16G$ for $0.1 \, M$ decylammonium bromide, and $35G$ for $0.5 \, M$ undecylammonium bromide.

Elaborate studies have been made on cetyltrimethylammonium bromide (C_{16}TAB) and chloride (C_{16}TAC) by measuring ^{79}Br, ^{81}Br, and ^{35}Cl signal widths at various concentrations and temperatures in H_2O and D_2O solvents (137). With increasing concentration of C_{16}TAB at 27°C, the ^{81}Br signal width changes rather stepwise: increasing rapidly in the range of 0 to 4% of C_{16}TAB by weight, staying constant at 4 to 7%, increasing gradually at 7 to 15%, staying constant at 15 to 25%, decreasing abruptly at 25%, and increasing slightly in a still higher concentration region.

This behavior has been explained in terms of equation 36. Between 4 and 7% of C_{16}TAB, the counter-ion binding occurs onto spherical micelles whose size and shape remain essentially constant in this region. At concentrations between 7 and 14%, the sphere-to-rod transition (138, 139) is taking place accompanied with a charge–density increase on the micellar surface, which in turn causes an increase in P_{bound}. The sudden drop of signal width at 25% is ascribed to the formation of a liquid–crystalline phase with a hexagonal arrangement of rod-shaped micelles (140). The signal in this mesophase is split into three peaks, owing to static quadrupolar effects.

The change in ^{35}Cl relaxation for C_{16}TAC in the corresponding concentration range is very small compared with the results for C_{16}TAB. This implys that no change in micellar shape occurs with $C_{16}TAC$, in accord with the conclusion obtained from x-ray data (139). The effects of temperature and of additives upon the sphere-to-rod transition of $C_{16}TAB$ micelle have also been discussed.

Rubidium caprylate and caproate solutions show a constant [85]Rb signal width at concentrations lower than the CMC (141), in contrast to the case of alkylammonium bromides mentioned above (135). A simple model assuming a constant degree of counter-ion binding is consistent with the data obtained at moderate concentrations above the CMC. For rubidium caprylate, ΔH_{bound} is estimated to be 0.7 to $0.8G$; $\Delta H_{bound}/\Delta H_{free}$ is about 3, in contrast to 23 or 50 for Br^- in alkylammonium bromides. A possible explanation for this is a more extensive hydration of the rubidium ions. The data at very high concentrations indicate promoted binding of the counterions.

Robb (142) has measured T_1 and T_2 of water protons in aqueous solutions of sodium dodecyl sulfate in the presence of the paramagnetic ions Mn^{2+}, Gd^{3+}, and Cu^{2+}. Comparing these results with the ion-pairing abilities of these ions and of the sodium ion, he arrived at the conclusion that the rotation rate of Na^+ will decrease only a little, less than 35%, when transferred from the bulk water onto the micellar surface. He also measured T_1 and T_2 (found to be equal) of [23]Na as a function of the surfactant concentration, and deduced that this transfer decreases T_1 by a factor of 3 to 5, assuming the ratio of Na^+ binding on the micelle = ca. $\frac{2}{3}$. Thus, the major factor in determining the T_1 of a counter ion was concluded to be the change in q rather than the change in τ_c; this is in contrast with the view of Eriksson et al. (134) mentioned previously.

Spin–lattice relaxation rates of [7]Li, [23]Na, [87]Rb, and [133]Cs have been measured for appropriate decyl sulfates and decanoates (143). The increase in relaxation rate when adsorbed from the bulk of solution into the Stern layer was estimated to be 1.5 to 1.7 times for Li^+, 4 for Na^+, 9 to 15 for Rb^+, and 20 to 30 for Cs^+ in the case of decyl sulfates, and 2.5 to 3 both for Na^+ and Rb^+ in the case of decanoates. The results are consistent with the view that the change in q is mainly responsible for the increase of relaxation rate caused by counter-ion binding.

The ternary system cetyltrimethylammonium bromide/n-hexanol/water gives rise to two regions with isotropic solutions, one in the vicinity of the soap–water axis (L_1, ordinary aqueous micellar solution) and the other extending from the hexanol corner (L_2) of the phase diagram (see Figure 34 for reference). The L_2 solutions are believed to contain the so-called reversed micelles with a water core surrounded by amphiphile molecules (144). The signal widths of [81]Br have been measured at various compositions (145). The results can be interpreted as showing that except for high hexanol contents the bromide ions are located in the water core of a reversed micelle. At the highest hexanol content (95%), the relaxation rate is very fast, owing to ion-pair formation. The binding of sodium ions in the L_2 solution (146) and in the lamellar mesophase (147) of the sodium caprylate/caprylic acid/water system has been discussed on the basis of ^{23}Na signal width data.

Future developments are expected in two directions. In mesomorphic phases, counter-ions with a non-zero nuclear quadrupole moment sometimes exhibit a signal splitting of the first and/or second order, since the environment around the ions is usually anisotropic on a time scale that is long compared to the reciprocal of quadrupole splitting expressed in hertz units. The possibilities of obtaining useful information from quadrupole-splitting measurements have been discussed (148, 149). It is expected that the chemical shifts of counter-ions will be affected by the binding onto micelles or colloidal particles, thus giving us some information on the mode of the interaction. An approach along this line has recently been made with promising prospects (150).

B. Bibliographies

In order to supplement the text, Table 21 has been prepared: Here accumulated references available to the authors are classified according to their intended purpose, the system examined, and the measurement performed. The abbreviations used are summarized at the bottom of the table, among which the following are rather unusual: ad = additive, bing = binding, c-ion = counter-ion, cs = chemical shift, idn = identification, intn = interaction, meso = mesomorphic phase, mic = micelle or micellar, motn = motion, n = none or nothing, pt = phase transition, ss = signal shape, str = structure, sw = signal width, w = water. The letter s, meaning plural, is frequently omitted.

TABLE 21

Collection of nMr data

Scheme	System	Measurement	Ref.
Review including nMr			151
Population & exchange of monom & mic surft	$\phi\text{-(CH}_2)_{4,\,5,\,6,\,8,\,10,\,12}\text{NMe}_3\text{Br}$ /D$_2$O	^{1}H cs, ss	22 24
Effect of mic formn on nMr ss	Me(CH$_2$)$_{7,\,9}$N$^+$Me$_2$CH$_2\phi$Me · Cl$^-$ /D$_2$O/NaCl	^{1}H cs, ss	63
Str of w in surft soln	Na C2, 4, 6, 8, 12yl sulfate/w	^{1}H cs of w	20
Chem shift of w in alk$_4$NBr solns	Me$_4-$, Et$_4-$, n–Pr$_4-$, n-Bu$_4$-NBr, CTAB/w	^{1}H cs	152

TABLE 21 (*Continued*)

Scheme	System	Measurement	Ref.
Mic formn & str, CMC	Promethazine/D_2O	^1H cs	122
Mic formn & str, CMC	Phenothiazines/D_2O, H_2O	^1H cs	123
Mic formn	$Me_3Si(CH_2)_3SO_3Na/D_2O$/ paramagnetic ions	^1H cs, sw	126
Str of w in surft soln	Na C2, 4, 6, 8, 12yl sulfate/w	T_1 of w ^1H	32
Mol motn of & environment around surft mol	Na C2, 4, 6, 8, 10, 12yl sulfate/D_2O	T_1 of CH_2 ^1H	33
CMC, relaxn (mic, monom)	Me_2C8ylamine oxide/D_2O	T_1 of ^1H	35
States of mic core, POE shell, & w	$C_{10}H_{21}O$ $(CH_2CH_2O)_5H/H_2O$, D_2O	T_1 of ^1H	37
Str of w in isotropic soln & meso; pt	Alk POE glycol monoethers/w	^1H cs, T_1	21
Mic formn	Na C2, 3, 4, 5, 6 ates/D_2O/n, NaCl; Li, Na, K, Rb, Cs C6ates/D_2O	^1H cs, T_1	153
W in mic core & POE shell, mol motn in mic	p-$Me_3CCH_2CMc_2$-ϕ-POE/n, (H_2O, D_2O), dioxane, $HOCH_2CH_2OH$, ϕ	^1H(F–T) cs, T_1	34
Prop of mic soln	SDS, K C10ate/w	^1H & ^{13}C T_1	10
Mol motn of monom & mic	C6, 8, 16-NMe_3Br/w	T_1 of ^{13}C	6
Mic str, w in mic core	Na C8ate/w	T_1 of ^{13}C	8
Mol motn in mic	SDS/w/n, NaCl	T_1 of ^{13}C	9
Mic formn, CMC, w in mic core	$CF_3(CH_2)_{8,10,11}COONa$/ H_2O, D_2O/n, salt, $CF_3\phi$; Na C10, 12ate/w/$CF_3\phi$	^{19}F cs	127
CMC, w in mic core, temp & salt effect	$CF_3(CH_2)_{8,10}COONa$/w/n, salt	^{19}F cs	128
Mic formn, w in mic core	C_3F_7COOH, $C_7F_{15}COONa/H_2O$, D_2O	T_1, T_2 of ^{19}F	45
CMC, mic str, ad effect	$CF_3(CH_2)_{11}SO_4Na$/w/n, org ad	^{19}F cs	129
Aggrn Nr, thermodynamic parameters of mic formn	$CF_3(CH_2)_7$POE ether, CF_3-$(CH_2)_7$-SO-Me/w/n, org ad	^{19}F cs	130
CMC, aggrn Nr, w in mic core	Na^+, K^+, $Me_4N^+ \cdot C_7F_{15}COO^-$/w/n, org ad	^{19}F cs	131
CMC, 2nd CMC, mic prop	$CF_3(CH_2)_{9,11}NMe_3^+ \cdot Br^-$, F^-, Cl^-, OH^-/w/n, salt	^{19}F cs	132
CMC	C_2F_5COONa, $C_7F_{15}COONa$/w	^{19}F cs	154
Mic formn, CMC	C_3F_7COOH/w	^{19}F cs	155
CMC	C8, 10, 12, 14acid/95% H_2SO_4; C12amm C4, 8, 14ate/CCl_4	^1H cs	156

TABLE 21 (*Continued*)

Scheme	System	Measurement	Ref.
CMC, aggrn Nr, equilib constant for mic formn	C4, 6, 8, 10, 12amm propionate /ϕ, CCl$_4$	^{1}H cs	27
CMC, aggrn Nr, equilib constant for mic formn	C8amm C3, 4, 6, 9, 12, 14ate/ ϕ, CCl$_4$	^{1}H cs	28
CMC, aggrn Nr, mic str	C4, 6, 8, 12amm C3ate/CDCl$_3$, CH$_2$Cl$_2$, ϕCl, CH$_3$CON(CH$_3$)$_2$	^{1}H cs	29
Str of mixed mic	$\phi-$(CH$_2$)$_5$NMe$_3$Br + $\phi-$(CH$_2$)$_8$-NMe$_3$Br/D$_2$O	^{1}H cs, ss	54
Str of mixed mic	C12-POE ether + C8ϕSO$_3$Na/w	^{1}H cs	55
Str of mixed mic	C12-POE ether + C8ϕSO$_3$Na, C4ϕC4SO$_3$Na, ϕC8SO$_3$Na/w; C8ϕSO$_3$Na + C12-POE-SO$_4$Na/w	^{1}H cs	56
Str of viscoelastic mic soln, mol motn	C$_8$H$_{17}$NMe$_3$Br, C$_{16}$H$_{33}$N$^+$Me$_2$-(CH$_2$)$_3$OSO$_2^-$/SDS/D$_2$O	^{1}H sw T_1, T_2	38 50
Loc solubn	C16-Py-Cl/w/ϕ, ϕBr, CH$_3$I, cyclohexane	^{1}H cs, ss	65
Population & exchange of free & solubd ϕ	SDS, C12-NMe$_3$Cl/D$_2$O, H$_2$O/ ϕ, cyclohexane	^{1}H cs, ss	23
Mode & loc solubn	C16-NMe$_3$Br/w/ϕ, ϕNMe$_2$, ϕNO$_2$, iso-C3ϕ, cyclohexane	^{1}H cs, sw	64
Mode & loc solubn	SDS/w/ϕ(CH$_2$)$_n$OH, Meϕ(CH$_2$)$_n$OH, H(CH$_2$)$_n\phi$OH	^{1}H cs	62
Mode of solubn	SDS/H$_2$O + D$_2$O/ϕOH	^{1}H cs	157
Loc solubn	C16-POE ether/w/ϕOH	^{1}H cs	158
Intn of ϕOH with POE & POE–POP polymer	POE polymer, POE-POP block polymer/w/ϕOH	^{1}H cs	159
Loc solubn	C16-POE ether/w/ϕCHO & or C3yl gallate	^{1}H cs	160
Loc solubn	C16-POE ether/D$_2$O/ϕCOOH	^{1}H cs	161
W in mic core	C12-POE ether, SDS, CTAB, Na C10, 12ate/w/ϕ/n, NaCl	^{1}H cs	133
Loc solubn	C12-$^+$NMe$_2$-(CH$_2$)$_3$SO$_3^-$/w/ϕ, ϕNO$_2$, other aromatic ad	^{1}H cs	66
Mode & loc solubn	SDS/D$_2$O/*p*-xylene/Mn^{2+}, Gd^{3+}, Cu^{2+}, Co^{2+}	^{1}H sw	162
Solubn by reversed mic	C12yl amm C3ate/ϕ, CD$_3$Cl, CH$_2$Cl$_2$/org ad	^{1}H cs, sw	30 31
Mol motn of guest mol in Na desoxycholate	Na desoxycholate/w/ *p*-xylene, 2-Me-naphthalene	T_1 of ^{13}C	7

152

TABLE 21 (*Continued*)

Scheme	System	Measurement	Ref.
Exchange of emulsified & solubd w	Tallow 1,3-propylenediamine dioleate/cyclohexane /H_2O, D_2O	^{1}H ss	67
W str in emulsion	SDS/w/ϕ	^{1}H cs of w	68
Pt, str of meso	Various systems	^{1}H sw, ss	163
Pt, mol motn	Anhyd K C10, 12, 14, 16, 18ate; Na C18ate	^{1}H sw	69 70
Pt, mol motn	Anhyd Rb, Cs C18ates	^{1}H sw	71
Pt, mol motn	Anhyd K C6, 8ate	^{1}H sw	72
Pt, mol motn	Anhyd Na C12, 14, 16, 18ate	^{1}H sw, ss	74
Pt, mol motn	Anhyd C18acid, Li C18ate	^{1}H sw, ss	73
Mol motn of soaps	Na C11–19ate, Li C14~17ate	^{1}H T_1	80
Pt, mol motn in & str of meso	Me_2-C12amine oxide/D_2O	^{1}H, ^{2}H sw, ss	36
Str of meso	K oleate/D_2O	^{1}H ss	82
Mol motn in soap-w meso	$C_{11}D_{23}COOK/H_2O$	^{2}H ss	83
Mol motn in meso	C8amm Cl, K oleate/w	^{1}H ss	84
Mol motn	K C12ate/D_2O	^{1}H T_1, T_2	164
Mol motn of w & amm ion, ^{1}H exchange between them, w-surft intn	Amm perfluoro-C8ate/w; mesomorphic phase	T_1, T_2 of ^{1}H, ^{7}H	165
Chain motn of surft in meso	$CF_3CF_2CF_2(CH_2)_{10}COONa/D_2O$	T_1, T_2 of ^{19}F, ^{1}H	85
Ratio of mobile & rigid domains	Na C10, 12, 14, 16, 18ate, oleate, linolate/H_2O	^{1}H ss, sig area	166
Mol motn in meso	C8yl amine/w; SDS/w/C8acid	^{1}H ss	4
Pt, mol motn & intn in meso	$C10$-NMe_3Br/w/$CHCl_3$; 2-amino-2-Me-1-C3ol-, 2-amino-1-C4ol oleate/w/ϕ	^{1}H cs	86
Mol intn, mol motn	Na C8ate/H_2O/C10ol	^{1}H cs, ^{23}Na sw	87
Str & mol motn in meso	Na C8ate/H_2O, D_2O/C10ol	^{1}H, T_1, T_2	48
Effect of surfaces on str & prop of w	Na C8ate/w/C10ol	^{1}H T_1, T_2	88
Mol motn, prop of meso	$C_7F_{15}COONH_4/H_2O$, D_2O/n, org ad	^{1}H T_1, ^{19}F T_1, T_2	89
Na^+ bing in C8acid rich phase	Na C8ate/C8acid/n, w	^{23}Na sw	146

153

Table 21 (*Continued*)

Scheme	System	Measurement	Ref.
Phase separation, state of w	Aerosol OT/w/n-octane/ diamagnetic salts	^{1}H cs	167
Prop of microemulsion	K oleate/w/C6o1/ϕ	^{1}H cs, ss	94
State & mol motn of w & surft in microemulsion	K oleate/1-hexanol/H_2O, D_2O/ϕ, Cl_2CCCl_2	^{1}H cs, sw ^{2}H T_1, T_2	95
W str of microemulsion	K oleate/C6o1/$C_{16}H_{34}$/w	^{1}H cs, sw	97
Intn of surft & protein	SDS/phycocyanin/D_2O buffer	^{1}H cs, sw	101
Bing of surft on BSA	$CF_3C_7H_{14}\phi SO_3Na$/BSA/w/salt	^{19}F cs, sw	103
Bing of surft on BSA	$CF_3C_{11}H_{22}SO_4Na$, $CF_3C_{12}H_{24}SO_4Na$/BSA/w/salt	^{19}F cs, sw	104
Competitive bing of surft & Cl$^-$ on BSA	Na C6, 8, 10, 12yl sulfate/ BSA/w/NaCl	^{35}Cl sw	105
Anal of POE-, POP-surft	C9ϕ-POE (POP), C12-POP ether; POE–POP copolymer	^{1}H sig area	106
Anal of POE-surft	POE derivative of alk-ϕOH	^{1}H sig area	107
Anal of POE-surft	POE derivative of alk-ϕOH, n-alcohol, POP, etc.	^{1}H sig area	108
Anal of POE-surft	Alk-ϕ, alk-ϕOH, & their POE derivatives	^{1}H sig area	109
Anal of nonionic surft	Polyalkylene glycols, glycol polyesters	^{1}H	110
Mol weight distribution	Amm polyether surft	^{1}H	111
Anal of surft	Alk ethoxylates	^{1}H sig area	112
Determination of HLB	Nonionic surft	^{1}H sig area	113
Idn of o,p-isomers	Ethoxylated alk-ϕOH surft	^{1}H cs	114
Chem str of betaine surft	C8, 10, 12yl betaine, N-C8yl betaine/D_2O	^{1}H ss	115
Chem str of N-alk betaines	N-C10, 11, 12, 14, 16yl betaine/D_2O	^{1}H ss	116
NMR catalog of surft	Numerous surft	^{1}H	117
Idn of surft	. Numerous surft	^{1}H cs	118
nMr prop	25 Na alk sulfonates/D_2O	^{1}H	168
Anal	Alkene sulfonates/D_2O	^{1}H	169
Str of C8acid rich soln	Na C8ate/C8acid/n, w	^{1}H cs, sw	119
Intn of fatty acid & POE alk ether	C12-POE ether/C8acid/n, CCl_4	^{1}H cs, ss	120
C-ion bing in mic soln	C16-Py-Br, C8amine-HBr, CTAB/w	^{81}Br sw	134
C-ion bing in mic soln	Rb C6, 8ate/w	^{85}Rb sw	141

TABLE 21 (*Continued*)

Scheme	System	Measurement	Ref.
C-ion bing	Mono, di, trialk amm Br/w	^{79}Br sw	135
C-ion bing	ternary amphiphile-w systems	^{2}H, ^{23}Na sw, ss	147
C-ion bing	C8yl amm Cl/C10ol/w	^{35,37}Cl ss	149
C-ion bing, mic shape, loc solubn	CTAB/w/n, ϕ, ϕNMe$_2$, C6ol, cyclohexane	^{81}Br sw	137
Bing of alkali & halide ions to mic, CMC	Na C8ate, Cs C8ate, C8amm chloride/w	^{23}Na, ^{35}Cl, ^{133}Cs cs	150
C-ion bing on reversed mic	CTAB/C6ol/w	^{81}Br sw	145
C-ion bing	SDS/H$_2$O/Mn^{2+}, Gd^{3+}, Cu^{2+}, Na$^+$ salt	T_1, T_2 of ^{1}H, ^{23}Na	142
C-ion bing	(Li$^+$, Na$^+$, Rb$^+$, Cs$^+$). (C$_n$H$_{2n+1}$SO$_4^-$, C$_n$H$_{2n+1}$COO$^-$) /H$_2$O	T_1 of ^{7}Li ^{23}Na, ^{87}Rb, ^{113}Cs	143
Na$^+$ bing on cellular membranes	Lecithin/Na cholate/w (as membrane model)	^{23}Na ss, T_1	148
Bing of Mn^{2+} on mic	SDS/w/n, C12ol/n, MnSO$_4$	T_1, T_2 of w ^{1}H	170

Abbreviations used: ad = additive, aggrn = aggregation, alk = alkyl, amm = ammonium, anal = analysis, anhyd = anhydrous, bing = binding, BSA = bovine serum albumin, Bu = butyl, C8ol = octanol, C8yl = octyl, chem = chemical, c-ion = counterion, cs = chemical shift, CTAB = cetyltrimethylammonium bromide, equilib = equilibrium, Et = ethyl, formn = formation, HLB = hydrophilic lipophilic balance, idn = identification, intn = interaction, loc = locus or loci of, Me = methyl, meso = mesomorphic phase, mic = micelle or micellar, mol = molecule or molecular, monom = monomer or monomeric, motn = motion, n = none or nothing, Nr = number, org = organic, POE = polyoxyethylene, POP = polyoxypropylene, Pr = propyl, prop = property, pt = phase transition, Py = pyridinium, relaxn = relaxation, SDS = sodium dodecyl sulfate, sig = signal, soln = solution, solubd = solubilized, solubn = solubilization, ss = signal shape, str = structure, surft = surfactant, sw = signal width, temp = temperature, w = water, ϕ = benzene or benzene ring.

ABBREVIATIONS OF SURFACTANTS AND OTHER COMPOUNDS*

C$_n$AP   Alkylammonium propionate
C$_n$DAO   Alkyldimethylamine oxide
C$_n$(EO)$_p$ Alkyl polyoxyethylene ether

---

*n = the number of carbon atoms in alkyl chain; p = the number of oxyethylene units per molecule.

155

$C_n(EO)_pC_1$	Terminally methylated $C_n(EO)_p$
C_nTAB	Alkyltrimethylammonium bromide
DTO	Tallow-1,3-propylenediamine dioleate
$NaC_4\phi C_4S$	Sodium δ-(p-butylphenyl)-butyl sulfonate
$NaC_8\phi S$	Sodium p-octylbenzene sulfonate
$NaC_{12}(EO)_pS$	Sodium dodecyl polyoxyethylene sulfate
NaC_nS	Sodium alkyl sulfate or sulfonate
NaF_3C_nS	Sodium ω,ω,ω-trifluoroalkyl sulfate
$NaF_3C_8\phi S$	Sodium p-(8,8,8,-trifluoro) octylbenzene sulfonate
$Na\phi C_8S$	Sodium ω-phenyloctyl sulfonate
ϕC_nTAB	ω-Phenylalkyltrimethylammonium bromide
BSA	Bovine serum albumin
$C_1\phi C_nOH$	ω-(p-Methylphenyl)-alkyl alcohol
$C_n\phi OH$	p-Alkylphenol
ϕC_nOH	ω-Phenylalkyl alcohol

SYMBOLS

b	Solubility of benzene in g/ml surfactant solution
b_M	Saturation amount of solubilized benzene in g/g micellar surfactant
b_w	Solubility of benzene in g/ml water
C	(Total) concentration of surfactant
C	Molar ratio of two surfactants
C_0	Concentration of monomeric surfactant
CMC	Critical micelle concentration
D_b, D_f	Concentrations of bound and free surfactant
d	Thickness of bound-water layer of a microemulsion droplet (see Figure 38)
E	Activation energy
E_{act}	Activation energy associated with molecular motion
E_T	Solvent polarity parameter
e	Proton charge
f_B, f_F	Mole fractions of bound and free water
g	Nuclear g factor
h	Planck's constant
h	Arbitrary constant (equation 8)
h	Half-height-width of a resonance signal (equation 14)
$\hbar$	$= h/2\pi$
I	Nuclear spin number
i	ith oxyethylene unit

J_0	Bessel function
K	Equilibrium constant
L_1, L_2	Regions in a phase diagram (see Figure 34)
m	Aggregation number of a micelle
N	Number of moles of water protons (equation 4)
N	Number of hydrogens directly attached to a carbon atom (equation 33)
N	Number of protons (equation 17)
n	Number of methylene units in the hydrophobic part of a surfactant molecule
$\bar{n}$	Average number of surfacatant ions bound per BSA molecule
n_w/n_s	Molar ratio of water to surfactant in a microemulsion droplet
P_A, P_B	Fraction of protons in alkyl chains and that in water (equation 21)
P_{bi}	Mole fraction of ions bound to the ith kind of sites
P_{free}	Mole fraction of free ions
p	Number of oxyethylene units per molecule (for polyoxyethylene-type surfactants)
p	pth oxyethylene unit
p	Fractional population in an environment
Q	Quadrupole moment
q	Number of nonionic surfactant molecules influenced by one anionic surfactant molecule
q	Electric field gradient scalar
R	Gas constant per mole
$R(8/2)$	Ratio of signal width at $\frac{1}{8}$ height to that at $\frac{1}{2}$ height
r_{F-F}	Fluorine–fluorine internuclear distance
r_{jf}	Distance between proton j and nucleus f other than proton
r_{jk}	Distance between protons j and k
r_w	Radius of water droplet (see Figure 38)
S	Solubilizing capacity
S	Order parameter (equation 20)
S	Second moment of resonance spectrum
S_{rigid}	Rigid-lattice moment
T	Absolute temperature
T_1	Spin–lattice relaxation time
T_2	Spin–spin relaxation time
T_2'	Effective spin–spin relaxation time
t	Average residence time of a nucleus or molecule
$\bar{v}$	Partial specific volume of surfactant
X	Mole fraction

x	Shift from the water peak
Z	Parameter measuring the degree of hydrocarbon-like character (equation 32)
α	Angle between axis of reorientation and internuclear vector (equation 19)
α	Factor referring to broad-line nMr (equation 15)
γ	Magnetogyric ratio
γ	Angular amplitude of torsional oscillation (equation19)
ΔH	Signal width (gauss)
$\Delta H'$	Signal width before motional narrowing has taken place
$\Delta H''$	Signal width after motional narrowing has taken place
ΔH_{bi}	Signal width of ions bound to the ith kind of sites
ΔH_{bound}	Signal width of bound ions
ΔH_{free}	Signal width of free ions
ΔS	Synergistic effect in solubilization
$\Delta\nu$	Shift from a referential peak
$\Delta\nu$	Doublet spacings (Hz, see Figure 32)
$\Delta\nu, \ \Delta\nu_{1/2}$	Signal width
$\Delta\nu_{bi}$	Signal width of ions bound to the ith kind of sites
$\Delta\nu_{\text{free}}$	Signal width of free ions
δ	Chemical shift (ppm)
δ_b	Chemical shift of bound surfactant
δ_M	Chemical shift of micellar surfactant
δ_m	Chemical shift of monomeric surfactant
δ_r	Chemical shift of surfactant forming protein–surfactant complex AD_r
η	Asymmetry parameter
μ	Nuclear magnetic moment
μ_0	Nuclear magneton
ν	Observed chemical shift (Hz or ppm)
ν_0	NMR frequency
ν_a	Weight-averaged chemical shift of monomeric and micellar species
ν_M	Chemical shift of micellar surfactant
ν_m	Chemical shift of monomeric surfactant
ρ	$= S/S_{\text{rigid}}$ for torsional oscillation (equation 19)
τ	Average residence time of a nucleus or molecule
τ_0	Constant (equation 16)
τ_c	Correlation time
τ_{eff}	Effective correlation time
ϕ	Angle between external magnetic field and normal to lamellae of oriented crystallites (equation 20).

ϕ_{ab} Angle between axis of motion and radius vector connecting nuclei a and b (equation 18)

Subscript (unless otherwise specified above)

1	Refers to component "1"
2	Refers to component "2"
A, B	Refer to environments A and B
a, b	Refer to water protons near to and far from solute molecules (equation 4)
f	Refers to nucleus other than proton (equations 17, 18)
H_2O	Refers to water
H_A	Refers to H_A proton(s)
H_B	Refers to H_B proton(s)
i	Refers to ith oxyethylene unit
i, j	Refer to states i and j
inter	Intermolecular
intra	Intramolecular
M	Refers to micellar species
m	Refers to monomeric species
sat	Saturation value
w	Refers to water or aqueous phase

REFERENCES

1. F. M. Purcell, H. C. Torrey, and R. V. Pound, *Phys. Rev.*, **69**, 37 (1946).
2. F. Bloch, W. W. Hansen, and M. E. Packard, *Phys. Rev.*, **69**, 127 (1946).
3. R. D. Spence, H. A. Moses, and P. L. Lain, *J. Chem. Phys.*, **21**, 380 (1953); R. D. Spence, H. S. Gutowsky, and C. H. Holm, ibid., **21**, 1891 (1953); P. L. Jain, H. A. Moses, J. C. Lee, and R. D. Spence, *Phys. Rev.*, **92**, 844 (1953).
4. M. P. McDonald, *Arch. Sci. (Geneva)*, **12**, Fasc. spec., 141 (1959).
5. G. C. Levy and G. L. Nelson, *Carbon-13 Nuclear Magnetic Resonance for Organic Chemists*, Wiley, New York, 1972.
6. E. Williams, B. Sears, A. Allerhand, and E. H. Cordes, *J. Amer. Chem. Soc.*, **95**, 4871 (9173).
7. D. Leibfritz and J. D. Roberts, *J. Amer. Chem. Soc.*, **95**, 4996 (1973).
8. T. Drakenberg and B. Lindman, *J. Colloid Interface Sci.*, **44**, 184 (1973).
9. R. T. Roberts and C. Chachaty, *Chem. Phys. Lett.*, **22**, 348 (1973).
10. M. Alexandre, C. Foughet, and P. Rigny, *J. Chim. Phys.*, **70**, 1073 (1973).
11. J. A. Pople, W. C. Schneider, and H. J. Bernstein, *High-Resolution Nuclear Magnetic Resonance*, McGraw-Hill, New York, 1959.
12. C. P. Slichter, *Principles of Magnetic Resonance*, Harper & Row, New York, 1963.

13. A. Abragam, *Principles of Nuclear Magnetism,* Oxford, 1962.

14. A. Carrington and A. D. McLachlan, *Introduction to Magnetic Resonance*, Harper & Row, New York, 1967.

15. A. Kowalsky and M. Cohn, *Ann. Rev. Biochem.*, **33**, 481 (1964).

16. O. Jardetzky, *Advan. Chem. Phys.,* **7**, 499 (1964).

17. B. Sheard and E. M. Bradbury, *Progr. Biophys. Mol. Biol.*, **20**, 187 (1970).

18. A. S. V. Burgen and J. C. Metcalfe, *J. Pharm. Pharmacol.*, **22**, 153 (1970).

19. N. J. Shoolery and B. J. Alder, *J. Chem. Phys.*, **23**, 805 (1955).

20. J. Clifford and B. A. Pethica, *Trans. Faraday Soc.,* **60**, 1483 (1964).

21. J. M. Corkill, J. F. Goodman, and J. Wyer, *Trans. Faraday Soc.,* **65**, 9 (1969).

22. T. Nakagawa and H. Inoue, *Proc. Int. Cong. Surface Act. Substances*, Brussels, 1964, Vol. II, Sect. B, p. 569.

23. T. Nakagawa and K. Tori, *Kolloid-Z. Z. Polym.*, **194**, 143 (1964).

24. T. Nakagawa, H. Inoue, H. Jizomoto, and K. Horiuchi, *Kolloid-Z. Z. Polym.,* **229**, 159 (1969).

25. P. H. Elworthy, A. T. Florence, and C. B. Macfarlane, *Solubilization by Surface Active Agents*, Chapman and Hill, London, 1968.

26. A. Kitahara, "Micelle Formation of Cationic Surfactants in Nonaqueous Media," in *Cationic Surfactants*, E. Jungermen, Ed., Marcel Dekker, New York, 1970, p. 289.

27. J. H. Fendler, E. J. Fendler, R. T. Medary, and O. A. EL Seoud, *J. Chem. Soc. Faraday Trans. I.*, **69**, 280 (1973).

28. E. J. Fendler, J. H. Fendler, R. T. Medary, and O. A. El Seoud, *J. Phys. Chem.*, **77**, 1432 (1973).

29. O. A. El Seoud, E. J. Fendler, J. H. Fendler, and R. T. Medary, *J. Phys. Chem.*, **77**, 1876 (1973).

30. O. A. El Seoud, E. J. Fendler, and J. H. Fendler, *J. Chem. Soc. Faraday Trans. I,* **70**, 450 (1974).

31. O. A. El Seoud, E. J. Fendler, and J. H. Fendler, *J. Chem. Soc. Faraday Trans. I,* **70**, 459 (1974).

32. J. Clifford and B. A. Pethica, *Trans. Faraday Soc.*, **61**, 182 (1965).

33. J. Clifford, *Trans. Faraday Soc.*, **61**, 1276 (1965).

34. F. Podo, A Ray, and G. Nemethy, *J. Amer. Chem. Soc.*, **95**, 6164 (1973).

35. K. D. Lawson and T. J. Flautt, *J. Phys. Chem.*, **69**, 3204 (1965).

36. K. D. Lawson and T. J. Flautt, *J. Phys. Chem.*, **72**, 2066 (1968).

37. C. J. Clemett, *J. Chem. Soc. (A)*, 2251 (1970).

38. C. A. Barker, D. Saul, G. J. T. Tiddy, B. A. Wheeler, and E. Willis, *J. Chem. Soc. Faraday Trans. I,* **70**, 154 (1974).

39. J. A. Pople, W. G. Schneider, and H. S. Bernstein, *High-Resolution Nuclear Magnetic Resonance*, McGraw-Hill, New York, 1959, pp. 82, 83.

40. J. M. Corkill and K. W. Herrmann, *J. Phys. Chem.*, **67**, 935 (1963).

41. L. Benjamin, *J. Phys. Chem.* **68**, 3575 (1964).

42. J. R. Zimmerman and J. A. Lasater, *J. Phys. Chem.*, **62**, 1157 (1958).

43. J. G. Powles, *Ber.*, **67**, 328 (1963).

44. D. E. Woessner, *J. Chem. Phys.*, **41**, 84 (1964).

45. U. Henriksson and L. Ödberg, *J. Colloid Interface Sci.*, **46**, 212 (1974).

46. N. Bloembergen, E. M. Purcell, and R. V. Pound, *Phys. Rev.*, **73**, 679 (1948).

47. D. E. Woessner, *J. Chem. Phys.*, **36**, 1 (1962).

48. G. J. T. Tiddy, *J. Chem. Soc. Faraday Trans. I*, **68**, 369 (1972).

49. C. A. Barker, D. Saul, G. J. T. Tiddy, B. A. Wheeler, and E. Willis, *J. Chem. Soc. Faraday Trans. I*, **70**, 154 (1974).

50. D. Saul, G. J. T. Tiddy, B. A. Wheeler, P. A. Wheeler, and E. Willis, *J. Chem. Soc. Faraday Trans.* I, **70**, 163 (1974).

51. H. W. Hoyer and A. Marno, *J. Phys. Chem.*, **65**, 1807 (1961).

52. F. Tokiwa, *J. Colloid Interface Sci.*, **28**, 145 (1968).

53. F. Tokiwa and N. Moriyama, *J. Colloid Interface Sci.*, **30**, 338 (1969).

54. H. Inoue and T. Nakagawa, *J. Phys. Chem.*, **70**, 1108 (1966).

55. F. Tokiwa and K. Tsujii, *J. Phys. Chem.*, **75**, 3560 (1971).

56. F. Tokiwa and K. Tsujii, *J. Colloid Interface Sci.*, **42**, 343 (1972).

57. F. Tokiwa and K. Tsujii, *Bull. Chem. Soc. Japan*, **46**, 1338 (1973).

58. F. Tokiwa, *J. Phys. Chem.*, **72**, 1214 (1968); **72**, 4331 (1968).

59. S. I. Chan, M. P. Schweizer, P. O. P. Ts'o, and G. K. Helmkamp, *J. Amer. Chem. Soc.*, **86**, 4182 (1964).

60. E. S. Hand and T. Cohen, *J. Amer. Chem. Soc.*, **87**, 133 (1965).

61. T. Nakagawa, *J. Japan Oil Chem. Soc.*, **22**, 181 (1973); and references cited therein; *Colloid Polym. Sci.*, **252**, 56 (1974).

62. F. Tokiwa and K. Aigami, *Kolloid-Z. Z. Polym.*, **246**, 688 (1971).

63. R. E. Atkinson, G. E. Clint, and T. Walker, *J. Colloid Interface Sci.*, **46**, 32 (1974).

64. J. C. Eriksson and G. Gillberg, *Acta Chem. Scand.*, **20**, 2019 (1966).

65. J. C. Eriksson, *Acta Chem. Scand.*, **17**, 1478 (1963).

66. E. J. Fendler, C. L. Day, and J. H. Fendler, *J. Phys. Chem.*, **76**, 1460 (1972).

67. P. D. Cratin and B. K. Robertson, *J. Phys. Chem.*, **69**, 1087 (1965).

68. T. Okamura, I. Satake, and R. Matsuura, *Bull. Chem. Soc. Japan*, **39**, 1071 (1966).

69. R. F. Grant and B. A. Dunell, *Can. J. Chem.*, **38**, 1951 (1960); **38**, 2395 (1960); **39**, 359 (1961).

70. M. R. Barr and B. A. Dunell, *Can. J. Chem.*, **42**, 1098 (1964).

71. D. J. Shaw and B. A. Dunell, *Trans. Faraday Soc.*, **58**, 132 (1962).

72. W. R. Janzen and B. A. Dunell, *Trans. Faraday Soc.*, **59**, 1260 (1963).

73. T. J. R. Cyr, W. R. Janzen, and B. A. Dunell, *Advances in Chemistry Series*, No. 63, Amer. Chem. Soc. Washington, D. C., 1967, Chapter 2, p. 13.

74. K. D. Lawson and T. J. Flautt, *J. Phys. Chem.*, **69**, 4256 (1965).

75. H. S. Gutowsky and G. E. Pake, *J. Chem. Phys.*, **18**, 162 (1950).

76. H. Frohlich, *Theory of Dielectrics*, 2nd ed. Oxford Univ. Press, 1958, p. 128.

77. D. W. McCall and W. P. Slichter, *J. Polym. Sci.*, **26**, 171 (1957).

78. H. S. Gutowsky, G. B. Kistakowsky, G. E. Pake, and E. M. Purcell, *J. Chem. Phys.*, **17**, 972 (1950).

79. E. R. Andrew, *J. Chem. Phys.*, **18**, 607 (1950).

80. K. van Putte, *Trans. Faraday Soc.*, **66**, 523 (1970).

81. V. Luzzati, *Biological Membranes*, D. Chapman, Ed., Academic, New York, 1968.

82. J. J. DeVries and H. J. C. Berendsen, *Nature*, **221**, 1139 (1969).

83. J. Charvolin, P. Manneville, and B. Deloche, *Chem. Phys. Lett.*, **23**, 345 (1973).

84. C. Dijkema and H. J. C. Berendsen, *J. Magnet. Res.*, **14**, 251 (1974).

85. G. J. T. Tiddy, *J. Chem. Soc. Faraday Trans. I*, **68**, 670 (1972).

86. I. A. Zlochower and J. H. Schulman, *J. Colloid Interface Sci.*, **24**, 115 (1967).

87. G. Gillberg and P. Ekwall, *Acta Chem. Scand.*, **21**, 1630 (1967).

88. J. Clifford, J. Oakes, and G. J. T. Tiddy, *Spec. Discuss. Faraday Soc.*, **1**, 175 (1970).

89. G. J. T. Tiddy, *Symposia Faraday Soc.*, **111**, 150 (1971).

90. D. Gill and S. Meiboom, *Rev. Sci. Instrum.*, **9**, 101 (1959).

91. K. Fontell, L. Mandell, H. Lehtinen, and P. Ekwall, *Acta Polytech. Scand.*, 1968, Ch. 74. I–III.

92. J. H. Schulman, W. Stoeckenius, and L. Prince, *J. Phys. Chem.*, **63**, 1677 (1959).

93. P. Becher, *Emulsions: Theory and Practice*, 2nd ed., Reinhold, New York, 1966, p. 297.

94. G. Gillberg, H. Lehtinen, and S. Friberg, *J. Colloid Interface Sci.*, **33**, 40 (1970).

95. J. R. Hansen, *J. Phys. Chem.*, **78**, 256 (1974).

96. J. E. L. Bowcott and J. H. Schulman, *Z. Elektrochem.*, **59**, 283 (1955).

97. D. O. Shah and R. M. Hamlin, Jr., *Science*, **171**, 483 (1971).

98. F. W. Putnam and H. Neurath, *J. Biol. Chem.*, **159**, 195 (1945).

99. (a) C. Tanford, *The Hydrophobic Effect: Formation of Micelles and Biological Membranes*, Interscience, New York, 1973. (b) C. Tanford, *J. Amer. Chem. Soc.*, **84**, 4240 (1962).

100. See references 3–17 cited in ref. 101.

101. R. M. Rosenberg, H. L. Crespi, and J. J. Katz, *Biochim. Biophys. Acta*, **175**, 31 (1969).

102. R. M. Rosenberg, H. L. Crespi, and J. J. Katz, *Biochim. Biophys. Acta*, **175**, 34 (1969).

103. T. W. Johnson and N. Muller, *Biochem.*, **9**, 1943 (1970).

104. N. Muller and R. J. Mead, Jr., *Biochem.*, **12**, 3831 (1973).

105. J. A. Magnuson and N. S. Magnuson, *J. Amer. Chem. Soc.*, **94**, 5461 (1972).

106. H. Walz and H. Kirschnek, *Original Lectures in 3rd Int. Congr. Surface Activity C*, p. 92 (1960).

107. P. W. Flanagan, R. A. Greff, Jr., and H. F. Smith, *Spectrochim. Acta*, **18**, 891 (1962).

108. R. A. Greff, Jr., and P. W. Flanagan, *J. Amer. Oil Chem. Soc.*, **40**, 118 (1963).

109. M. M. Crutchfield, R. R. Irani, and J. T. Yoder, *J. Amer. Oil Chem. Soc.*, **41**, 129 (1964).

110. T. F. Page, Jr., and W. E. Bresler, *Anal. Chem.*, **36**, 1981 (1964).

111. N. Kalish, L. Weintraub, and M. Kowblansky, *J. Soc. Cosmet. Chemists*, **23**, 153 (1972).

112. C. K. Cross and A. C. Mackay, *J. Amer. Oil Chem. Soc.*, **50**, 249 (1973).

113. G. Ben-Et and D. Tatarsky, *J. Amer. Oil Chem. Soc.*, **49**, 499 (1972).

114. G. E. Stolzenberg, R. G. Zaylskie, and P. A. Olson, *Anal. Chem.*, **43**, 908 (1971).

115. K. Tori and T. Nakagawa, *Kolloid-Z. Z. Polym.*, **187**, 44 (1963); **188**, 47 (1963).

116. P. Molyneux, C. T. Rhodes, and J. Swarbrick, *Trans. Faraday Soc.*, **61**, 1043 (1965).

117. H. Koenig, *Z. Anal. Chem.*, **251**, 225 (1970).

118. H. Koenig, *Tenside*, **8**, 63 (1971).

119. S. Friberg, L. Mandell, and P. Ekwall, *Kolloid-Z. Z. Polym.*, **233**, 955 (1969).

120. B. Lincoln, S. Friberg, and S. Gravsholt, *Colloid Polym. Sci.*, **252**, 39 (1974).

121. D. Attwood, A. T. Florence, and J. M. N. Gillan, *J. Pharm. Sci.*, **63**, 988 (1974).

122. A. T. Florence and R. T. Parfitt, *J. Pharm. Pharmacol.*, **22**, 121S (1970).

123. A. T. Florence and R. T. Parfitt, *J. Phys. Chem.*, **75**, 3554 (1971).

124. M. P. Schweizer, S. I. Chan, and P. O. P. Ts'o, *J. Amer. Chem. Soc.*, **87**, 5241 (1965).

125. D. J. Blears and S. S. Danyluk, *J. Amer. Chem. Soc.*, **89**, 21 (1967).

126. B. R. Donaldson and J. C. P. Schwarz, *J. Chem. Soc. (B)*, 395 (1968).

127. N. Muller and R. H. Birkhahn, *J. Phys. Chem.*, **71**, 957 (1967).

128. N. Muller and R. H. Birkhahn, *J. Phys. Chem.*, **72**, 583 (1968).

129. N. Muller and T. W. Johnson, *J. Phys. Chem.*, **73**, 2042 (1969).

130. N. Muller and F. E. Platko, *J. Phys. Chem.*, **75**, 547 (1971).

131. N. Muller and H. Simsohn, *J. Phys. Chem.*, **75**, 942 (1971).

132. N. Muller, J. H. Pellerin, and W. W. Chen, *J. Phys. Chem.*, **76**, 3012 (1972).

133. J. E. Gordon, J. C. Robertson, and R. L. Thorne, *J. Phys. Chem.*, **74**, 957 (1970).

134. J. C. Eriksson, Å. Johansson, and L. Andersson, *Acta Chem. Scand.*, **20**, 2301 (1966).

135. B. Lindman, H. Wennerström, and S. Forsén, *J. Phys. Chem.*, **74**, 754 (1970).

136. M. Shirai and B. Tamamuahi, *Bull. Chem. Soc. Japan*, **29**, 733 (1956); **30**, 411 (1957).

137. G. Lindblom, B. Lindman, and L. Mandell, *J. Colloid Interface Sci.*, **42**, 400 (1973).

138. (a) P. Ekwall, L. Mandell, and P. Solyom, *J. Colloid Interface Sci.*, **35**, 519 (1971).
(b) K. G. Götz and K. Heckmann, *J. Colloid Sci.*, **13**, 206 (1958); *Z. Physik. Chem.*, **20**, 42 (1959). (c) E. Graber, J. Lang, and R. Zana, *Kolloid-Z. Z. Polym.*, **238**, 470 (1970).

139. F. Reiss-Husson and V. Luzzati, *J. Phys. Chem.*, **68**, 3504 (1964).

140. P. Ekwall, L. Mandell, and K. Fontell, *J. Colloid Interface Sci.*, **29**, 639 (1969).

141. B. Lindman and I. Danielsson, *J. Colloid Interface Sci.*, **39**, 349 (1972).

142. I. D. Robb, *J. Colloid Interface Sci.*, **37**, 521 (1971).

143. I. D. Robb and R. Smith, *J. Chem. Soc. Faraday Trans. I.*, **70**, 287 (1974).

144. P. Ekwall, *J. Colloid Interface Sci.*, **29**, 16 (1969).

145. G. Lindblom, B. Lindman, and L. Mandell, *J. Colloid Interface Sci.*, **34**, 262 (1970).

146. B. Lindman and P. Ekwall, *Kolloid-Z. Z. Polym.*, **234**, 1115 (1969).

147. N. Persson and Å. Johansson, *Acta Chem. Scand.*, **25**, 2118 (1971).

148. G. Lindblom, *Acta Chem. Scand.*, **25**, 2767 (1971).

149. G. Lindblom, H. Wennerström, and B. Lindman, *Chem. Phys. Lett.*, **8**, 487 (1971).

150. H. Gustavsson and B. Lindman, *J. C. S. Chem. Commun.*, 93 (1973).

151. P. A. Winsor, *Chem. Rev.*, **68**, 1 (1968).

152. A. LoSurdo and H. E. Wirth, *J. Phys.*, **76**, 130 (1972).

153. L. Oedberg, B. Svens, and I. Danielsson, *J. Colloid Interface Sci.*, **41**, 298 (1972).

154. R. Haque, *J. Phys. Chem.*, **72**, 3056 (1968).

155. R. E. Bailey and G. H. Cady, *J. Phys. Chem.*, **73**, 1612 (1969).

156. J. F. Yan and M. B. Palmer, *J. Colloid Interface Sci.*, **30**, 177 (1969).

157. J. J. Jacobs, R. A. Anderson, and T. R. Watson, *J. Pharm. Pharmacol.*, **23**, 148 (1971).

158. J. J. Jacobs, R. A. Anderson, and T. R. Watson, *J. Pharm. Pharmacol.*, **23**, 786 (1971).

159. J. J. Jacobs, R. A. Anderson, and T. R. Watson, *J. Pharm. Pharmacol.*, **24**, 586 (1972).

160. J. J. Jacobs and G. A. Groves, *J. Pharm. Pharmacol.*, **25**, 844 (1973).

161. M. Donbrow and C. T. Rhodes, *J. Pharm. Pharmacol.*, **18**, 424 (1966).

162. K. K. Fox, I. D. Robb, and R. Smith, *J. Chem. Soc. Faraday Trans. I.*, **68**, 445 (1972).

163. T. J. Flautt and K. D. Lawson, *Advances in Chemistry Series*, No. 63, Amer. Chem. Soc. Washington, D. C., 1967, Chapter 3, p. 26.

164. J. Charvolin and P. Rigny, *Mol. Cryst. Liq. Cryst.*, **15**, 211 (1971); *J. Magn. Resonance*, **4**, 40 (1971).

165. G. J. T. Tiddy, *J. Chem. Soc. Faraday Trans. I.*, **68**, 653 (1972).

166. S. Noguchi, Nippon Nogei Kagaku Kaishi (*J. Agri. Chem. Soc. Japan*), **34**, 416 (1960). In Japanese.

167. S. G. Frank, Y. Shaw, and N. C. Li, *J. Phys. Chem.*, **77**, 238 (1973).

168. S. Nakanishi and F. Yoshimura, *Kogyo Kagaku Zasshi (J. Chem. Soc. Japan, Ind. Chem. Sect.)*, **73**, 1134 (1970). In Japanese.

169. T. Nagai, I. Tamai, S. Hashimoto, I. Yamane, and A. Mori, *Kogyo Kagaku Zasshi (J. Chem. Soc. Japan, Ind. Chem. Sect.)*, **74**, 32 (1971). In Japanese.

170. J. Oakes, *J. Chem. Soc. Faraday Trans. II.*, **69**, 1321 (1973).

3

Micellar Aspects of Casein

D.G. Schmidt and T.A.J. Payens

Netherlands Institute for Dairy Research
Ede, The Netherlands

I. INTRODUCTION

This review deals with the micellar properties of casein, the major protein constituent of milk. As is demonstrated by Figure 1, in milk casein largely

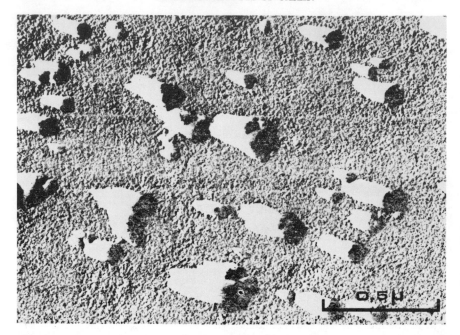

Fig. 1.   Electron micrograph of casein micelles in fresh skim milk. Shadow-casting with platinum-carbon. Note the triangular shadows, characteristic of collapsed spherical particles.

occurs as roughly spherical particles of colloidal dimensions, which give the milk its white appearance. The diameter of these so-called casein micelles varies from 20 to 300 mn. As we shall see, apart from the protein, the micelles also contain a small (about 5 %) but essential proportion of inorganic constituents, such as Ca, Mg, and phosphate and citrate.

The colloidal stability of casein micelles is remarkable: They survive the severe heat treatments during pasteurization or sterilization of food products and can easily be redispersed after the complete drying of milk to milk powder. Two factors have been recognized to be of primary importance for the stability of the micelle. The first of these is that the stability is clearly electrocratic in nature. Thus casein can be precipitated from milk by simple acidification to pH 4.6, that is the isoelectric point of casein (1). It has also been observed (2–4) that the curdling of milk by milk-clotting enzymes, which forms the basic process of the cheese industry, is accompanied by a drastic decrease of the zeta potential of the particles. Both observations have gained wide application with the winning of casein, either on a laboratory or industrial scale (5). The other factor governing casein stability goes back to Linderstrøm Lang's important discovery in 1929 (6) that casein is not a homogeneous protein, but can be separated into Ca-sensitive and Ca-insensitive fractions. In particular Lang was able to demonstrate that the latter can stabilize the former against flocculation by the calcium ions, which normally are present in the milk. This led Lang (6) to the introduction of another colloid chemical concept; the protective colloid theory of the micelle. Though in recent years the presence of a protective layer of the Ca-insensitive fraction on the surface of the micelle has been rejected for good experimental reasons (7–9), the very existence of a Ca-stable casein has been firmly re-established. Starting with the rediscovery of this so-called κ-casein in 1956 by Waugh and Von Hippel (10), numerous investigators (11–16) have confirmed its stabilizing action towards the Ca-sensitive casein fractions, α_{s1}- and β-casein. Interaction between these and κ-casein leads to protein complexes of decreased calcium sensitivity (see Section III). An example is given in Figure 2, from which it is seen that the relative amount of calcium-sensitive α_{s1}-casein dispersed in $CaCl_2$ solution increases with the concentration of κ-casein. The key position that κ-casein holds with respect to the stability of the micelle is further demonstrated by the fact that the curdling of milk is initiated by the specific proteolysis of this protein by milk-clotting enzymes (17). This reaction forms the molecular explanation of the decrease of the zeta potential of clotting micelles, as will be later explained.

It has been shown by means of electron microscopy (18–20) that casein micelles are not compact particles but consist of several thousands of

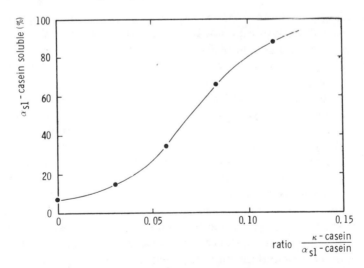

Fig. 2.   Solubilization of α_{s1}-casein in a 0.02 M $CaCl_2$ solution by increasing amounts of κ-casein. Experimental conditions: initial α_{s1}-casein concentration 3×10^{-3} g/ml; temperature 30°C; pH 6.7. (After C.A. Zittle (11), by permission of the American Dairy Association.)

Fig. 3.   Electron micrograph of casein micelles in fresh skim milk. Specimen prepared by the freeze-fracturing technique; shadowing with platinum-carbon. Note the submicellar structure.

"submicelles," which give the particle a strawberrylike appearance (see Figures 1 and 3). Dialysis experiments (Section II.A) have revealed that these submicellar particles in the micelle are held together by Ca -or Ca-phosphate linkages. The submicelles proper resist exhaustive dialysis against distilled water and apparently owe their existence to strong protein–protein interactions. These interactions are also reflected in the tendency of the caseins to form large associated polymers and complexes *in vitro*. In fact, the tendency to complex formation is so large that it was about a century before the heterogeneity of casein was recognized (21). During recent decades much effort has been made to evaluate the thermodynamic parameters involved in these casein associations in order to get some insight into the energetics of micelle formation. As we shall see, there is now overwhelming chemical and physical evidence that the caseins interact mainly through hydrophobic bonding, which is also seen in the formation of other biological particles of colloidal dimensions, for instance, the tobacco mosaic virus particle (22), insulin fibrils (23), or antigen–antibody complexes (24).

   Our understanding of the contribution of the inorganic constituents to the stability of the micelle is less satisfactory than that of the protein moiety. Strong binding of calcium to casein was already indicated by the early studies of

Chanutin et al. (25). Most probably, calcium ions act through the formation of calcium bridges between the carboxylic and phosphate groups on the protein surface. Casein micelles, however, also contain about 1.4 % of inorganic phosphorus in the form of so-called colloidal calcium phosphate. There has been much speculation about the type of Ca-phosphate involved and the way in which it is bound to the micelle. There is now some evidence (26, 27) that this calcium phosphate is a tricalcium phosphate. Whether it plays a cementing role in the micelle, as has been claimed repeatedly (15, 19, 28–32), remains still to be proved.

The enormous practical significance of the subject treated here can hardly be overestimated: For instance in 1971 in the then six EEC countries about 2 million tons of casein were worked up in a large number of widely different food articles. In addition to this 3.10^4 tons were produced for nonnutritive purposes, such as the manufacturing of paints, glues, paper, and textile or were used as a natural emulsifier. Similar figures can be cited for other countries. In practice, the texture of dairy products and important technological aspects of their preparation, such as the heat stability of milk, its curdling, or the gelation of evaporated milks on storage, depend sensitively on the state of dispersion, the solvation, and the charge of the micelle. Apart from biochemical and nutritional aspects, this explains the basic interest in the colloidal properties of casein, described in the next section.

II. COLLOIDAL PROPERTIES OF CASEIN MICELLES

A. The Role of the Inorganic Constituents

As mentioned in the introduction, casein occurs in milk as nearly spherical particles, which are composed of a large number of submicelles, (cf. Figures 1 and 3). The chemical composition of these micelles is presented in Table 1. The protein moiety, which constitutes the major part of the micelles, will be discussed in Section III. In this section we shall restrict ourselves to the inorganic components of the micelles.

The calcium of the micelles is partly incorporated into a calcium phosphate –citrate complex (26), which in some way is linked to the protein. The remaining part of the calcium in the micelles is directly bound to carboxylic or serine-esterified phosphate groups of casein.

Much work on the composition of the calcium phosphate and its binding to the protein moiety of the micelles has been done by Pyne and McGann (26, 28, 29). After acidifying milk and subsequently dialyzing it against unaltered milk to restore the original ionic conditions, they observed that the calcium phosphate content of the micelles had decreased irreversibly

TABLE 1.

Approximate composition of casein micelles in cow's milk[a]

Component	Concentration (wt. %)
α-Casein	42
β-Casein	28
κ-Casein	14
Minor casein components	10
Calcium	2.9
Magnesium	0.1
Organic phosphorus	0.8
Inorganic phosphorus	1.4
Citrate (as citric acid)	0.5
Sialic acid	0.3
Galactose	0.2
Galactosamide	0.2

[a] Averaged from different sources.

depending on the degree of acidification (see Figure 4). In contrast to the original milk the resulting colloidal phosphate-free milk (CPF milk) had a transluscent appearance caused by desintegration of the micelles into smaller particles, clearly demonstrated by a comparison of the micrographs of

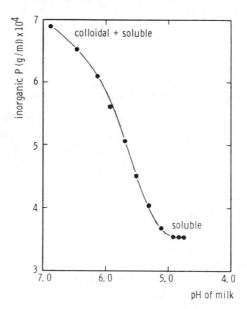

Fig. 4.   Effect of acidification and subsequent dialysis on the inorganic phosphate content of milk. (After G.T. Pyne and T.C.A., McGann (26), by permission of Cambridge University Press.)

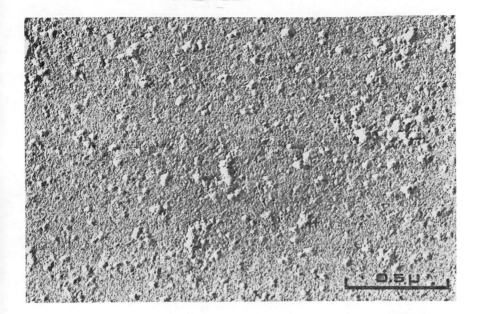

Fig. 5. Electron micrography of partly desintegrated casein micelles in colloidal phosphate-free milk (CPF-milk). Shadow-casting with platinum-carbon.

Figures 1 and 5. Apparently, calcium phosphate acts as a cementing agent for the submicelles. Complete desintegration of micelles may also be accomplished by dialysis of milk against calcium-free solvents as has been demonstrated by Schmidt and Buchhcim (19).

The composition of the colloidal calcium phosphate has been determined by Pyne and McGann (26) by chemical analysis and oxalate titration of milk and of the corresponding CPF milk. From this Pyne and McGann inferred that the colloidal calcium phosphate is a tricalcium phosphate, containing also some citrate, and is chemically bound to casein.

These results concerning the composition of the colloidal calcium phosphate were recently confirmed by Jenness (27), who titrated CPF milk and unaltered milk with HCl to pH 2. The difference between the amounts of HCl required in both cases is caused by the absence of colloidal phosphate in the CPF milk. This difference was equal to the value calculated from the phosphorus and citrate contents of both milks, assuming these components to be present as tricalcium phosphate and tricalcium citrate, respectively.

The presence of tricalcium phosphate in the micelles is remarkable, since in protein-free solutions trivalent phosphate is present in significant amounts only at pH values higher than 8 (33). This indicates the favored formation of

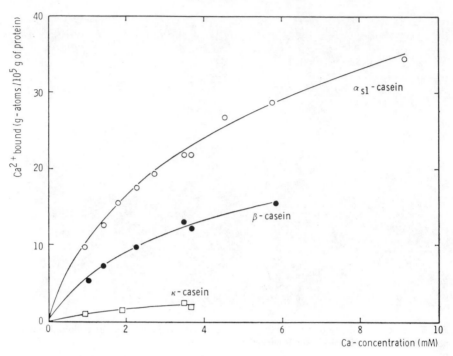

Fig. 6.   Calcium binding of α_{s1}-, β-, and κ-casein at pH 7.4 and ionic strength 0.1.(After I.R. Dickson and D.J. Perkins (40), by permission of the Biochemical Society.)

tricalcium phosphate in the presence of casein, as demonstrated by Termine and co-workers (34, 35).

Although the calcium binding of casein itself has been studied repeatedly (25, 36–40), we restrict ourselves to the recent investigations of Dickson and Perkins (40) concerning the calcium binding of the different casein components. By equilibrium dialysis using radioactive [47]Ca Dickson and Perkins established that the calcium-binding capacity of the various caseins decreased in the order α_{s1}-casein > β-casein > κ-casein (see Figure 6). The calcium binding appeared to be fully reversible, increasing with pH and decreasing with ionic strength. The latter effect was ascribed to a lowering of the calcium ion activity rather than to competition for the binding sites by the other ions.

Dickson and Perkins (40) compared the calcium binding of native and dephosphorylated caseins. Calcium binding is disproportionally decreased in the latter (Table 2), suggesting predominant binding to the phosphate groups. They established an apparent pK for binding of -2.6, which is comparable to that of low molecular weight serine-O-(dihydrogen phosphate).

Since calcium may form bridges between different casein molecules it

TABLE 2.

Comparing calcium binding to native and dephosphorylated α_{s1}-casein at pH 7.4[a]

	Number of $PO_4^{2-}/10^5$ g protein	Number of $COO^-/10^5$ g protein	Number of $Ca^{2+}/10^5$ g protein
Native	34	136	36
Dephosphorylated	0	136	5

[a]After Dickson and Perkins (40).

is clear that the extent of binding may sensitively affect the size of the micelles. This has indeed been observed repeatedly (41–43). On the other hand, changes in the casein concentration have much less effect (44).

Much speculation has been presented about the nature of the binding between calcium, the colloidal calcium phosphate, and the casein (31, 45–53). Table 3 lists several theoretically acceptable types of covalent binding, which could easily account for the variations in the Ca/P ratio observed with different milks (26, 29, 31). Calcium phosphate bridges linking molecules of α_{s1}-, β- and κ-casein may be, of type 1, but binding without bridge formation may also occur (cf. Table 3, lines 2–4). Lines 5 and 6 of the table account for the incorporation of minor quantities of citrate and hydroxyl groups. On the

TABLE 3.

Possible linkages in the calcium phosphate–citrate–casein complex.

1. casein $-\boxtimes-$ $\left(Ca-PO_4\!\!\begin{array}{c}Ca\\ \diagup \diagdown \\ Ca\end{array}\!\!PO_4-\right)_k$ $-Ca-\boxtimes-$casein

2. casein $-\boxtimes-$ $\left(Ca-PO_4\!\!\begin{array}{c}Ca\\ \diagup \diagdown \\ Ca\end{array}\!\!PO_4-\right)_k$ $-H$

3. casein $-\boxtimes-$ $\left(Ca-PO_4\!\!\begin{array}{c}Ca\\ \diagup \diagdown \\ Ca\end{array}\!\!PO_4-\right)_k$ $-Ca-OH$

4. casein $-\boxtimes-$ $\left(Ca-PO_4\!\!\begin{array}{c}Ca\\ \diagup \diagdown \\ Ca\end{array}\!\!PO_4-\right)_k$ $-Ca-PO_4=Ca$

$k = 0, 1, 2, ...$

5. casein $-\boxtimes-$ $Ca-PO_4-Ca-\boxtimes-$casein
        $|$
        Ca-citrate $=Ca$

6. casein $-\boxtimes-$ $Ca-PO_4-Ca-\boxtimes-$casein
        $|$
        $Ca-OH$

$\boxtimes$ may be either a carboxylic group or a phosphate group.

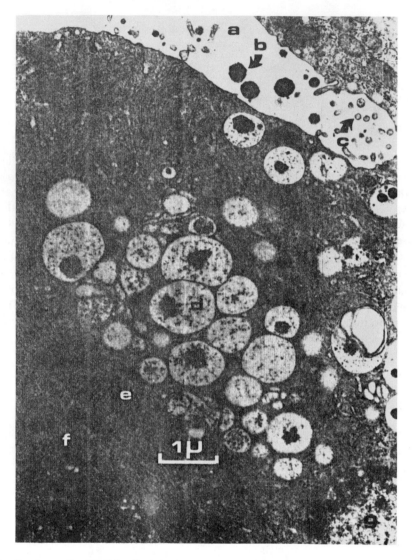

Fig. 7.   Apex of epithelial cell of the lactating mammary gland. (*a*) glandular lumen; (*b*) casein micelle in the lumen; (*c*) cross-sectioned microvillus of epithelia cell; (*d*) Golgi vacuole with submicelles which are already partly aggregated; (*e*) rough endoplasmic reticulum; (*f*) mitochondrium; (*g*) cellular nucleus. (After W. Buchheim and U. Welsch (58).)

distribution curve, but instead observed a gradually increasing number of micelles with decreasing diameter (Figure 8). Although the particles smaller casein moiety the functional groups consist of carboxylic groups or phosphate groups, esterified to serine or threonine residues. The use of apatite for the chromatographic separation of protein (54) suggests, however, that also relatively weak secondary bonds might exist between the calcium phosphate and the protein.

B. Substructure and Size Parameters

As mentioned in the introduction, electron microscopy has revealed a substructure in the micelle, which appears to be composed of a large number of spherical particles (18–20) (cf. Figure 3). The presence of such a substructure was already indicated earlier by the electron-microscopical studies of Shimmin and Hill (55, 56) and of Calapaj (57). Reported estimates of the diameter of these submicellar particles range from 11 to 20 nm, with the lower values found for the freeze-fractured specimen. Buchheim and Welsch (58) pointed out that in such preparations the diameter of small particles appear less than the actual one, because the protein particles differ markedly in their elastic properties from the surrounding ice; even at $-210°C$ they are inelastically deformed and torn apart during fracturing (59, 60). This led Buchheim and Welsch to conclude that the diameter of the submicelles most probably is 15 to 20 nm.

The formation of the micelles from submicelles has been most clearly demonstrated by Carroll et al. (7, 9, 61) and by Buchheim and Welsch (58) by electron microscopy of ultrathin sections of mammary gland tissue of lactating rats (Figure 7). After the biosynthesis of the polypeptide chains of the caseins on the ribosomes of the rough endoplasmatic reticulum, they interact to form submicelles, which are expelled into the vacuoles of the Golgi apparatus. In these Golgi vacuoles the submicelles aggregate to micelles under the influence of calcium ions and calcium phosphate, and finally are secreted into the alveolar lumen.

The size of the micelles has been investigated repeatedly by a large variety of methods (Table 4), although in recent years mostly by electron microscopy. For the same purpose Lin et al. (76) used inelastic light scattering to determine the diameter of ultracentrifugally fractionated micelles.

In nearly all investigations concerning the size distribution of the micelles a maximum was found to occur in the number frequency at a micellar diameter of 80 to 120 nm (41, 42, 66, 67, 70–75). It should be noted, however, that, as a consequence of the low resolution of the earlier electron microscopical techniques the numbers of small micelles and free submicelles certainly were underestimated. Therefore Schmidt et al. (20) reinvestigated the size distribution by electron microscopy using the freeze-fracturing technique in combination with an improved shadow-casting method, which resulted in highly improved resolution. These authors could not verify the maximum in the size

TABLE 4.

Size of casein micelles as determined by various techniques.

Method	Investigator and Reference	Size (nm)
Ultrafiltration	H. Bechold (62)	>40
Ultramicroscopy	G. Wiegner (63, 64)	5–100
Ultracentrifugation	T. Svedberg et al. (65)	20–140
	J.B. Nichols (66)	10–200
	M. Carić and J. Djordjević (67)	53–100
Light scattering	W. Lotmar and H. Nitschmann (68)	<120
	A. Leviton and H.S. Haller (69)	80
	P.F. Dyachenko (70)	86–123
	D.G. Schmidt et al. (44)	450
Electron microscopy	H. Nitschmann (41)	40–120
	H. Hostettler and K. Imhof (42)	30–200
	P.F. Dyachenko et al. (70, 71)	40–160
	E. Knoop and A. Wortmann (72)	10–300
	Z. Saito and Y. Hashimoto (73)	20–200
	D. Rose and J.R. Colvin (74)	25–150
	R.J. Carroll et al. (75)	50–300
	D.G. Schmidt et al. (20)	20–300
Inelastic light scattering	S.H.C. Lin et al. (76)	80–440

distribution curve, but instead observed a gradually increasing number of micelles with decreasing diameter (Figure 8). Although the particles smaller than 20 nm, being the free submicelles, outnumbered the micelles, their total mass constituted only a few percent of the whole. It was shown that the size distribution is well approximated by a modified log-normal distribution having an upper limit (Figure 9). This may be expected for particles built up from a large number of submicelles (77).

The particle weight of the micelles has been determined in various ways. Bloomfield and co-workers (78, 79) measured the intrinsic viscosity, $[\eta]$, of the micelles and determined the particle weight M from the Einstein–Stokes relation (80)

$$M = \frac{10\pi N_0 R_h^3}{3\,[\eta]} \tag{1}$$

in which N_0 is Avogadro's number. The hydrodynamic radius R_h was calculated from the diffusion coefficient, determined by inelastic light scattering, as 94 nm. Bloomfield et al. thus arrived at a value of 5.20×10^8 daltons for the weight-average particle weight of the micelles. Schmidt et al. (44), by conventional light scattering, found an average value of 11×10^8 (SD = 2×10^8; $\varphi = 8$) daltons for the weight-average particle weight of the micelles in nine different milks (cf. Figure 10). These values differ considerably, which

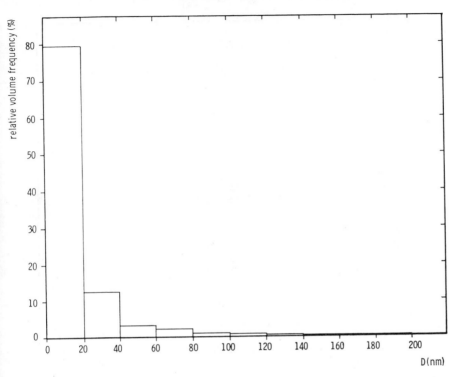

Fig. 8. Number frequency distribution of casein micelles in milk as determined by electron microscopy (% by number per 20 nm size class) (20).

suggests real differences in the average micellar size of the milks investigated. However, both results are much lower than the estimate arrived at earlier by Nitschmann (41) using electron microscopy. This discrepancy should be ascribed to the high water content of the micelles (cf. Section II.C), which was not accounted for by Nitschmann, and to size distortions during specimen preparation for electron microscopy.

By light scattering it is also possible to determine the value of the light-scattering average diameter, $\{\langle D^2 \rangle_{ls}\}^{1/2}$, defined by (81)

$$\langle D^2 \rangle_{ls} \equiv \frac{\sum_i c_i M_i D_i^2}{\sum_i c_i M_i} \qquad (2)$$

where M_i is the particle weight of the micelles with diameter D_i and concentration c_i. For unfractionated micelles Schmidt et al. (44) found $\{\langle D^2 \rangle_{ls}\}^{1/2}$ to be 450 nm, which value is approximately two times higher than that calculated from the electron microscopical data of Schmidt et al. (20) when it is assumed that all micelles have the same density, that is, $M_i \sim D_i^3$. Schmidt

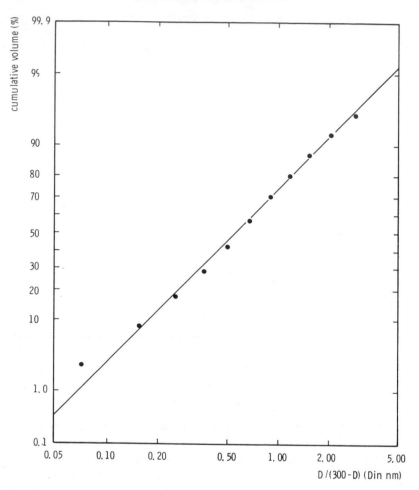

Fig. 9. Cumulative volume as a function of $D/(300-D)$ of casein micelles in bulk milk on a logarithmic scale (20).

et al. (44) demonstrated that this discrepancy could, at least qualitatively, be explained by the presence of a small amount of very big micelles, that are easily overlooked in the electron microscope. In this connection it should be reminded that with the latter technique the number of particles which is measured is 8 to 9 orders of magnitude smaller than with a light-scattering experiment.

C. Voluminosity and Micelle Hydration

The voluminosity of the micelles has been determined by means of electron

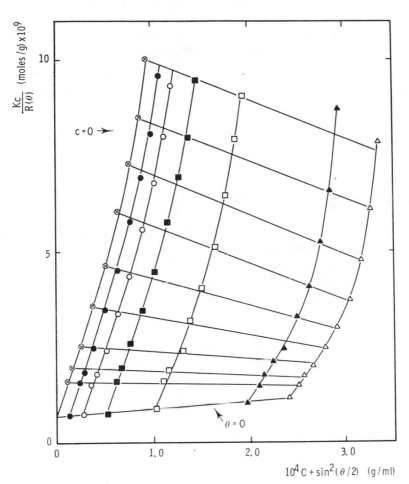

Fig. 10.   Typical Zimm plot of casein micelles redispersed in milk dialysate. (By permission of Elsevier Scientific Publishing Company, ref. 44.)

microscopy, viscometry, and from the moisture content of sedimented casein pellets. The electron microscopical method involves the determination of the fractional area of the cross sections of the micelles in freeze-fractured preparation of milk. This fractional area is equal to the volume fraction of the micelles, provided corrections are made in case the thickness of the sections is not negligible compared to the size of the micelles (82).

Compared to the other methods, electron microscopy yields the lowest values for the voluminosity as demonstrated in Table 5. We would ascribe this to the presence of an immobilized water layer surrounding the micelles, rather than to shrinkage during specimen preparation. Such a shrinkage

TABLE 5.

Volume fraction, φ, and voluminosity, v, of casein micelles at room termperature as determined by different authors.

Technique	φ (ml/ml)	v(ml/g)	Author and Reference
Viscosity	0.088	3.0	H. Eilers et al. (83)
	0.072–0.096	3–4	C.H. Whitnah and W.D. Rutz (84)
	0.106	4.4	R.K. Dewan et al. (85)
Moisture content of sediment	0.038–0.053	1.6–2.2	M.P. Thompson et al. (86)
	0.066	2.7	B.Ribadeau Dumas and J. Garnier (87)
Electronomicroscopy	0.037–0.038	1.5–1.6	D.G. Schmidt et al. (20)

would be more than has ever been mentioned (88), in particular with the freeze-fracturing techniques.

Viscometry is the most reliable method to determine the voluminosity. For a suspension of noninteracting spherical particles we have the well-known Einstein expression

$$\eta_r = 1 + 2.5\varphi \tag{3}$$

where φ is the volume fraction of the particles and η_r their relative viscosity. Defining the voluminosity v by $\varphi = cv$ with c the mass concentration, we find from equation 3

$$v = 0.4 \, [\eta] \tag{4}$$

in which the intrinsic viscosity $[\eta]$ is given by

$$[\eta] = \lim_{c \to 0} \frac{\eta_r - 1}{c} \tag{5}$$

Equation 4 is used by Dewan et al. (85). Whitnah and Rutz (84) have used, instead, the equation

$$v = 0.4 \, \frac{d\eta_r}{dc} \tag{6}$$

which again is valid only at low concentrations. The trouble of evaluating data at different concentrations was circumvented by Eilers et al. (83) with the empirical expression

$$\eta_r = \left\{ 1 + \frac{2.5\varphi}{1 - 1.35\varphi} \right\}^2 \tag{7}$$

The values of the voluminosity determined from the moisture content of ultracentrifugal sediments seem less reliable than those determined by viscometry on account of the presence of unknown quantities of residual solvent in the pellets. Also the extent of squeezing, which occurs under the hydrostatic pressure during centrifugation, is not known. Such uncertainties may account for the discrepancies between the voluminosities determined by various methods of Table 5.

From the measured voluminosity we may estimate the hydration of the micelles, δ, with the expression (80)

$$v = \bar{v}_2 + \delta v_1^0 \tag{8}$$

where $\bar{v}_2$ and v_1^0 are the partial specific volume of the micelles and the specific volume of the water, respectively. For instance with the voluminosity $v =$ 4.4 ml/g reported by Dewan et al. (85) and with $\bar{v}_2 = 0.7$ ml/g (80), we find for the hydration of micelles $\delta = 3.7$ g/g dry protein. This value is of an order 10 higher than that normally observed with compact proteins (89). No doubt, this extraordinarily large hydration should be ascribed to a large amount of interstitial water inside a spongelike particle.

The loose, spongy structure is also evident from the collapse of the micelles during dehydration as observed electron-microscopically with air-dried preparations of micelles, contrasted by shadow-casting (cf. Figure 1). The triangular shadows are characteristic of spherical particles that are partly collapsed on the grid during drying. Further evidence for an open structure has been obtained from a comparison of the extent of interaction of reagents such as dansyl chloride (mol.wt. = 270) and carboxypeptidase A (mol.wt. = 34.600) with molecularly dispersed sodium caseinate and with intact micelles. Since the reactivities of both types of casein are comparable, Ribadeau Dumas and Garnier (87) inferred that these reagents were able to penetrate the interior of the micelles, which therefore must possess sites easily accessible to these substances. This porous structure should not only be ascribed to the interstices between the submicelles but also to the porosity of the submicelles themselves. The porosity of the latter is demonstrated by its comparatively low molecular weight, approximately 250,000 (ref. 90 and D.G. Schmidt and B.W. van Markwijk, unpublished results), which is an order of magnitude smaller than expected for a similarly sized globular protein (80)

III. CASEIN ASSOCIATIONS IN RELATION TO MICELLE STABILITY

A. Introduction

It has been explained in the foregoing section that casein submicelles remain intact after exhaustive dialysis against distilled water. Their particle weight of about 2.5×10^5 daltons (D. G. Schmidt and B. W. van Markwijk, unpublished results, and ref. 90), is about 10 times the minimum molecular weights of the casein components, to be described below. Obviously then, the stability of these submicelles is due to strong protein–protein interactions, rather than to the formation of calcium bridges. Indeed it has been suspected since the earliest electrophoretic studies of casein heterogeneity (91–94) that the abnormalities of the electrophoretic pattern should be explained by complex formation between the casein components. It has also been mentioned before that in fact the complex formation is so pronounced that it took almost a century before the heterogeneity of casein was convincingly demonstrated (6, 95). For the very same reason, the separation and purification of the various casein components has proven to be an enormous task, which has become practicable only after the introduction of urea or alcohol as a dissociating agent by Hipp et al. in 1952 (96).

This section, after surveying the heterogeneity of casein, will deal with the self-associations and complex formations of the caseins, and describe some methods in use to evaluate the thermodynamic parameters involved in these interactions.

B. Characterization of Casein Components

By the method of free electrophoresis Mellander in 1939 (95) observed three different moving boundaries, which he designated as α-, β-, and γ-casein in order of decreasing mobility. The relative amounts were estimated as 75, 22, and 3 % respectively. These figures should be regarded with care, however, on account of the strong interaction between the caseins during electrophoresis, as will be explained below.

It soon became clear that Mellander's α-casein did not represent a homogeneous protein, but contains at least one other protein with properties reminiscent of the protective colloid predicted by Linderstrøm Lang (cf. Section I). One piece of evidence for this is given by zonal electrophoresis of total casein in the presence of relatively high proportions of urea (97). As is seen from Figure 11, urea- addition leads to the fair separation of three fractions, which differ strikingly in stability against flocculation by calcium ions or the milk-clotting enzyme chymosin.* It is not difficult to imagine the second, Ca-stable but chymosin-susceptible fraction of Figure 11, to be identical to the protective colloid of Linderstrøm Lang (6) and of Waugh and Von Hippel (10). The first and third fraction were identified as

---

*Chymosin (EC 3.4. 23.4), a milk-clotting enzyme from calf stomachs, often is also referred to as rennin. To avoid confusion with renin, a blood pressure-regulating substance from the kidney, we prefer the first name.

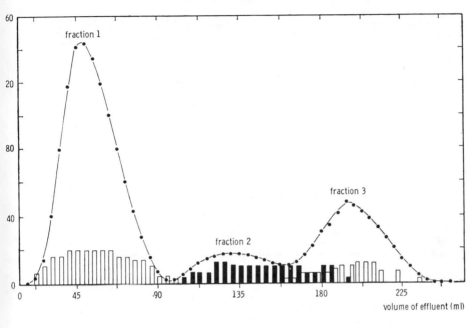

Fig. 11. Zonal electrophoresis of isoelectric casein in a column of powdered cellulose. Experimental conditions: column size 105 × 3 cm, barbiturate buffer of pH 7.5, and ionic strength 0.035 with 4 *M* urea, electrophoresis for 72 hr at 5 V/cm and 30 °C. White columns: fractions flocculated by calcium ions, but not by chymosin. Black columns: calcium-stable, but chymosin-sensitive fractions. (By permission of Elsevier Scientific Publishing Company, ref. 97.)

Ca-sensitive α- and β-casein, repectively. The separation achieved is not sufficient enough to show the presence of γ- and some other minor caseins.

It is worthwhile to mention that none of the fractions of Figure 11 can withstand the criticism of heterogeneity on closer inspection. This was convincingly demonstrated by Wake and Baldwin in 1961 (98), who, by application of starch gel electrophoresis in a urea-containing buffer, detected about 20 different casein components. These numerous components can be grouped into three families corresponding to the fractions of Figure 11, best characterized as α-, κ-, and β-caseins. Thus four genetic variants have been described of α-casein (99). Following the recommendation of the Committee for Nomenclature of the Milk Proteins (100), these α-caseins will be designated hereafter as α_{s1}-casein. Further six genetic variants are known of β-casein (99) and two of κ-casein (17). The heterogeneity of the latter is further enlarged by a variable amount of esterified carbohydrate (17). The study of the amino acid sequences of the genetic variants of γ-casein has

```
                              10                                    20
H.Arg-Pro-Lys-His-Pro-Ile-Lys-His-Gln-Gly-Leu-Pro-Gln┤Glu-Val-Leu-Asn-Glu-Asn-Leu-
                                                      deletion of A-variant
```

```
                              30                                    40
Leu-Arg-Phe-Phe-Val-Ala┤Pro-Phe-Pro-Gln-Val-Phe-Gly-Lys-Glu-Lys-Val-Asn-Glu-Leu-
```

```
                              50                                    60
Ser-Lys-Asp-Ile-Gly-Ser-Glu-Ser-Thr-Glu-Asp-Gln┤Ala┤Met-Glu-Asp-Ile-Lys-Gln-Met-
              |            |
              P            P                       ThrP   (D-variant)
```

```
                              70                                    80
Glu-Ala-Glu-Ser-Ile-Ser-Ser-Ser-Glu-Glu-Ile-Val-Pro-Asn-Ser-Val-Glu-Gln-Lys-His-
              |         |   |   |                              |
              P         P   P   P                              P
```

```
                              90                                   100
Ile-Gln-Lys-Glu-Asp-Val-Pro-Ser-Glu-Arg-Tyr-Leu-Gly-Tyr-Leu-Glu-Gln-Leu-Leu-Arg-
```

```
                             110                                   120
Leu-Lys-Lys-Tyr-Lys-Val-Pro-Gln-Leu-Glu-Ile-Val-Pro-Asn-Ser-Ala-Glu-Glu-Arg-Leu-
                                                              |
                                                              P
```

```
                             130                                   140
His-Ser-Met-Lys-Glu-Gly-Ile-His-Ala-Gln-Gln-Lys-Glu-Pro-Met-Ile-Gly-Val-Asn-Gln-
```

```
                             150                                   160
Glu-Leu-Ala-Tyr-Phe-Tyr-Pro-Glu-Leu-Phe-Arg-Gln-Phe-Tyr-Gln-Leu-Asp-Ala-Tyr-Pro-
```

```
                             170                                   180
Ser-Gly-Ala-Trp-Tyr-Tyr-Val-Pro-Leu-Gly-Thr-Gln-Tyr-Thr-Asp-Ala-Pro-Ser-Phe-Ser-
```

```
                             190                                   199
Asp-Ile-Pro-Asn-Pro-Ile-Gly-Ser-Glu-Asn-Ser┤Glu┤Lys-Thr-Thr-Met-Pro-Leu-Trp.OH
                                             Gly   (C-variant)
```

a

Fig. 12.   Primary structure of the three main casein components. (*a*) α_{s1}-casein B (after J.C. Mercier et al., ref. 104); page 185: (*b*) β-casein A² (after B. Ribadeau Dumas et al., ref. 105); page 186: (*c*) κ-casein B (after J.C. Mercier et al., ref. 106). Note the amino acid substitutions in the different genetic variants and the fragmentary β-casein molecules constituting γ- and other minor caseins. Note also the chymosin labile phenylalanine–methionine bond (residus 105–106) in the κ-casein sequence.

revealed that these are major fragments of the β-casein molecule, in which the first 28 residues of the polypeptide chain are lacking (101). A similar explanation was given recently for the origin of a number of other minor caseins (TS-A[2], TS-B, R- and S-casein), which in turn proved to be fragments of γ-casein (102). As a consequence of their related composition, one may, therefore, expect the properties of these minor caseins not to be very different from those of β-casein. The relative amounts of these minor fractions do not exceed 2% of whole casein. For a more complete description of the heterogeneity of casein the reader is referred to the above recommendations for nomenclature (100) and the recent surveys of Thompson (99) and Mackinlay and Wake (17).

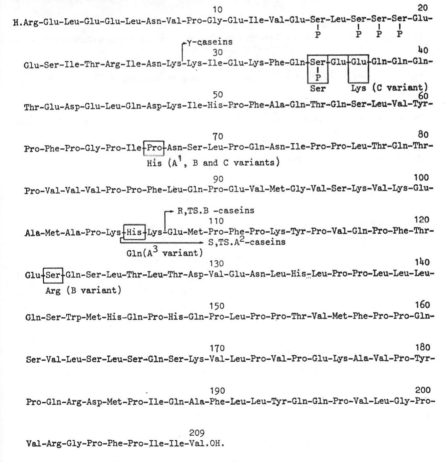

b

```
   1                                        10                                        20
PyroGlu-Glu-Gln-Asn-Gln-Glu-Gln-Pro-Ile-Arg-Cys-Glu-Lys-Asp-Glu-Arg-Phe-Phe-Ser-Asp-


                                          30                                        40
     Lys-Ile-Ala-Lys-Tyr-Ile-Pro-Ile-Gln-Tyr-Val-Leu-Ser-Arg-Tyr-Pro-Ser-Tyr-Gly-Leu-


                                          50                                        60
     Asn-Tyr-Tyr-Gln-Gln-Lys-Pro-Val-Ala-Leu-Ile-Asn-Asn-Gln-Phe-Leu-Pro-Tyr-Pro-Tyr-


                                          70                                        80
     Tyr-Ala-Lys-Pro-Ala-Ala-Val-Arg-Ser-Pro-Ala-Gln-Ile-Leu-Gln-Trp-Gln-Val-Leu-Ser-


                                          90                                        100
     Asp-Thr-Val-Pro-Ala-Lys-Ser-Cys-Gln-Ala-Gln-Pro-Thr-Thr-Met-Ala-Arg-His-Pro-His-

                       ↓
                                         110                                        120
     Pro-His-Leu-Ser-Phe-Met-Ala-Ile-Pro-Pro-Lys-Lys-Asn-Gln-Asp-Lys-Thr-Glu-Ile-Pro-


                                         130                                        140
     Thr-Ile-Asn-Thr-Ile-Ala-Ser-Gly-Glu-Pro-Thr-Ser-Thr-Pro-Thr─Ile─Glu-Ala-Val-Glu-
                                                                  Thr  (variant A)

                                         150                                        160
     Ser-Thr-Val-Ala-Thr-Leu-Glu─Ala─Ser-Pro-Glu-Val-Ile-Glu-Ser-Pro-Pro-Glu-Ile-Asn-
                                  Asp   (variant A)
                                         169
     Thr-Val-Gln-Val-Thr-Ser-Thr-Ala-Val.OH
```

c

The amino acid composition and sequence of the different caseins have been described during the last decade (103–107). Figure 12 shows the primary structures of α_{s1}-, β-, and κ-casein. It should be noticed from this figure that, when compared to other proteins (23), the caseins have an unusually high proportion of apolar residues, such as valine, alanine, leucine, isoleucine, phenylalanine, tryptophane, tyrosine, and proline. The abundancy of proline is striking. The influence of this residue on the structural organization of the peptide chain is, as well known, to prevent the formation of ordered helical structures (108). Indeed it has been concluded from optical rotatory dispersion and hydrodynamic behavior (109–111), that in solution the caseins are best described as a random coil.

1. Molecular Weights

The molecular weights of the various caseins, which follow from the sequence studies referred to above are for α_{s1}-, β-, and κ-casein 23,615, 23,983, and 19,023, respectively. These values are in excellent agreement with the

results obtained from physical studies like light scattering and ultracentrifugation (H.J. Vreeman, B.W. van Markwijk, and J.A. Brinkhuis, unpublished
results and refs. 109, 111, 112). Divergent results may be found especially in
earlier reports (113–117). They are explained by the strong tendency of the
caseins to associate (cf. Section III.D) and nonideality effects, which interfere
with the proper extrapolation of the experimental data obtained at finite
concentrations.

2. Hydrophobicities

It has been mentioned above that the caseins are characterized by a
relatively high proportion of nonpolar amino acid residues. Tanford (118)
has demonstrated that the hydrophobicity of a protein usefully can be
expressed in terms of the average free enthalpy change accompanying
the transfer of the amino acid side-chains from water to ethanol. Thermodynamically, the standard free enthalpy of transfer of an amino acid,
$H_3N^{\oplus}-C-\underset{R}{\overset{H}{|}}-COO^{\ominus}$, is related to the ratio of the solubilities in both media
by

$$\Delta G^{\ominus} = -RT \ln\left(\frac{S_e}{S_w}\right) \tag{9}$$

where S_e and S_w stand for the solubilities in ethanol and water, respectively,
and the other symbols have their usual meaning. Since $\Delta G^{\ominus}$ will be roughly
additive with respect to composition, one may obtain the side-chain contribution of a particular amino acid by substracting from $\Delta G^{\ominus}$ the correspond-

TABLE 6.

Amino acid side-chain hydrophobicities according to Tanford (118).

Amino Acid	$-\Delta G^{\ominus}$ (kcal/mole)
Tryptophan	3.00
Isoleucine	2.97
Tyrosine	2.87
Phenylalanine	2.65
Proline	2.60
Leucine	2.42
Valine	1.69
Lysine	1.50
Methionine	1.30
Cysteine	1.00
Alanine	0.73
Arginine	0.73
Threonine	0.44
Glycine	0

ing quantity for glycine for which R = H. This leads to the significant side-chain hydrophobicities listed in Table 6.

The average hydrophobicity of a protein is then defined as the sum of the constituent residue hydrophobicities divided by the number of residues in that molecule:

$$\frac{\sum_i N_i \left(\Delta G_i^{\ominus} - \Delta G_{\text{glycine}}^{\ominus} \right)}{\sum_i N_i}$$

where N_i is the number of residues of type i. For α_{s1}-, β-, and κ-casein Hill and Wake (119) thus calculated hydrophobicities of 1.24, 1.32, and 1.20 kcal/mole, respectively. If these figures are compared to the tabulated hydrophobicities of a large number of other proteins (120), it appears that they are very high. This is especially true for β-casein. It was shown by Fisher and Bigelow (120, 121) that the combination of such high hydrophobicities with a molecular weight as reported for the caseins prevents the formation of a globular structure in which the nonpolar groups are completely buried inside the protein's interior. Rather, one expects a part of the molecular surface to be covered by nonpolar residues, which will give rise to strong hydrophobic interactions. This expectation is wholly fulfilled with the caseins, as will be demonstrated in the following sections.

C. Methods to Study Protein Interactions

In principle, any property that is a function of the degree of association, can be used to evaluate the thermodynamic parameters involved in protein interactions. Thus, restricting ourselves to casein, uv difference spectroscopy, optical rotatory dispersion, and polarization of fluorescent light (122–124) have been applied to study the association. Most directly, however, the relevant information is obtained from the concentration dependence of the apparent molecular weight and from the anomalies observed in transport experiments such as sedimentation, electrophoresis and gelchromatography.

1. Molecular Weight Approach

For the analysis of self-association reactions, any technique, such as osmometry, sedimentation equilibrium, or light scattering, yielding a well-defined average of the molecular weight, may do the job. In practice it is found, however, that the applicability of osmometry is very limited on account of the low precision of the data obtained in the low-concentration region. The other two techniques have proved extremely useful for the study of self-associating proteins (125, 126). Equations have also been derived (127, 128) to analyze complex formations, which, however, demand at present unattainable accuracy of experimental data. This is, no doubt, the reason why complex reactions have until now been investigated by transport methods mainly.

The analysis of self-associating systems was first dealt with by Steiner (129). This author considers a two-component solution in which the protein solute behaves ideally. Its self-association can be represented by a number of consecutive equilibria:

$$k_i = \frac{c_i}{c_1 c_{i-1}} \quad (i = 2, 3, ...) \tag{10}$$

and its weight-average molecular weight, $\bar{M}_w$, rewritten as

$$\frac{\bar{M}_w c}{M_1 c_1} = 1 + \sum_{i=2} i c_1{}^{i-1} \prod_{j=2}^{i} k_j \tag{11}$$

where $c = \sum_i c_i$ is the total protein concentration (g/ml) and M_1 is the subunit molecular weight. For the relative concentration of the latter Steiner derived

$$\ln\left(\frac{c_1}{c}\right) = \int_0^c \left\{ \frac{(M_1/\bar{M}_w) - 1}{c} \right\} dc \tag{12}$$

which shows that c_1 can be obtained by graphical integration of the plot of $[(M_1/\bar{M}_w) - 1]/c$ versus c. In practice it is found that the outcome of this procedure depends heavily on the accuracy of the experimental data at low concentrations (130), and more so when the molecular weight of the subunit, M_1, is not known beforehand and must be obtained by extrapolation of the weight-average molecular weight to zero concentration. Equation 11 shows that, once c_1 is known, the dimerization constant k_2 may be obtained from the limiting slope at zero concentration of the graph of $\bar{M}_w c/M_1 c_1$ versus c_1; k_3 is found from the plot of $[(\bar{M}_w c/M_1 c_1) - k_2 c_1]$ versus c_1^2; and so on. Once again, the importance of accurate data in the low-concentration range becomes obvious.

With proteins, being macromolecular polyampholytes, the assumption of ideal behavior appears to be rather unrealistic. Modifications of Steiner's treatment have, therefore, been proposed (127, 131), to include the effects of nonideality. In the case of light scattering, the experimental quantity available is the apparent molecular weight, M_a, which is related to the weight-average by (132)

$$\frac{1}{M_a} = \frac{1}{\bar{M}_w} + 2\bar{B}c + 0\,(c^2) \tag{13}$$

The second virial coefficient, $\bar{B}$, in multicomponent systems was calculated by Kirkwood and Goldberg (133) and Stockmayer (134) as

$$\bar{B} = \frac{\sum_i \sum_j c_i c_j M_i M_j B_{ij}}{\sum_i \sum_j c_i c_j M_i M_j} \tag{14}$$

where the interaction parameter B_{ij}, of the species i and j in sufficiently dilute solution is found by differentiation towards c_j of (135)

$$\ln y_i = M_i \sum_j B_{ij} c_j + 0\,(c^2) \tag{15}$$

where y_i is the activity coefficient of component i.
For the consecutive association constants we further have

$$k_i = \frac{y_i c_i}{y_1 c_1 y_{i-1} c_{i-1}} \quad (i = 2, 3, ...) \tag{16}$$

It is seen from equations 13 and 14 that the dependence of the apparent molecular weight on concentration is the result of the counteracting influences of the concentration on $\bar{M}_w$ and $\bar{B}$. This interferes with the straightforward determination of the consecutive association constants by the simple Steiner treatment outlined above. Several proposals have been made to overcome this difficulty. The simplest and widely accepted solution is to assume all interaction parameters B_{ij} to be identical (136–138). Equation 16 is then seen to reduce to equation 10 again. Further, the analysis of the concentration dependence of the apparent molecular weight then comes to finding an appropriate set of $(i + 1)$ equations to be solved for the i association constants plus the one universal interaction parameter B. These are given by direct or indirect experimental quantities such as M_a, the apparent number-average molecular weight

$$\bar{M}_{n,a} = \int_0^c M_1 \frac{dc}{M_a}$$

and so on. The merits and drawbacks of this approach have been amply discussed by Adams and associates (126, 139). Again, the experimental precision of the data in the low-concentration region appears to be a serious problem. Also, as with Steiner's treatment, it is not clear in how far intermediate association steps can be neglected without affecting the veracity of the analysis. The assumption of a universal interaction parameter itself has also been severly criticized (131, 135, 140). More satisfying solutions have been proposed in relation with the self-associations of α_{s1}- and β-casein (cf. Section III.D).

Strictly, the above treatment of associating systems is only valid for two-component systems. If, as is usually the case, the protein is dissolved in a buffer solution, preferential interaction with salt ions should be taken into account. In practice, this requires that the protein be dialyzed against the buffer prior to the experiment (141–143). If then the polyampholyte, P, is redefined as PCl–zNaCl, with z its effective valency, the equations given above remain formally unchanged. For details the reader is referred to the original literature.

2. Migration Experiments

Moving boundary experiments such as electrophoresis, sedimentation velocity, and gel chromatography constitute powerful complementary techniques to study the interaction of proteins (125, 135, 144, 145). For caseins, two extreme cases can be distinguished. The first is that the rate of re-equilibration during the process of migration is very small compared with the rate of separation of the various species present. Obviously, if this is the case the velocities and equilibrium concentrations of the reactants can be measured directly.

In the opposite case of instantaneous re-equilibration Tiselius (146) already recognized that migration could never lead to the complete separation of the reactants and that a single boundary would be observable, the median* of which moves with the weight-average velocity of the species present. Tiselius' conclusion has often been interpreted as the appearance of just one peak in the schlieren pattern (148, 149). During the last two decades, however, much evidence has been gathered that the single boundary does not necessarily imply the absence of partially separated peaks. In fact, as was convincingly demonstrated by Gilbert (150) and Gilbert and Jenkins (151), association makes the boundary spread in a manner typical of the type of association involved.

Boundary spreading during electrophoresis or sedimentation generally takes place for the following reasons (135):

a. Diffusion, which causes the boundary to spread in proportion to the square root of the time.

b. A polydispersity insufficiently large to bring about complete separation of the migrating species. This effect, for obvious reasons, will not be considered here.

c. Protein interactions of the type found with the caseins. The latter two effects give rise to a spreading proportional to the time itself.

In general, the different contributions to the boundary spreading are difficult to distinguish. Baldwin (135) and Gilbert (150), however, afforded a practical solution to this problem by neglecting the diffusional contribution. In view

---

*In the ultracentrifuge, owing to the sectorial shape of the cell, the median is to be replaced by the second moment of the gradient curve (147).

of the time dependence of the various effects mentioned above, this practically comes to extrapolation of the shape of the boundary to infinite time.

For rapidly re-equilibrating, self-associating systems the continuity equation then reduces to (150, 152)

$$\frac{\partial \sum_i c_i v_i}{\partial \sum_i c_i} = \frac{dx}{dt} \tag{17}$$

where $dx/dt = -\partial c/\partial t/\partial c/\partial x$ represents the velocity of a section of constant concentration through the boundary and v_i that of species i. Combination of equations 17 and 10 and subsequent integration yields a relation between the local concentration and velocity as a function of the various k_i's and v_i's. This might be differentiated to yield a "reduced" schlieren pattern in which the concentration gradient is plotted against x/t. Since the velocities v_i usually are not known, their value must be estimated. For the case of sedimentation of globular proteins Gilbert made use of the Stokes–Einstein relation and accordingly

$$v_i = v_1 \cdot i^{2/3} \tag{18}$$

where v_1 is the limiting velocity at zero concentration. For the caseins a model of sedimenting coils might be more appropriate (109–111) and therefore (153)

$$v_i = v_1 \cdot i^{1/2} \tag{19}$$

Considering the above approximations, it is surprising to find how often Gilbert's theory settles matters in the resolve interpretation of moving boundary experiments on associating systems. Its success has convincingly been demonstrated with:

a. The unmasking of the misleading likeness of the boundary spreading due to polydispersity and association.

b. The interpretation of the maximum gradient velocities and apparent percentages of schlieren patterns.

c. The estimation of association constants.

It is worthwhile to summarize some of Gilbert's most important conclusions pertinent to the self-associations of the caseins.

1. In the case of simple dimerization ($i = 2$) or multiple associations ($i = 2, 3, \ldots$) the migrating peak in the sedimenting or descending electrophoretic boundary will be asymmetric with a trailing edge as shown in Figure 13*a*.

2. In the case of a single association step with $i > 2$, two partially separated peaks will be observable beyond a certain minimum concentration (Figure

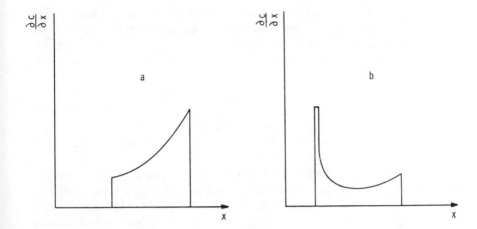

Fig. 13. Schematic picture of the sedimentation (or descending electrophoretic) pattern of self-associating systems, neglecting diffusion. (After G.A. Gilbert (150).) (a) Dimerization or a series of consecutive association steps; (b) discrete polymerization with a degree polymerization larger than two.

13b). Most importantly, Gilbert found that the area under the slower peak remains constant on further increase of the concentration. This result, which has been verified repeatedly (154, 155), usefully distinguishes between an associating and a polydisperse or slowly re-equilibrating system, since with the latter the size of the slower peak necessarily increases with concentration. A related consequence is that peak area's should not be identified with the percentages of the reactants present. Equations have also been given (125) by which the association constant can be calculated from the area of the slower peak and the degree of association. The latter quantity is readily obtained from the relative position of the minimum gradient, which according to Gilbert is given by

$$\frac{[(x/t)_{min} - v_1]}{(v_i - v_1)} = \frac{(i - 2)}{3(i - 1)} \tag{20}$$

Such relations allow as yet an estimation of the reactant concentrations to be made.

3. If the velocities of the associated species are larger than that of the subunit, the ascending electrophoretic boundary will become hypersharp due to progressive dissociation through the boundary.

The same approach of neglected diffusional spreading may be used to analyze complex formations between different species, although the mathematics involved is much more complicated. Gilbert and Jenkins (151) (see also Nichol and Ogston (156)) have treated the anomalous electrophoretic and sedimentation behavior for a number of typical cases, which may be very helpful for diagnostic purposes. Instead of summarizing their theory completely, we shall restrict ourselves to the particular example of Figure 14. This figure shows the electrophoretic patterns of complex-forming α_{s1}- and β-casein.

Complex formation can easily be recognized from the following features.

First, apart from the stationary δ- and ε-boundaries, the ascending limb shows three distinct peaks, the descending side only two. This cannot be explained by the Dole–Longsworth theory (152) for the electrophoresis of two independently migrating components.

Second, only the mobilities of the fastest rising and slowest descending peaks correspond to those of pure α_{s1}- and β-casein. The other peaks of intermediate mobilities are called the reaction boundaries (151).

Third, the relative areas of the α_{s1}- and β-peaks are distinctly less than would be expected from the amounts of these components put into the reaction mixture.

Such anomalies are successfully explained by the theory of Gilbert and Jenkins, provided a complex of intermediate mobility between α_{s1}- and β-casein is formed.

As justified above, the effect of diffusion on the spreading of migrating boundaries has not yet been considered. Because it might easily be imagined that diffusion could blur small peaks or shoulders, a number of authors (144,

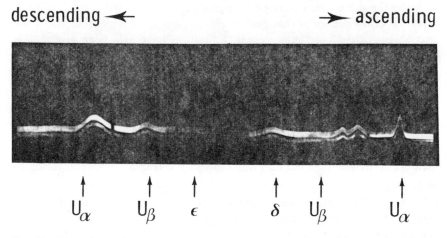

descending ←   → ascending

U_α U_β ϵ δ U_β U_α

Fig. 14. Free electrophoresis of complex-forming α_{s1}- and β-casein at a total protein concentration of 12×10^{-3} g/ml. Experimental conditions: mixing ratio $\alpha_{s1}/\beta = 50/50$; barbiturate buffer of pH 6.6 and ionic strength 0.10. Electrophoresis for 4800 sec at 3.05 V/cm and 2°C. (Ref. 160 and by permission of Elsevier Scientific Publishing Company, Ref. 185.)

157–159) have solved the complete conservation-of-mass equation by numerical methods. Cann and Goad in particular (144) solved the continuity equation by considering the contributions of the individual species to the fluxes due to velocity and diffusional transport. The moving boundary is simulated by dividing it into a large number of small segments, and the flux at a particular segment edge is related to the average concentrations in neighboring boxes. After each short transport interval, chemical equilibrium should be reestablished. Goad proved that, if the velocity flux is restricted to transfer of material from one box to the next, a diffusionlike error is introduced. This author, therefore, minimized this error by taking into account the contribution of the contents of several neighboring boxes to the flux and calculated the diffusional flux separately. Nijhuis and Payens (159, 160), however, made use of the diffusionlike error to simulate the effect of the diffusion. This leads to a drastic reduction of the necessary computer time, whereas it can be shown that this procedure is identical to the countercurrent analog method developed earlier by Bethune and Kegeles (161, 162) to account for the diffusional spreading. We shall demonstrate the usefulness of this approach when discussing the complex formation between α_{s1}- and β-casein (cf. Section III.D).

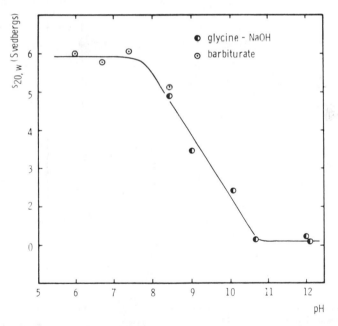

Fig. 15.   The sedimentation coefficient, $s_{20,w}$, of α_{s1}-casein as a function of pH. Protein concentration 7.0×10^{-3} g/ml. Sedimentation in barbiturate–HCl or glycine–NaOH buffers of ionic strength 0.2 at 2°C. (By permission of Academic Press, ref. 112).

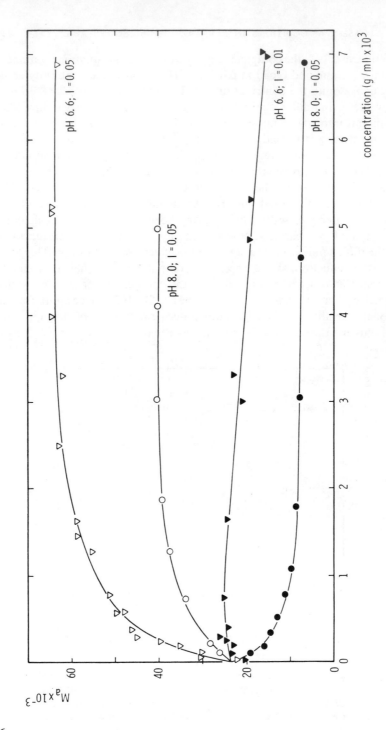

Fig. 16. Dependence of the apparent molecular weight of α_{s1}-casein B on concentration at different values of pH and ionic strength, I. Experimental conditions: imidazole–HCl–NaCl Buffers; 21°C. (131).

D. Casein Polymerizations and Complex Formations

1. Self-Association of α_{s1}-Casein

Hawler (163), in his early light-scattering studies, concluded that α_{s1}-casein is subject to aggregation in the presence of low molecular weight electrolyte, impeding the proper extrapolation of molecular weight measurements. This difficulty was circumvented by Schmidt et al. (112), who studied the protein at alkaline pH values. As is evident from the falling-off of the sedimentation coefficient with increasing pH (cf. Figure 15), the protein is completely dissociated beyond pH 10.5, apparently as a result of strong electrostatic repulsion. The subunit molecular weight was established at pH 12 by light scattering and sedimentation equilibrium as 23,400 and 22,600, respectively. These values are in good agreement with the molecular weight resulting from the amino acid sequence (see Section III.B). In further studies at more neutral pH, Schmidt and Van Markwijk (131, 164, 165) confirmed the above molecular weight by extrapolation of light-scattering data at very low protein concentrations. Some examples are collected in Figure 16, which shows that, independent of pH and ionic strength, all data extrapolate to the same molecular weight of about 23,000. This figure also demonstrates the influence of the degree of ionization and the screening effect of the supporting electrolyte upon the association. Evidently, electrostatic interactions cannot be disregarded when considering the self-association of this protein. This led Schmidt to analyze the concentration dependence of the apparent molecular weight (cf. equations 11, 13) in terms of association constants by successive approximations. In the first instance, nonideality effects were completely ignored and the apparent molecular weight data at concentrations below 10^{-3} g/ml analyzed by the ideal Steiner relations, equations 11 and 12. This yields the consecutive association constants and polymer concentrations in first approximation. Next, the second virial coefficient (equation 14) was estimated, accepting the interaction parameter B_{ii} between identical particles to be given by (166):

$$B_{ii} = \left(\frac{2\pi N_0}{M_i^2}\right) \int_0^\infty \left[1 - \exp\left(\frac{-V_r}{kT}\right)\right] r^2 \, dr \tag{21}$$

where V_r is the potential energy of two particles a distance r apart, M_i their molecular weight, and N_0 Avogadro's number.

The interaction parameter B_{ij} between dissimilar particles was assumed to be given by

$$B_{ij} = \frac{(B_{ii} + B_{jj})}{2} \tag{22}$$

TABLE 7

Association constants of α_{s1}-casein B at pH 6.6 in imidazole–HCl–NaCl buffers of different ionic strengths, I, at 21°C. Values of the association constants expressed in (ml/g) $\times$ 10^{-3}.

I		k_2	k_3	k_4
0.010	(first approximation)	0.13	—	—
0.010	(second approximation)	0.44	—	—
0.05	(first approximation)	2.70	2.28	5.19
0.05	(second approximation)	2.83	3.13	4.25
0.10		3.35	5.45	4.67
0.20		15.9	20.8	14.7
0.50		36.6	23.7	47.3

It is seen from the relations 14, 21, and 22 that, once V_r is known, we are able to estimate the second virial coefficient $\bar{B}$. The latter can be calculated from the electrophoretically determined ζ potential (167) using the well-known theory of Verwey and Overbeek (168).

It should further be kept in mind that, according to equation 16, non-ideality corrections are also necessary for the association constants. It can easily be shown that, subject to the approximation 22,

$$\frac{y_i}{y_1 y_{i-1}} = \exp\left\{M_1 c\left[iB_{ii} - B_{11} - (i-1)B_{i-1,\,i-1}\right]\right\} \tag{23}$$

with c the total protein concentration.

In second approximation, $\bar{M}_w$ is found after correcting M_a for the non-ideality contribution whereupon Steiner's analysis is repeated. As expected, the successive approximations converge rapidly at low protein concentrations. Some results of such an analysis for the genetic variant α_{s1}-casein B are collected in Table 7 and Figure 17.* One should note the strong influence of electrostatic interaction on the association, especially at ionic strengths below 0.05. It is further obvious from the dotted line of Figure 17 that a universal interaction parameter B, as has been proposed repeatedly (128, 136, 169, 170), certainly cannot account for the observed concentration dependence of the apparent molecular weight.

Until now it has tacitly been assumed that α_{s1}-casein indeed undergoes consecutive association. The rationale for this is provided by the shape of the sedimenting boundary, which, in agreement with Gilbert's prediction, shows

---

*The data in Table 7 and the calculated curves in Figure 17 have been corrected for an underestimation of nonideality effects in the previous publications (131, 140, 165).

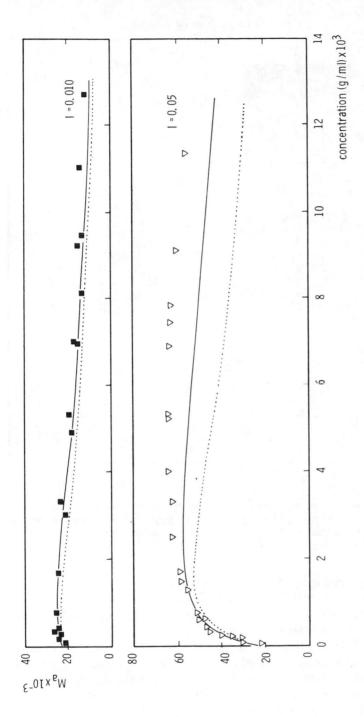

Fig. 17. Comparison between theoretical and experimental apparent molecular weights of α_{s1}-casein B at 21°C in imidazole–HCl-NaCl buffers of pH 6.6 and different ionic strengths. I. Full line: curve calculated by successive approximations as outlined in the text. Dotted line: curve calculated with a universal interaction parameter B.

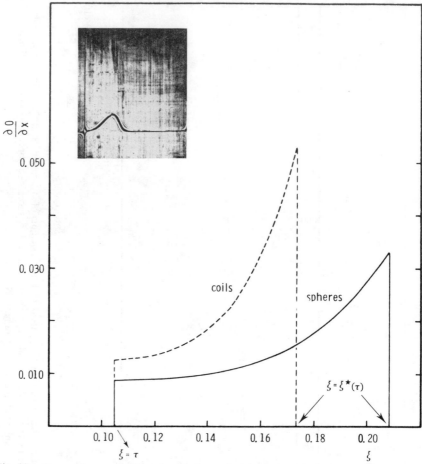

Fig. 18. Comparing experimental and theoretical boundary spreading of sedimenting α_{s1}-casein B. Experimental conditions: protein concentratin 5.72×10^{-3} g/ml, barbiturate buffer of pH 6.6 and ionic strength 0.2. Centrifugation for 60 min at 130,000 g and 2°C. (By permission of Academic Press, ref. 153).

a trailing edge (see Figure 18). The theoretical sedimentation patterns for this case, which were calculated after correcting the Gilbert theory for the non-uniformity of the ultracentrifugal field and sectorial dilution (135, 153), show the same type of boundary spreading.

We conclude by mentioning that the association of α_{s1}-casein is found to be nearly temperature independent between 8 and 30°C, which suggests it to be entropy driven. Schmidt (131) has evaluated in particular the thermodynamic parameters of the dimerization step and shown that it can be explained by the predominant formation of hydrophobic bonds (171, 172).

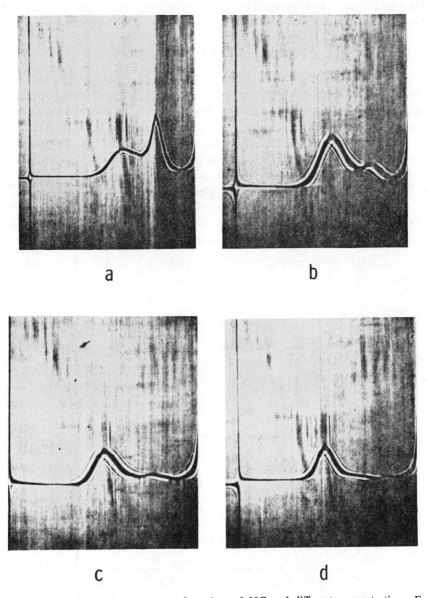

Fig. 19. Sedimentation patterns of β-casein at 8.5 °C and different concentrations. Experimental conditions: barbiturate buffer of pH 7.5 and ionic strength 0.20. Centrifugation in a synthetic boundary cell at 143.000 g. (a) Concentration 20×10^{-3}g/ml, after 100 min; (b) 10×10^{-3}g/ml, after 65 min; (c) 8×10^{-3}g/ml, after 80 min; (d) 3.3×10^{-3}g/ml, after 25 min. Sedimentation to the right. (By permission of Elsevier Scientific Publishing Company, ref. 109.)

The likelihood of hydrophobic interactions among the caseins was indicated when we discussed their amino acid composition and hydrophobicity indices.

2. The Association of β-Casein

The association of β-casein differs from that of α_{s1}-casein in several respects. Firstly, it was observed by Sullivan et al. (173) that the sedimentation pattern consists of two well-separated peaks, indicating that the association is of the discrete type (cf. Figure 19). Secondly, Hawler (163) found that, in contrast to α_{s1}-casein, the rate of association is low. This was confirmed by Payens and Van Markwijk (109), who noticed that the area of the slower peak of the sedimentation pattern increases with concentration, in contrast with the predictions of Gilbert for rapidly re-equilibrating systems.

It was suggested in Section III.B that β-casein, on account of its exceptionally high hydrophobicity, is subject to strong hydrophobic interactions. This was demonstrated by the sedimentation studies of Sullivan et al. (173), of Von

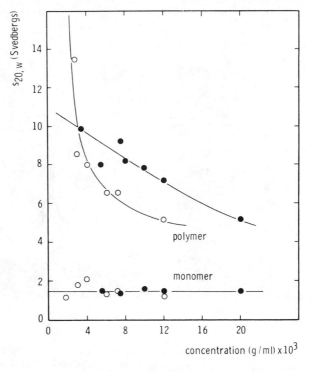

Fig. 20. Concentration dependence of the sedimentation coefficients, $s_{20,w}$ of β-casein monomers and polymers at different temperatures. Experimental conditions: barbiturate buffer of pH 7.5 and ionic strength 0.20. Full circles: experiments at 8.5°C; open circles: experiments at 13.5°C. (By permission of Elsevier Scientific Publishing Company, ref. 109.)

Hippel and Waugh (113), and of Payens and Van Markwijk (109), which showed endothermic, and therefore entropy-driven, association. At temperatures below 5°C Payens and Van Markwijk found the protein to be completely dissociated with a molecular weight in agreement with that calculated from its amino acid composition (cf. Section III.B). An increase in temperature of only a few degrees leads to the sudden appearance of a rapidly sedimenting peak, the sedimentation coefficient of which is extremely concentration dependent, as shown in Figure 20. Obviously, β-casein polymers behave very nonideally even under conditions of suppressed electrostatic interaction. This is also evident from the maximum in the graph of the apparent molecular weight versus concentration as shown in Figure 21. The separation of the contributions of nonideality and association to the apparent molecular weight was achieved by Payens and Van Markwijk on the results obtained by the Archibald technique in the ultracentrifuge (109), and by Payens et al. (174) using the more accurate light-scattering results of Figure 21. The sedimentation coefficients of the two peaks of the sedimentation pattern and their concentration dependence, shown in Figure 20, suggest that

$$M_p \gg M_1$$
$$\text{and} \quad B_{pp} \gg B_{11} \tag{24}$$

where M_p is the molecular weight of the β-casein polymers and B_{pp} their interaction parameter. Therefore, in equation 14, the average light-scattering virial coefficient, $\bar{B}$, reduces to B_{pp} and consequently, by equation 13, the apparent molecular weight becomes

$$\frac{1}{M_a} = \left(\frac{1}{\bar{M}_w} \right) + 2B_{pp}c \tag{25}$$

It can further be shown that the condition of the maximum of the apparent molecular weight leads to

$$(M_a)_{\max} = \left(\frac{\bar{M}_w^2}{\bar{M}_z} \right)_{\max} \tag{26}$$

with $\bar{M}_z$ the z-average molecular weight. With the concentrations of monomer and polymer given by the areas of the sedimenting peaks, equation 25 can easily be solved for the degree of polymerization i. Some results have been collected in Table 8 together with the values calculated for the weight-average molecular weight and the virial coefficient B_{pp}. Compared to other associating proteins (125), β-casein polymers appear to be characterized by an unusually high degree of association and virial coefficients typical for random coils or rodlike particles (175). The latter configuration, however, must be rejected in view of the absence of any significant dissymmetry of the scattered light.

The above results suggest the similarity of β-casein polymers and soap

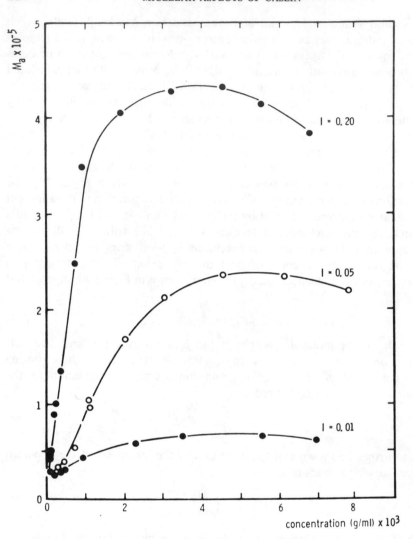

Fig. 21. Concentration dependence of the apparent molecular weight of β-casein as measured by light scattering at 21 °C in phosphate buffers of pH 7.0 and different ionic strengths, I. (Ref. 174 and by permission of Academic Press, ref. 140.)

micelles, the more so since the shape of the sedimentation pattern suggests that the polymers formed have a narrow size-distribution. Indeed we have been able to demonstrate the existence of a critical micelle concentration (140). At low protein concentrations, where nonideality effects can be neglected, the weight-average molecular weight can be rewritten as

TABLE 8

Experimental and computed data of β-casein polymerization
at two ionic strengths, I (174)

I	$(M_a)_{max}$ $\times 10^{-5}$	c_{max} (g/ml) $\times 10^3$	c_1^b (g/ml) $\times 10^3$	c_p^b (g/ml) $\times 10^3$	n	$(\bar{M}_w)_{max}$ $\times 10^{-5}$	$2B$ (ml mole/g^{-2}) $\times 10^4$
0.05	2.38	5.3	2.44	2.86	41	4.8	4.08
0.2	4.32	4.0	1.7	2.3	50	7.25	2.5

[a]From the plots of Figure 21.
[b]From the areas under the sedimenting peaks, neglecting the Johnston–Ogston effect.

$$\bar{M}_w = M_1 \left\{ \frac{[i - (i - 1)]}{(1 + k_i c_1^{i-1})} \right\} \tag{27}$$

which shows that for concentrations corresponding to $k_i c_1^{i-1} \ll 1$, $\bar{M}_w$ should be equal to the monomer molecular weight M_1. As soon as $k_i c_1^{i-1}$ becomes of order unity, equation 27 predicts an abrupt increase of $\bar{M}_w$. The concentration at which this takes place is comparable to the critical micelle concentration found with soap solutions. An example of this behavior is shown in Figure 22. One should notice that the shift of the critical micelle concentration with ionic strength is as expected for electrostatic interaction interfering with the micellization. The electrostatic interaction is also obvious from the data presented in Table 8.

It is worthwhile to mention that light-scattering measurements at elevated pressures have shown that α_{s1}- and β-casein polymers dissociate under pressure (140, 176). Again, this effect is most readily explained by the occurrence of hydrophobic bonds, the partial specific volume of which is positive (172). The quantitative interpretation of such dissociation in terms of the number of hydrophobic bonds involved is rather hazardous, however, since it is not known in how far other bond types (hydrogen and electrostatic bonds) contribute to the phenomenon, and since the pressure effects are reversed beyond 1500kg/cm^2 (177).

3. Associations of κ-Casein

Virtually nothing has been published about the self-association of this important casein component. Early attempts to determine a subunit molecular weight have yielded a value for this quantity of 28,000 (116). Comparison with the molecular weight (19,023) calculated from sequence analysis (106) suggests that this value must be explained by the polymerization of κ-casein. This has indeed been confirmed recently by Vreeman et al. (H.J. Vreeman, B.W. van Markwijk, and J.A. Brinkhuis, unpublished results), who studied the molecular fragments of κ-casein obtained by the action of the

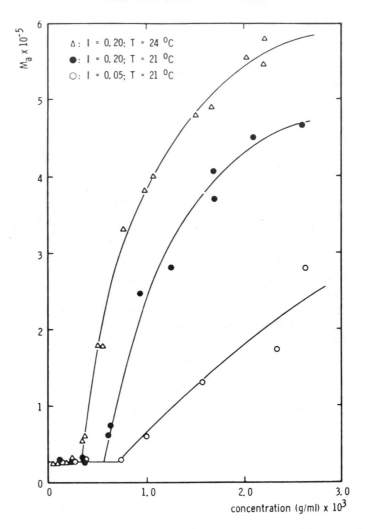

Fig. 22. Concentration dependence of the apparent molecular weight M_a and critical micelle concentration as determined by light scattering in dilute solutions of β-casein A in phosphate buffers of pH 7.0 at different ionic strengths, I, and temperatures. (By permission of Academic Press, ref. 140).

milk-clotting enzyme, chymosin. This enzyme splits κ-casein in so-called *para*-κ-casein, which is insoluble at neutral pH and a soluble macropeptide, which may have varying amounts of carbohydrate attached to it.

Molecular weight determinations of these fragments, as well as of whole κ-casein, are hampered by the presence of readily oxidizable cysteine residues

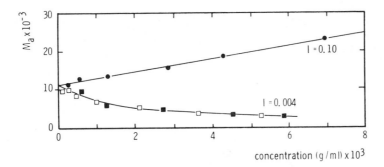

Fig. 23.   Concentration dependence of the apparent molecular weight M_a of *para-κ*-casein B as measured by light scattering. Experimental conditions: HCl–NaCl solutions of pH 2.5 and different ionic strengths, I, at 20°C. (H.J. Vreeman and B.W. van Markwijk, unpublished results.)

and a strong tendency to noncovalent polymerization. It has, therefore, become practice to study the protein and its fragments in the presence of cystine-reducing mercapto-ethanol and urea or to block the reduced sulf-hydryl groups (17). Vreeman et al. studied *para-κ*-casein at pH 2.5 using light scattering and sedimentation equilibrium. From their results, which are shown in Figures 23 and 24, molecular weight values of 11,500 and 12,500 are derived. It was also demonstrated that *para-κ*-casein undergoes extensive dimerization, which probably should be ascribed to a dipolar interaction.

The macropeptide part of *κ*-casein was freed from fractions containing carbohydrate by ion-exchange chromatography. Its molecular weight, by sedimentation equilibrium, was found to be 6700 (cf. Figure 25). Together the above values yield a molecular weight of about 19,000 for the intact *κ*-casein, in good agreement with the sequence study of Mercier et al. (106).

The sedimentation pattern of whole *κ*-casein suggests a *β*-casein-like as-

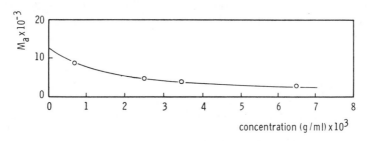

Fig. 24.   Concentration dependence of the apparent molecular weight of *para-κ*-casein B as determined by sedimentation equilibrium. Experimental conditions: HCl solution of pH 2.5 and ionic strength 0.004 at 20°C. (H.J. Vreeman and J.A. Brinkhuis, unpublished results.)

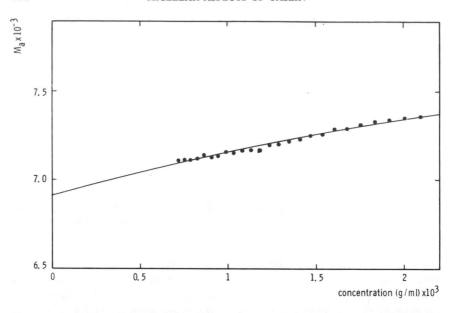

Fig. 25.   Concentration dependence of the apparent molecular weight M_a of the macro-peptide of κ-casein A as determined by sedimentation equilibrium. Experimental conditions: glycine–NaOH–NaCl buffer of pH 11 and ionic strength 0.6 at 20°C. (H.J. Vreeman and J.A. Brinkhuis, unpublished results.)

sociation. The parameters characterizing this polymerization have not been established as yet.

It was mentioned in the introduction that κ-casein forms calcium-stable complexes with α_{s1}- and β-casein. These complex formations (cf. Figure 2 and refs. 11 and 12), which seem to be of primary importance for the solubilization of casein in its natural surroundings, have not yet been studied in detail. The dispersing capacity of κ-casein with respect to the other caseins is evident from the κ-casein content of the micelles. This was first demonstrated by Sullivan et al. (178), who, by means of ultracentrifugation, isolated differently sized micelles, in which the sialic acid was determined as a measure of the κ-casein content. The results, shown in Figure 26, clearly demonstrate that the smaller micelles contain more κ-casein than the bigger ones. These results, which have been confirmed by others (51, 179–181) affirm that κ-casein is the component responsible for the colloidal dispersion of casein.

4. Complex Formation between the Different Casein Components

The importance of interaction between different caseins is clearly indicated by the formation of calcium-stable complexes between α_{s1}- and β-casein with

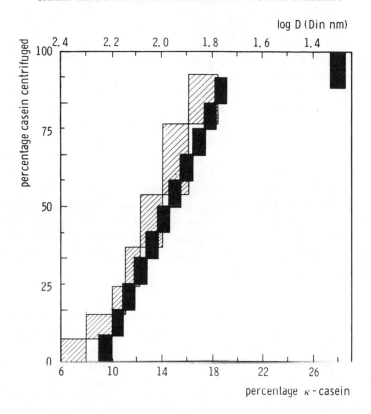

Fig. 26.   Relation between κ-casein content and amount of centrifugable casein micelles. The diameter D of the micelles was estimated from the size distribution as obtained by differential centrifugation. (By permission of MacMillan Journals Ltd., ref. 178.)

κ-casein, as discussed above. Such complex formation also explains the stability of the casein submicelles in the absence of calcium. Several attempts have been made to study these interactions quantitively. We shall summarize here three different approaches: polarization of fluorescence, difference spectroscopy, and the study of migration behavior. Though these studies yield quantitatively different results, they all agree insofar that casein interactions are found to be highly endothermic, which suggests again the importance of intermolecular hydrophobic bonding.

Clarke and Nakai (124) have applied polarization of fluorescence to study the complex formation between α_{s1}- and κ-casein after dansylating the α_{s1}- component.

The polarization was related to the fluorescent intensities F_0 and F_3 of unreacted α_{s1}-casein and complex by the relation

$$\frac{(\bar{P} - P_1)}{(P_3 - P_1)} = \frac{F_3}{F_0} \tag{28}$$

where $\bar{P}$, P_3, and P_1 represent the polarizations of the equilibrium mixture, the complex, and the dansylated α_{s1}- casein, respectively. The polarization of the complex, P_3, was measured at 37°C, where all α_{s1}-casein was assumed to be converted into complex and, therefore, F_3 equal to F_0, the fluorescent intensity of α_{s1}-casein. At intermediate temperatures then the equilibrium concentrations are readily calculated once a stoichiometric ratio of α_{s1}- and κ-casein in the complex is accepted. Clarke and Nakai provided some evidence that this ratio would be 1 : 1, but claimed at the same time that their results would not be influenced seriously if other stoichimetries would occur. This would indicate that the method used is rather insensitive, though the sign of the thermodynamic parameters obtained from the equilibrium constants and their temperature dependence can be established with certainty. Some of their results are collected in Table 9, which confirm the endothermic nature of the complex formation.

The above assumption of a 1 : 1 stoichiometry of α_{s1}- and κ-casein in the complex is conflicting with earlier results of Waugh (182) and Garnier (123), which suggest molar ratio's of 3 and 4, respectively. Garnier in particular obtained his result by application of the method of continuous variation to the difference spectra of solutions of α_{s1}- and κ-casein of varying mixing ratios. In principle this should yield the stoichiometry directly, but Garnier et al. (183) admitted that the quantitative interpretation is hazardous on account of the possible self-association of the caseins. Also in this study the endothermicity of the complex formation was clearly demonstrated.

A 1 : 1 stoichiometric ratio of the α_{s1}/κ-complex was also presumed by Slattery and Evard (32) and by Chun (184), who analyzed ultracentrifugal patterns of mixtures of α_{s1}- and κ-casein with the aid of the Gilbert theory. It may be clear from what has been stated above that there is no a priori justification for such a presumption.

The presence of higher complexes in the reaction mixture of α_{s1}- and β-casein was convincingly demonstrated by Nijhuis and Payens (160, 185) who

TABLE 9

Changes in free enthalpy, enthalpy, and entropy for the complex
formation between α_{s1}- and κ-casein[a]

Temperature (°C)	$\Delta G^{\ominus}$ kcal/mole	$\Delta H^{\ominus}$ kcal/mole	$\Delta S^{\ominus}$ (eu)
24	−5.71	3.33	30.4
30	−6.19	4.67	35.8

[a]After R. Clarke and S. Nakai (124).

studied the anomalies of electrophoretic patterns, such as those, presented in Figure 14. As has been discussed above, β-casein at the temperature of free electrophoresis, that is 2°C, is completely dissociated, whereas α_{s1}-casein is subject to consecutive association. In principle therefore, we should have the following multiple equilibria:

$$\left. \begin{aligned} i\alpha &\leftrightarrows \alpha_i \\ j\alpha + k\beta &\leftrightarrows \alpha_j\beta_k \end{aligned} \right\} \quad (i, j, k = 1, 2, \ldots) \tag{29}$$

The evidence for the occurrence of higher complexes, gathered by Nijhuis and Payens, was threefold. Firstly, they observed in the sedimentation pattern a rapid peak, the velocity of which was far too high to be explained on the basis of a 1 : 1 complex of α_{s1}- and β-casein subunits. Secondly, they applied the conservation-of-mass condition to the displacement of the reaction boundaries of Figure 14. This leads to the following expression for the concentration c_α reflected by the area of the leading ascending peak

$$c_\alpha = \frac{\bar{c}_\alpha (\bar{u}_\alpha - \bar{u}_\beta)}{(u_\alpha - \bar{u}_\beta)} \tag{30}$$

where u_α is the mobility of pure α_{s1}-casein and $\bar{c}_\alpha$ its constituent concentration defined as

$$\bar{c}_\alpha = \sum_i c_{\alpha_i} + \sum_j \sum_k \left[\frac{jM_\alpha}{(jM_\alpha + kM_\beta)} \right] c_{\alpha_j\beta_k} \tag{31}$$

Further in equation 30 $\bar{u}_\alpha$ is the constituent mobility of α_{s1}-casein, defined as

$$\bar{u}_\alpha = \frac{\left(\sum_i c_{\alpha_i} u_{\alpha_i} + \sum_j \sum_k c_{\alpha_j\beta_k} u_{\alpha_j\beta_k} \right)}{\bar{c}_\alpha} \tag{32}$$

A similar expression may be written down for $\bar{u}_\beta$. These constituent mobilities can be measured from the displacement of the medians bisecting the reaction boundaries (186). Their relatively high values as well as that of c_α were explained by Nijhuis and Payens by the predominant formation of complexes with mobilities approaching that of pure α_{s1}-casein, u_α, (cf. equations 30 and 32). This implies that complexes $\alpha_j\beta_k$ with $j/k > 1$ will prevail in the equilibrium mixture.

This conclusion was confirmed by computer simulation of the electrophoretic reaction boundaries of Figure 14. Nijhuis and Payens (159, 160) simulated the formation of a reaction boundary by subdividing the electrophoretic channel into a large number of small boxes and repeating alternating cycles of transport and re-equilibration through the boxes. All velocities were taken relative to that of the slowest component, that is, β-casein. They further showed that, provided a proper ratio of box dimension to mobility

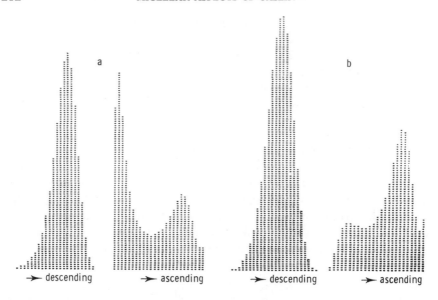

→ descending   → ascending   → descending   → ascending

Fig. 27. Simulated reaction boundaries for complex formation between α_{s1}- and β-casein during electrophoresis under the same conditions as with Fig. 14 (ref. 160). (a) Mixing ratio α_{s1}/β = 50/50; (b) mixing ratio α_{s1}/β = 70/30. (By permission of Elsevier Publishing Company, ref. 185.)

was made, it is possible to simulate the effect of diffusional spreading on the boundary. The equilibrium constants of the reaction presented in equation 29 and the mobilities of the complexes must be introduced as variable parameters in the computations. All other parameters can be obtained from the self-association studies of pure α_{s1}- and β-casein discussed above. Fortunately, the computations proved to be rather insensitive to the actual values chosen for the complex mobilities. Therefore the latter were estimated by linear interpolation between the pure component mobilities. Systematic search for a set of satisfying equilibrium constants then showed that the anomalies of the electrophoretic patterns could only be reproduced if the existence of higher complexes was taken into account. This is in agreement with the high constituent mobility of α_{s1}-casein discussed above. Nijhuis and Payens obtained a satisfying set of equilibrium constants by letting all polymers of α_{s1}-casein interact with the β-monomer. Though this seems reasonable from the physical point of view, one should realize that this requires the introduction of no less than six different parameters. The set of equilibrium constants obtained, therefore, can never be claimed to represent a unique solution. The point of interest is, however, that only the introduction of higher complexes leads to satisfying agreement between the observed and calculated reaction boundaries. Some of the results of the simulation

TABLE 10

Comparing simulated and experimental reaction boundaries for
complex-forming α_{s1}- and β-casein during electrophoresis (160, 185)

Mixing Ratio α_{s1}/β		Ascending Boundary			Descending Boundary	
		Velocity (cm/sec) $\times 10^5$		Relative Area	Velocity (cm/sec) $\times 10^5$	Relative Area
50/50	Fig. 27	10.8	20.0	63.6	17.7	64.2
	experimental (Fig. 14)	12.7	16.9	68	18.5	72
70/30	Fig. 27	13.8	20.9	51.9	18.1	86.2
	experimental	13.9	18.9	47	20.2	86

are represented in Figure 27 and Table 10. The agreement is as good as one may expect with the uncertainties discussed above.

Similar anomalies are observed with the sedimentation and electrophoresis patterns of mixture of κ- and α_{s1}- or κ- and β-casein (T.A.J. Payens, unpublished results, and ref. 184). Apparently all caseins have pronounced tendencies to associate and form complexes, even in the absence of calcium ions. From what has been explained above, it will be clear that such interactions are predominantly hydrophobic in nature and can account for the stability of the submicelles in calcium free solution.

IV. STRUCTURAL ORGANIZATION OF THE MICELLE

A. Factors Affecting Micelle Reassembly

It has been mentioned in the introduction that the presence of κ-casein enables the dispersion of α_{s1}-casein in $CaCl_2$ solutions (cf. Figure 2). The reformation of casein micelles in this process is suggested by the milky appearance of the casein–$CaCl_2$ dispersion. This has led a number of investigators to study the nature of such artificial micelles and the parameters that influence their formation.

Noble and Waugh (187, 188) prepared artificial micelles with different mixtures of α_{s1}- and κ-casein and increasing amounts of $CaCl_2$. Some of their results are shown in Figure 28, from which it is seen that the micellization of casein over the range of $CaCl_2$ concentrations investigated becomes complete at a κ/α_{s1} ratio of about 0.2. Noble and Waugh observed a considerable path dependence of this micelle formation. Thus these authors found that rapid addition of $CaCl_2$ resulted in the formation of stable micelles. Slow incremental addition, on the other hand, led to increasing precipitation of calcium caseinate, although some micellization did occur. It was, therefore, concluded

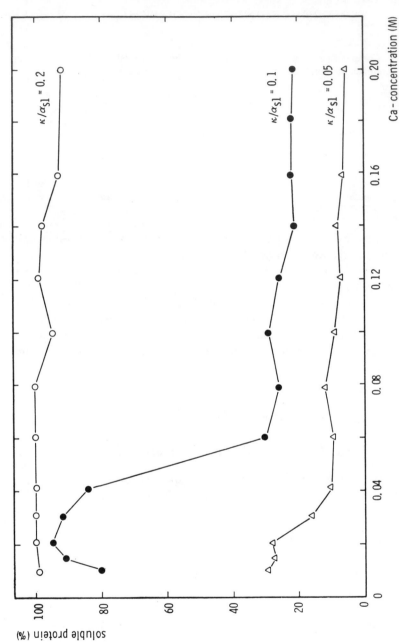

Fig. 28. Soluble protein for mixtures of α_{s1}- and κ-casein as a function of the CaCl$_2$ concentration. Experimental conditions: 0.07 M imidazole–KCl buffer of pH 7.1 at 37°C. Each solution initially contained 1 × 10⁻³ g/ml κ-casein. The different initial concentration of α_{s1}-casein are: △ for 20 × 10⁻³g/ml; ● for 10 × 10⁻³ g/ml; ○ for 5 × 10⁻³ g/ml. (Redrawn from data of D.F. Waugh and R.W. Noble Jr. (188), by permission of the American Chemical Society.)

that micelle formation was not an equilibrium process. This conclusion was further supported by the observation that, inversely, the addition of κ-casein to calcium α_{s1}-caseinate only resulted in a limited redispersion of the latter.

Schmidt et al. (16) also performed experiments on the formation of artificial micelles by addition of a fixed amount of $CaCl_2$ to mixtures of α_{s1}- and κ-casein or of α_{s1}-, β-, and κ-casein, in which the proportion of κ-casein was varied. After removal of the coarse precipitate of nondispersed casein by centrifugation, the amount of protein dispersed and the extent of micelle formation in the supernatant were estimated from the optical densities at 278 and 360 nm, respectively. An example is given in Figure 29. The maxi-

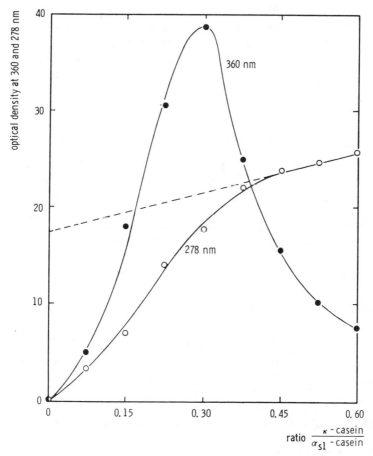

Fig. 29. Dependence of the optical density at 360 and 278 nm on the κ/α_{s1}-ratio in a solution initially containing 13×10^{-3} g/ml α_{s1}-casein, 9×10^{-3} g/ml β-casein, 30 mM $CaCl_2$ and increasing amounts of κ-casein. The dashed line represents the theoretical optical density if no precipitation would occur. Experimental conditions: 0.07 M NaCl solution of pH 6.6 at 37°C. (By permission of Volkswirtschaftlicher Verlag GmbH, ref. 16.)

mum in the supernatant turbidity in this figure does not imply a maximum in micellar size. It is due to the fact that without κ-casein all α_{s1}-casein is precipitated by calcium. Subsequent addition of κ-casein results in the micellization of increasing amounts of α_{s1}-casein, until at the optimal κ/α_{s1} ratio of 0.045 all α_{s1}-casein remains in dispersion. Further addition of κ-casein then leads to a decrease of the micellar size. This dispersing effect of κ-casein is confirmed by the electron micrographs of Figure 30 and by the volume frequency distributions given in Figure 31. From these it is clear that artificial micelles become smaller the higher the proportion of κ-casein. This is reminescent of the observations made with natural micelles (refs. 51, 178–181; see also Figure 26).

In mixtures of β- and κ-casein, β-casein was also found to be stabilized against calcium flocculation. The stable product formed, however, did not resemble the original micelles from milk, but rather appeared to consist of irregularly shaped aggregates which were small in number (16). Contrary to what was reported by Noble and Waugh (187, 188), Schmidt et al. (16) observed that slow addition of $CaCl_2$ did yield micelles. Actually, the slow

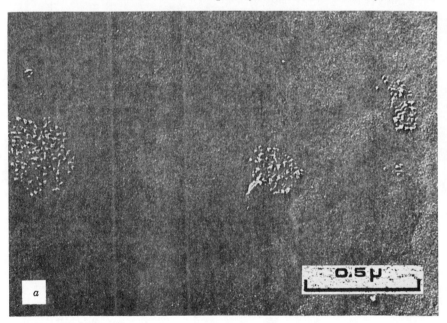

Fig. 30.   Electron micrographs of artificial micelles prepared by mixing α_{s1}-, β-, and κ-casein and $CaCl_2$. Experimental conditions: 0.07 M NaCl solution of pH 6.6 at 37°C containing 30 mM $CaCl_2$. Initial concentrations of α_{s1}- and β-casein 13 × 10^{-3} and 9 × 10^{-3} g/ml, respectively; κ-casein contents 2 × 10^{-3}, 4 × 10^{-3}, and 6 × 10^{-3} g/ml, respectively, for pictures a, b, and c.

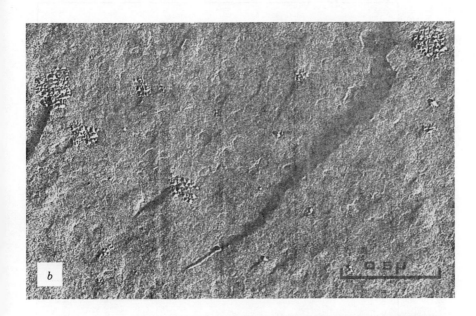

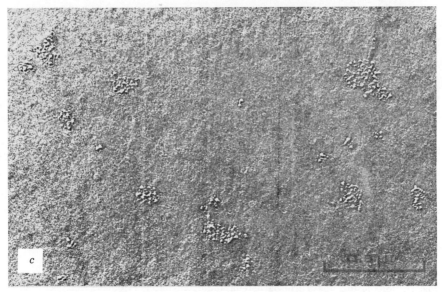

217

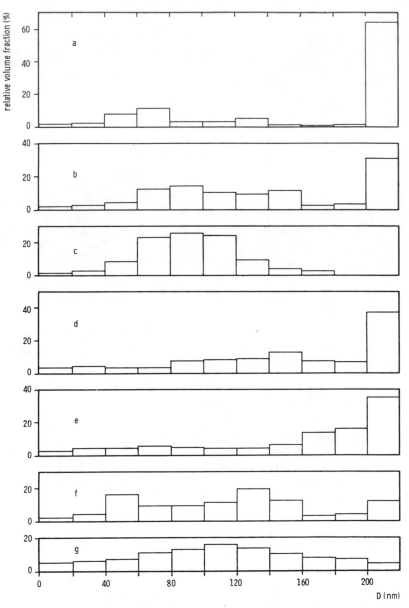

Fig. 31. Size histograms of artificial and natural casein micelles from electron microscopy. Experimental conditions with the artificial micelles: 0.07 M NaCl solution of pH 6.6 at 37°C, contaning 30 mM CaCl$_2$. (a) 13 × 10^{-3} g/ml α_{s1}-casein and 2 × 10^{-3} g/ml κ-casein; (b) 13 × 10^{-3} g/ml α_{s1}-casein and 4 × 10^{-3} g/ml κ-casein; (c) 13 × 10^{-3} g/ml α_{s1}-casein and 6 × 10^{-3} g/ml κ-casein; (d) 13 × 10^{-3} g/ml α_{s1}-casein, 9 × 10^{-3} g/ml β-casein, and 2 × 10^{-3} g/ml κ-casein; (e) 13 × 10^{-3} g/ml α_{s1}-casein, 9 × 10^{-3} g/ml β-casein, and 4 × 10^{-3} g/ml κ-casein; (f) 13 × 10^{-3} g/ml α_{s1}-casein, 9 × 10^{-3} g/ml β-casein, and 6 × 10^{-3} g/ml κ-casein; (g) natural micelles. (By permission of Volkswirtschaftlicher Verlag GmbH, ref. 16.)

addition was found to be necessary to obtain reproducible results. With rapid addition local excesses of calcium cannot be avoided and the bigger aggregates thus formed did not reduce in size with time, apparently as a consequence of their poor equilibrium properties.

The addition of β-casein seems also to influence the size of artificial micelles of α_{s1}- and κ-casein, as demonstrated by the size distributions represented in Figure 31. In the case of natural micelles Schmidt et al. (16) observed that a lowering of their β-casein content, accomplished by prolonged cooling (74, 189-192), likewise results in smaller micelles. They found that the number-average diameter of the micelles in a particular milk decreased from 27 to 23 nm upon 3 days' cooling at 4°C. When the free submicelles were not taken into account the number-average diameter decreased from 56 to 46 nm. Obviously, this reduction in micellar size should be ascribed to the release of a part of β-casein, which is linked to the micelle by hydrophobic, but not by calcium bonds.

From the above it is clear that artificial micelles from calcium ions, α_{s1}- and κ-casein may be prepared, the electronmicroscopical appearance of which is not different from that of the natural micelle. β-Casein can influence the micellar size, but seems not to be required for micelle stability.

B. The Structural Organization of the Micelle

The evidence gathered in the foregoing sections permit a tentative interpretation of the electronmicroscopical picture of the natural micelles (cf. Figure 3). They seem to be composed of a large number of more or less spherical submicelles having an average molecular weight of approximately 250,000 (Schmidt and Van Markwijk, unpublished results, and ref. 90) and a diameter of 15 to 20 nm (58). From a comparison with the molecular weights of the casein subunits (cf. Section III.B), the submicelles thus appear to consist of about 12 molecules of α_{s1}-, β-, and κ-casein. Apparently, they are not held together by calcium bonds, since they survive dialysis against calcium-free buffers (19). From what has been explained in Section III, it is clear that hydrophobic bonds are mainly responsible for their integrity.

It is useful to compare the association of the casein subunits into submicelles with the formation of soap micelles, the structural organization of which is determined by the hydrophile–lipophile balance of the constituting soap molecules. According to this concept the core of the submicelles would consist of an accumulation of hydrophobic groups, screened from the solvent by the polar moiety of the protein, notably its carboxylic and phosphate groups.

In the micelle several thousands of submicelles are linked together via their phosphate or carboxylic groups by calcium or magnesium ions as well as by apatitelike calcium phosphate, as schematized in Table 3. This leads

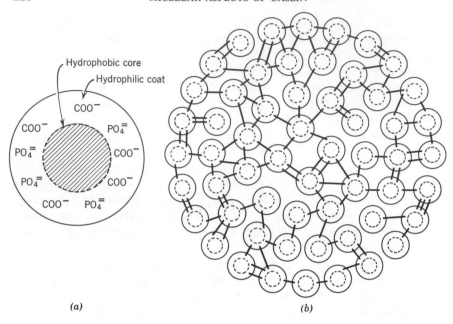

Fig. 32.   Schematic picture of submicelles (*a*) forming the casein micelle (*b*).–: inorganic bonds as outlined in Table 3; hatched area: region of predominantly hydrophobic nature.

to the schematic picture of the micelles, presented in Figure 32. Schmidt et al. (20) have suggested that ultimately the size of the micelles is limited by the size of the Golgi vacuoles of the mammary gland, where the bioassembly of the micelles takes place (7, 9, 58, 61, 193).

The interstitial space in the micelle explains its ready accessibility to enzyme molecules such as chymosin and carboxypeptidase A (87). It is worth realizing that at least part of the micellar surface must be occupied by κ-casein to explain its specific reaction with huge molecules such as those of insolubilized papain (194, 195), denatured β-lactoglobulin (196), and κ-carrageenan (197).

C. Epilogue

In the last two decades various models of the casein micelle have been proposed. Since they have already been reviewed or refuted on experimental grounds (7, 9, 30, 198, 199), we shall restrict ourselves to a short discussion of the most important alternatives.

Various so-called coat–core models have been described in which a coat of κ-casein completely covers a core of α_{s1}- and β-casein, thereby protecting the micelles against flocculation by calcium ions (15, 200, 201). Such models, which are based on the early protective colloid concept of Linderstrøm-Lang (6), have been discussed at length by Waugh and co-workers (15). On the basis of his solubility studies of α_{s1}- and κ-casein in the presence of calcium, as discussed in section IV. A, Waugh developed a model of the micelle to explain the relationship between κ-casein content and micellar size (cf. Figure 26). The core–coat concept, however, is in contradiction with the proteolytic studies of Ashoor (194, 195) using insolubilized papain. This author observed that glutaraldehyde-polymerized papain degraded micellar casein at nearly the same rate as molecularly dispersed sodium caseinate. From this he concluded that the components of sodium caseinate are evenly distributed over the micellar surface.

In general, the molecular structure of the micelle, proposed by Waugh, suffers from overinterpretation of experimental data and cannot account for the observed porosity of the micelles discussed in Section II.C.

So-called internal structure models have been discussed by Rose (30) and by Garnier and Ribadeau Dumas (198, 202). Rose proposed a framework of β-casein polymers to which molecules of α_{s1}- and κ-casein are bound by hydrophobic forces. In the micelle such structures are crosslinked by calcium phosphate. This model, however, cannot account for the low-temperature dissociation of β-casein (109, 173, 189–192) without desintegration of the micelle. In the model of Garnier and Ribadeau Dumas, trimers of κ-casein act as polyfunctional nodes for a network of α_{s1}- and β-casein. Though such a model accounts for the high porosity of the micelle, reliable evidence for a crosslinking role of κ-casein is not available.

The arguments in favor of a model with a submicellar structure are provided by the electronmicroscopic work of Shimmin and Hill (55), of Carroll et al. (7, 9, 61), and of Buchheim and Welsch (58), and from the ultracentrifugation studies by Morr (203) and Slattery and Evard (32). Most convincingly, however, the existence of submicelles is demonstrated by application of the freeze-etching technique in electronmicroscopy, as Figure 3 shows. Further evidence that the submicelles are the building units of the micelle has been obtained by Schmidt and co-workers (19, 20) from particle counting and the electronmicroscopical appearance of the colloidal casein before and after dialysis against a calcium-free medium. Knoop et al. (204) demonstrated by electronmicroscopy that the calcium phosphate was distributed through the whole micelle, which seems to be in agreement with its postulated cementing role.

Our knowledge about the biosynthesis of the casein micelles is poor. On the basis of his electron microscopical studies, Farrell (9) hypothesized that, after synthesis on the ribosomes, the polypeptide chains are transferred to the Golgi apparatus where they interact to form submicelles (cf. Figure 7). Subsequently, phosphorylation and cementing by calcium was supposed to take place. Deposition of calcium phosphate would finally yield the mature micelles that are secreted into the alveolar lumen.

Carroll et al. (7, 9, 61) observed strings of beads of submicelles in the Golgi vacuoles, but this could not be confirmed by Buchheim and Welsch (58), who reported only the direct aggregation of the submicelles into the micelles.

To what extent the size distribution of the micelles is an equilibrium has not been established. Schmidt et al. (44) observed, however, that narrow-sized micellar fractions, obtained by density gradient centrifugation showed re-equilibration. Re-equilibration did not take place after previous fixation of the micelles with a crosslinking agent like glutaraldehyde.

One may wonder wether the stability of the micelle can be explained on the basis of the well-known DLVO theory (168), especially since the stability seems to be governed by the surface potential of the micelle. Thus it is found that the clotting of milk caused by chymosin is accompanied by a drastic reduction of the ζ potential of the micelles (2–4). For instance, Green and Crutchfield (2) report a decrease from 27 to 11–14 mV upon the action of chymosin. However, the magnitude of the Van der Waals attraction between the micelles is difficult to estimate. The atomic composition of the casein micelle suggests them to be comparable with polystyrene particles, for which recently a Hamaker constant of 3.5×10^{-14} erg has been computed (205). The combination of the latter value and the above-mentioned value of the ζ potential results in a potential energy barrier of about 14 kT at an interparticle ditance of about 5 Å. It is clear, however, that such considerations are rather hazardous, although they could afford an explanation for the instability of clotting micelles (2–4).

ACKNOWLEDGMENT

We are grateful to Dr. P. Walstra, Agricultural University, Wageningen, the Netherlands, for critical reading of the manuscript.

SYMBOLS

δ	Hydration
$[\eta]$	Intrinsic viscosity
η_r	Relative viscosity
φ	Volume fraction
$\bar{B}$	Mean second virial coefficient
B_{ij}	Interaction parameter between the species i and j
c_i	Concentration of species i
$\bar{c}_i$	Constituent concentration of species i

D_i	Diameter of a particle i
$\{\langle D^2 \rangle_{ls}\}^{1/2}$	Light-scattering average diameter
F	Fluorescent intensity
$\Delta G^{\ominus}$	Change in standard free enthalpy
i	Degree of polymerization
k	Boltzmann's constant
k_i	Equilibrium constant for the ith association step
M_a	Apparent molecular weight
M_i	Molecular weight of species i
$\bar{M}_n$	Number-average molecular weight
$\bar{M}_w$	Weight-average molecular weight
$\bar{M}_z$	z-average molecular weight
M_1	Molecular weight of associating subunit
N_i	Number of amino acid residues of type i in a protein
N_0	Avogadro's number
P	Polarization of fluorescence
r	Interparticle distance
R_h	Hydrodynamic radius
S	Solubility
T	Absolute temperature
u_i	Electrophoretic mobility of species i
$\bar{u}_i$	Constituent electrophoretic mobility of species i
v	Voluminosity
v_i	Electrophoretic velocity of species i
v_1^0	Specific volume of water
$\bar{v}$	Partial specific volume
V_r	Potential energy of two particles a distance r apart
y_i	Activity coefficient of species i

REFERENCES

1. H. A. McKenzie, in *Milk Proteins*, H. A. McKenzie, Ed., Academic Press, New York, 1971, Vol. II, Chapter 10.

2. M. L. Green and G. Crutchfield, *J. Dairy Res.*, **38**, 151 (1971).

3. O. Kirchmeier, *Neth. Milk Dairy J.*, **27**, 191 (1973).

4. M. L. Green, *Neth. Milk Dairy J.*, **27**, 278 (1973).

5. R. Beeby, R. D. Hill, and N. S. Snow, in *Milk Proteins*, H. A. McKenzie, Ed., Academic Press, New York, 1971, Vol. II, Chapter 17.

6. K. Linderstrøm-Lang, *Compt. Rend. Trav. Carlsberg*, **17** (9), 1 (1929).

7. M. P. Thompson and H. M. Farrell Jr., *Neth. Milk Dairy J.*, **27**, 220 (1973).

8. J. Garnier, *Neth. Milk Dairy J.*, **27**, 240 (1973).

9. H. M. Farrell Jr., *J. Dairy Sci.*, **56**, 1195 (1973).

10. D. F. Waugh and P. H. Von Hippel, *J. Amer. Chem. Soc.,* **78**, 4576 (1965).

11. C. A. Zittle, *J. Dairy Sci.,* **44**, 2101 (1961).

12. C. A. Zittle and M. Walter, *J. Dairy Sci.,* **46**, 1189 (1963).

13. D. G. Schmidt, P. Both, and P. J. de Koning, *J. Dairy Sci.,* **49**, 776 (1966).

14. M. P. Thompson, W. G. Gordon, R. T. Boswell, and H. M. Farrell Jr., *J. Dairy Sci.,* **52**, 1166 (1969).

15. D. F. Waugh, in *Milk Proteins,* H. A. McKenzie, Ed., Academic, New York, 1971, Vol. II, Chapter 9.

16. D. G. Schmidt, C. A. van der Spek, W. Buchheim, and A. Hinz, *Milchwissenschaft,* **29**, 455 (1974).

17. A. G. Mackinlay and R. G. Wake, in *Milk Proteins,* H. A. McKenzie, Ed., Academic, New York, 1971, Vol. II, Chapter 12.

18. H. Eggmann, *Milchwissenschaft,* **24**, 479 (1969).

19. D. G. Schmidt and W. Buchheim, *Milchwissenschaft,* **25**, 596 (1970).

20. D. G. Schmidt, P. Walstra, and W. Buchheim, *Neth. Milk Dairy J.,* **27**, 128 (1973).

21. T. L. McMeekin, in *Milk Proteins,* H. A. McKenzie, Academic Press, New York, 1970, Vol. I, Chapter 1.

22. D. L. D. Caspar, *Advan. Prot. Chem.,* **18**, 37 (1963).

23. D. F. Waugh, *Advan. Prot. Chem.,* **9**, 325 (1954).

24. S. J. Singer and D. H. Campbell, *J. Amer. Chem. Soc.,* **77**, 3499 (1955).

25. A. Chanutin, S. Ludewig, and A. V. Masket, *J. Biol. Chem.,* **143**, 737 (1942).

26. G. T. Pyne and T. C. A. McGann, *J. Dairy Sci.,* **27**, 9 (1960).

27. R. Jenness, *Neth. Milk Dairy J.,* **27**, 251 (1973).

28. T. C. A. McGann and G. T. Pyne, *J. Dairy Res.,* **27**, 403 (1960).

29. G. T. Pyne, *J. Dairy Res.,* **29**, 101 (1962).

30. D. Rose, *Dairy Sci. Abstr.,* **31**, 171 (1969).

31. M. Boulet, A. Yang, and R. R. Riel, *Can. J. Biochem.,* **48**, 816 (1970).

32. C. W. Slattery and R. Evard, *Biochim, Biophys. Acta,* **317**, 529 (1973).

33. A. Chughtai. R. Marshall, and G. H. Nancollas, *J. Phys. Chem.,* **72**, 208 (1968).

34. J. D. Termine and A. S. Posner, *Arch. Biochem. Biophys.,* **140**, 307 (1970).

35. J. D. Termine, R. A. Peckauskas, and A. S. Posner, *Arch. Biochem. Biophys.,* **140**, 318 (1970).

36. F. C. McLean and A. B. Hastings, *J. Biol. Chem.,* **108**, 135 (1935).

37. E. G. Weir and A. B. Hastings, *J. Biol. Chem.,* **114**, 397 (1936).

38. C. A. Zittle, E. S. Della Monica, R. K. Rudd, and J. H. Custer, *Arch. Biochem. Biophys.,* **76**, 342 (1958).

39. C. Ho and D. F. Waugh, *J. Amer. Chem. Soc.,* **87**, 889 (1965).

40. I. R. Dickson and D. J. Perkins, *Biochem. J.,* **124**, 235 (1971).

41. H. Nitschmann, *Helv. Chim. Acta,* **32**, 1258 (1949).

42. H. Hostettler and K. Imhof, *Landw. Jb. Schweiz,* **66**, 307 (1952).

43. S. H. C. Lin, S. L. Leong, R. K. Dewan, V. A. Bloomfield, and C. V. Morr, *Biochemistry,* **11**, 1818 (1972).

44. D. G. Schmidt, P. Both, B. W. van Markwijk, and W. Buchheim, *Biochim. Biophys. Acta*, **365**, 72 (1974).

45. G. T. Pyne, *Biochem. J.*, **28**, 940 (1934).

46. G. S. de Kadt and G. van Minnen, *Rec. Trav. Chim.*, **62**, 577 (1943).

47. P. van der Burg, *Neth. Milk Dairy J.*, **1**, 1 (1947).

48. C. J. Schipper, Thesis, Wageningen, 1961.

49. S. A. Visser, *J. Dairy Sci.*, **45**, 710 (1962).

50. M. G. ter Horst, *Neth. Milk Dairy J.*, **17**, 185 (1963).

51. D. Rose, *J. Dairy Sci.*, **48**, 139 (1965).

52. D. Rose and J. R. Colvin, *J. Dairy Sci.*, **49**, 351 (1966).

53. K. Yamauchi, Y. Yoneda, Y. Koga, and T. Tsugo, *Agr. Biol. Chem.*, **6**, 907 (1969).

54. A. Tiselius, S. Hjertén, and Ö. Levin, *Arch. Biochem. Biophys.*, **65**, 131 (1956).

55. P. D. Shimmin and R. D. Hill, *J. Dairy Res.*, **31**, 121 (1964).

56. P. D. Shimmin and R. D. Hill, *Austr. J. Dairy Technol.*, **20**, 119 (1965).

57. G. G. Calapaj, *J. Dairy Res.*, **35**, 1 (1968).

58. W. Buchheim and U. Welsch, *Neth. Milk Dairy J.*, **27**, 163 (1973).

59. H. W. Meyer and H. Winkelman, *Protoplasma*, **68**, 253 (1969).

60. W. Buchheim and U. Welsch, *Z. Zellforsch.*, **131**, 429 (1972).

61. R. J. Carroll, M. P. Thompson, and H. M. Farrell Jr., *Proc. 28th EMSA Meeting*, 1970, p. 150.

62. H. Bechold, *Z. Phys. Chem.*, **60**, 257 (1907).

63. G. Wiegner, *Kolloid Z.*, **8**, 227 (1911).

64. G. Wiegner, *Kolloid Z.*, **15**, 105 (1914).

65. T. Svedberg, *Kolloid Z.*, **51**, 10 (1930).

66. J. B. Nichols, E. D. Baily, G. E. Holm, G. R. Greenbank, and E. F. Deysher, *J. Phys. Chem.*, **35**, 133 (1931).

67. M. Carić and J. Djordjević, *Z. Lebensmittel-Untersuch. Forsch.*, **150**, 24 (1972).

68. W. Lotmar and H. Nitschmann, *Helv. Chim. Acta*, **24**, 242 (1941).

69. A. Leviton and H. S. Haller, *J. Phys. Coll. Chem.*, **51**, 460 (1947).

70. P. F. Dyachenko, *Proc. Fed. Sci. Res. Inst. Dairy Ind.*, No. 19, Moscow (1959).

71. P. Dyachenko, E. Zhdanova, and Yu. Polukarov, *Mol. Prom.*, **5**(5), 35 (1955).

72. E. Knoop and A. Wortmann, *Milchwissenschaft*, **15**, 273 (1960).

73. Z. Saito and Y. Hashimoto, *J. Facult. Agr. Hokkaido Univ.*, **54**(1), 17 (1964).

74. D. Rose and J. R. Colvin, *J. Dairy Sci.*, **49**, 1091 (1966).

75. R. J. Carroll, M. P. Thompson and G. C. Nutting, *J. Dairy Sci.*, **51**, 1903 (1968).

76. S. H. C. Lin, R. K. Dewan, V. A. Bloomfield, and C. V. Morr, *Biochemistry*, **10**, 4788 (1971).

77. R. R. Irani and C. F. Callis, *Particles Size: Measurement, Interpretation, and Application*, Wiley, New York 1963, Chapter 4.

78. V. A. Bloomfield and C. V. Morr, *Neth. Milk Dairy J.*, **27**, 103 (1973).

79. R. K. Dewan and V. A. Bloomfield, *J. Dairy Sci.*, **56**, 66 (1973).

80. C. Tanford, *Physical Chemistry of Macromolecules*, Wiley, New York, 1961, Chapter 6.

81. E. P. Geiduschek and A. Holtzer, *Advan. Biol. Med. Phys.*, **6**, 431 (1958).

82. E. R. Weibel, in *Principles and Techniques of Electronmicroscopy*, M. A. Hayat, Ed., Van Nostrand Reinhold, New York, 1973, Vol. III, Chapter 6.

83. H. Eilers, R.N.J. Saal, and M. van der Waarden, *Chemical and Physical Investigations on Dairy Products*, Elsevier, New York, 1947.

84. C.H. Whitnah and W.D. Rutz, *J. Dairy Sci.*, **42**, 227 (1959).

85. R.K. Dewan, V.A. Bloomfield, A. Chudgar, and C.V. Morr, *J. Dairy Sci.*, **56**, 699 (1973).

86. M.P. Thompson, R.T. Boswell, V. Martin, R. Jenness, and C. Kiddy, *J. Dairy Sci.*, **52**, 796 (1969).

87. B. Ribadeau Dumas and J. Garnier, *J. Dairy Res.*, **37**, 269 (1970).

88. M.A. Hayat, *Principles and Techniques of Electronmicroscopy*, Van Nostrand Reinhold, New York, 1970, Vol. I, Chapter 2.

89. J.L. Oncley in *Proteins, Amino Acids, and Peptides*, E.J. Cohn, and J.T. Edsall, Ed., Reinhold, New York, 1943, Chapter 22.

90. L. Pepper, *Biochim. Biophys. Acta*, **278**, 147 (1972).

91. L.E. Krejci, R.K. Jennings, and L.D. Smith, *J. Franklin Inst.*, **232**, 592 (1941).

92. L.E. Krejci, *J. Franklin Inst.*, **234**, 197 (1942).

93. R.C. Warner, *J. Amer. Chem. Soc.*, **66**, 1725 (1944).

94. H. Nitschmann and H. Zürcher, *Helv. Chim. Acta*, **33**, 1698 (1950).

95. O. Mellander, *Biochem. Z.*, **300**, 240 (1939).

96. N.J. Hipp, M.L. Groves, J.H. Custer, and T.L. McMeekin, *J. Dairy Sci.*, **35**, 272 (1952).

97. T.A.J. Payens, *Biochim. Biophys. Acta*, **46**, 411 (1961).

98. R.G. Wake and R.L. Baldwin, *Biochim. Biophys. Acta*, **47**, 225 (1961).

99. M.P. Thompson, in *Milk Proteins*, H.A. McKenzie, Ed., Academic Press, New York, 1971, Vol. II, Chapter 11.

100. D. Rose, J.R. Brunner, E.B. Kalan, B.L. Larson, P. Melnychyn, H.E. Swaisgood, and D.F. Waugh, *J. Dairy Sci.*, **53**, 1 (1970).

101. W.G. Gordon, M.L. Groves, R. Greenberg, S.B. Jones, E.B. Kalan, R. F. Peterson, and R.E. Townend, *J. Dairy Sci.*, **55**, 261 (1972).

102. M.L. Groves, W.G. Gordon, E.B. Kalan, and S.B. Jones, *J. Dairy Sci.*, **56**, 558 (1973).

103. P.J. de Koning, Thesis, Amsterdam, 1967.

104. J.C. Mercier, F. Grosclaude, and B. Ribadeau Dumas, *Eur. J. Biochem.*, **23**, 41 (1971).

105. B. Ribadeau Dumas, G. Brignon, F. Grosclaude, and J.C. Mercier, *Eur. J. Biochem.*, **25**, 505 (1972).

106. J.C. Mercier, G. Brignon, and B. Ribadeau Dumas, *Eur. J. Biochem.*, **35**, 222 (1973).

107. F. Grosclaude, M.F. Mahé, and B. Ribadeau Dumas, *Eur. J. Biochem.*, **40**, 323 (1973)

108. A.G. Szent-Gyorgi and C. Cohen, *Science*, **126**, 697 (1957).

109. T.A.J. Payens and B.W. van Markwijk, *Biochim. Biophys. Acta*, **71**, 517 (1963).

110. T.T. Herskovits, *Biochemistry*, **5**, 1018 (1966).

111. M. Noelken and H. Reibstein, *Arch. Biochem. Biophys.*, **123**, 397 (1968).

112. D.G. Schmidt, T.A.J. Payens, B.W. van Markwijk, and J.A. Brinkhuis, *Biochem. Biophys. Res. Commun.*, **27**, 448 (1967).

113. P.H. Von Hippel and D.F. Waugh, *J. Amer. Chem. Soc.*, **77**, 4311 (1955).

114. D.F. Waugh, H.L. Ludwig, J.M. Gillespie, B. Melton, M. Foley, and E.S. Kleiner, *J. Amer. Chem. Soc.*, **84**, 4929 (1962).

115. D.G. Schmidt and T.A.J. Payens, *Biochim. Biophys. Acta*, **78**, 492 (1963).

116. H.E. Swaisgood, J.R. Brunner, and H.A. Lillevik, *Biochemistry*, **3**, 1616 (1964).

117. C. Ho and A.H. Chen, *J. Biol. Chem.*, **242**, 551 (1967).

118. C. Tanford, *J. Amer. Chem. Soc.*, **84**, 4240 (1962).

119. R.J. Hill and R.G. Wake, *Nature*, **221**, 635 (1969).

120. C.C. Bigelow, *J. Theoret. Biol.*, **16**, 187 (1967).

121. H.F. Fisher, *Proc. Nat. Acad. Sci. Wash.*, **51**, 1285 (1964).

122. J. Garnier, *J. Mol. Biol.*, **19**, 586 (1966).

123. J. Garnier, *Biopol.*, **5**, 473 (1967).

124. R. Clarke and S. Nakai, *Biochemistry*, **10**, 3353 (1971).

125. L.W. Nichol, J.L. Bethune, G. Kegeles, and E.L. Hess, in *The Proteins*, H. Neurath, Ed., 2nd ed., Academic Press, New York, 1964, Vol. II, Chapter 9.

126. E.T. Adams Jr., *Fractions*, No. 3, Spinco Division of Beckmann Instruments Inc., Palo Alto, Calif., 1967.

127. E.T. Adams Jr., A.H. Pekar, D.A. Soucek, L.H. Tang, G. Barlow, and J.L. Armstrong, *Biopol.*, **7**, 5 (1969).

128. R.F. Steiner, *Biochemistry*, **9**, 4268 (1970).

129. R.F. Steiner, *Arch. Biochem. Biophys.*, **39**, 333 (1952).

130. P.G. Squire and S. Benson, *Protides of the Biological Fluids*, **14**, 441 (1966).

131. D.G. Schmidt, Thesis, Utrecht, 1969.

132. C. Tanford, *Physical Chemistry of Macromolecules*, Wiley, New York, 1961, Chapter 5.

133. J.G. Kirkwood and R.J. Goldberg, *J. Chem. Phys.*, **18**, 54 (1950).

134. W.H. Stockmayer, *J. Chem. Phys.*, **18**, 58 (1950).

135. H. Fujita, *Mathematical Theory of Sedimentation Analysis*, Academic Press, New York, 1962.

136. E.T. Adams Jr. and H. Fujita, in *Ultracentrifugal Analysis in Theory and Experiment*, J.W. Williams, Ed., Academic Press, New York, 1963, p. 119.

137. M. Derechin, *Biochemistry*, **8**, 921 (1969).

138. P.W. Chun and S.J. Kim, *Biochemistry*, **8**, 1633 (1969).

139. J. Visser, R.C. Deonier, E.T. Adams Jr., and J.W. Williams, *Biochemistry*, **11**, 2634 (1972).

140. D.G. Schmidt and T.A.J. Payens, *J. Colloid Interface Sci.*, **39**, 655 (1972).

141. G. Scatchard, *J. Amer. Chem. Soc.*, **68**, 2315 (1946).

142. E.F. Casassa and H. Eisenberg, *Advan. Prot. Chem.*, **19**, 287 (1964).

143. E.T. Adams Jr., *Biochemistry*, **4**, 1646 (1965).

144. J.R. Cann and W.B. Goad, *Interacting Macromolecules*, Academic Press, New York, 1970.

145. L.W. Nichol and D.J. Winzor, *Migration of Interacting Systems*, Clarendon Press, Oxford, 1972.

146. A. Tiselius, *Nova Acta Regiae Soc. Sci. Upsaliensis, IV*, **7** (4), 1 (1930).

147. R.J. Goldberg, *J. Phys. Chem.*, **57**, 194 (1953).

148. A.E. Alexander and P. Johnson, *Colloid Science*, Oxford University Press, Oxford, 1949.

149. E.O. Field and A.G. Ogston, *Biochem. J.*, **60**, 661 (1955).

150. G.A. Gilbert, *Proc. Roy. Soc.*, **A250**, 377 (1959).

151. G.A. Gilbert and R.C.L. Jenkins, *Proc. Roy. Soc.*, **A253**, 420 (1959).

152. L.G. Longsworth, in *Electrophoresis*, M. Bier, Ed., Academic Press, New York, 1959, Chapter 3.

153. T.A.J. Payens and D.G. Schmidt, *Arch. Biochem. Biophys.*, **115**, 136 (1966).

154. E.L. Smith, J.R. Kimmel, and D.M. Brown, *J. Biol. Chem.*, **207**, 533 (1954).

155. R. Townend, R.J. Winterbottom, and S.N. Timasheff, *J. Amer. Chem. Soc.*, **82**, 3161 (1960).

156. L.W. Nichol and A.G. Ogston, *Proc. Roy. Soc.*, **B163**, 343 (1965).

157. J.L. Bethune, *J. Phys. Chem.*, **74**, 3837 (1970).

158. D.J. Cox, *Arch. Biochem. Biophys.*, **146**, 181 (1971).

159. H. Nijhuis and T.A.J. Payens, *Biochem. Biophys. Acta*, **336**, 213 (1974).

160. H. Nijhuis, Thesis, Wageningen, 1974.

161. J.L. Bethune and G. Kegeles, *J. Phys. Chem.*, **65**, 1755 (1961).

162. J.L. Bethune and G. Kegeles, *J. Phys. Chem.*, **65**, 1761 (1961).

163. M. Halwer, *Arch. Biochem. Biophys.*, **51**, 79 (1954).

164. D.G. Schmidt and B.W. van Markwijk, *Biochim. Biophys. Acta*, **154**, 613 (1968).

165. D.G. Schmidt, *Biochim. Biophys. Acta*, **207**, 130 (1970).

166. W.G. McMillan and J.E. Mayer, *J. Chem. Phys.*, **13**, 276 (1945).

167. J. Th. G. Overbeek and P.H. Wiersema, in *Electrophoresis*, M. Bier, Ed., Academic Press, New York, 1967, Vol. II, Chapter 1.

168. E.J.W. Verwey and J.Th.G. Overbeek, *Theory of the Stability of Lyophobic Colloids*, Elsevier, Amsterdam, 1948.

169. E.T. Adams Jr. and J.W. Williams, *J. Amer. Chem. Soc.*, **86**, 3454 (1964).

170. D.A. Albright and J.W. Williams, *Biochemistry*, **7**, 67 (1968).

171. W. Kauzmann, *Advan. Prot. Chem.*, **14**, 1 (1959).

172. G. Némethy and H.A. Scheraga, *J. Phys. Chem.*, **66**, 1773 (1962).

173. R.A. Sullivan, M.M. Fitzpatrick, E.K. Stanton, R. Annino, G. Kissel, and F. Palermiti, *Arch. Biochem. Biophys.*, **55**, 455 (1955).

174. T.A.J. Payens, J.A. Brinkhuis, and B.W. van Markwijk, *Biochim. Biophys. Acta*, **175**, 434 (1969).

175. C. Tanford, *Physical Chemistry of Macromolecules*, Wiley, New York, 1961, Chapter 4.

176. T.A.J. Payens and K. Heremans, *Biopol.*, **8**, 335 (1969).

177. K. Suzuki and, Y. Taniguchi, *Proc. 26th Symp. Soc. Exp. Biol.*, 103, (1972).

178. R.A. Sullivan, M.M. Fitzpatrick, and E.K. Stanton, *Nature*, **183**, 616 (1959).

179. N.S. Snow and A.K. Hay, *Proc. 17th International Dairy Congress*, Munich, B, 105 (1966).

180. D. Rose, D.T. Davies, and M. Yaguchi, *J. Dairy Sci.*, **52**, 8 (1969).

181. L.K. Creamer, J.V. Wheelock, and D. Samuel, *Biochim. Biophys. Acta*, **317**, 202 (1973).

182. D.F. Waugh, *Discuss. Faraday Soc.*, **25**, 186 (1958).

183. J. Garnier, G. Mocquot, B. Ribadeau Dumas, and J.L. Maubois, *Ann. Nutr. Aliment.*, **22**, B495 (1968).

184. P.W.L. Chun, Thesis, Missouri, 1965.

185. T.A.J. Payens and H. Nijhuis, *Biochim. Biophys. Acta*, **336**, 201 (1974).

186. L.G. Longsworth, in *Electrophoresis*, M. Bier, Ed., Academic Press, New York, 1959, Chapter 4.

187. R.W. Noble Jr. and D.F. Waugh, *J. Amer, Chem. Soc.*, **87**, 2236 (1965).

188. D.F. Waugh and R.W. Noble, Jr., *J. Amer. Chem. Soc.*, **87**, 2246 (1965).

189. C.W. Kolar, Jr. and J.R. Brunner, *J. Dairy Sci.*, **50**, 941 (1967).

190. D. Rose, *J. Dairy Sci.*, **51**, 1897 (1968).

191. W.K. Downey and R.F. Murphy, *J. Dairy Res.*, **37**, 361 (1970).

192. W.K. Downey, *Neth. Milk Dairy J.*, **27**, 218 (1973).

193. K.E. Beery, L.F. Hood, and S. Patton, *J. Dairy Sci.*, **54**, 911 (1971).

194. S.H. Ashoor, Thesis, Wisconsin, 1971.

195. S.H. Ashoor, R.A. Sair, N.F. Olson, and T. Richardson, *Biochim. Biophys. Acta*, **229**, 423 (1971).

196. H.A. McKenzie, in *Milk Proteins*, H.A. McKenzie, Ed., Academic Press, New York, 1971, Vol. II, Chapter 14.

197. T.A.J. Payens, *J. Dairy Sci.*, **55**, 141 (1972).

198. J. Garnier, *Neth. Milk Dairy J.*, **27**, 240 (1973).

199. H.M. Farrell, Jr., and M.P. Thompson, in *Fundamentals of Dairy Chemistry*, B. Webb., Ed., 2nd ed. Avi Publishing Co, Westport, Conn., 1973.

200. T.A.J. Payens, *J. Dairy Sci.*, **49**, 1317 (1966).

201. T. Nagasawa, M. Kuboyama, and M. Tsuda, *Proc. 18th Int. Dairy Congr.*, Sydney, **1E**, 23 (1970).

202. J. Garnier and B. Ribadeau Dumas, *J. Dairy Res.*, **37**, 493 (1970).

203. C.V. Morr, *J. Dairy Sci.*, **50**, 1744 (1967).

204. A.M. Knoop, E. Knoop, and A. Wiechen, *Neth. Milk Dairy J.*, **27**, 121 (1973).

205. H. Krupp, W. Schnabel, and G. Walter, *J. Colloid Interface Sci.*, **39**, 421 (1972).

4

The Adsorption of Gases on Porous Solids

S. J. GREGG AND K. S. W. SING

School of Chemistry, Brunel University,
Uxbridge, Middlesex, England.

I. General 232
    A. Introduction 232
        1. The Origin of Pores 234
        2. Practical Importance 235
        3. Adsorption Isotherms 235
    B. Surface Forces 237
        1. "Specific" and "Nonspecific" Adsorption 242
        2. Application to the Adsorption Isotherm....................... 244
II. Adsorption on Nonporous Solids 245
    A. Introduction 245
        1. Experimental Findings 245
    B. Some Mathematical Expressions for the Type II Isotherm 247
        1. The BET Equation 247
        2. The Equations of Frenkel, Hill, and Halsey 250
        3. The Virial Approach 252
    C. The Packing of Molecules in the Monolayer.
        The Molecular Area a_m 253
        1. The BET Method for Determination of Specific Surface.
            The Role of Parameter c.................................... 254
    D. The Concept of a Standard Isotherm 255
        1. Deviations from the Standard Isotherm. 258
            a. The t-plot 258
            b. The α_s-plot 259
            c. The Effect of Mesoporosity 260
            d. The Effect of Microporosity 263
III. Adsorption on Mesoporous Solids 263
    A. Porosity and the Type IV Isotherm 263
        1. Types of Hysteresis Loop 268
        2. The Surface Area of Mesoporous Solids 268
    B. Derivation of the Kelvin Equation 269

I. GENERAL

A. Introduction

The uptake of vapors by solids has long been associated with the presence of pores. As early as 1777, both Fontana (1) and Scheele (2) had described the occlusion of gases by the porous solid, charcoal. It was later realized that gases could also be taken up on the free surface of a solid, and Kayser (3) in 1881 distinguished between "adsorption" on the surface and dissemination throughout the mass of the solid. In 1906 Ostwald (4) further distinguished between *adsorbed* and *capillary condensed* liquid. Shortly afterwards, in 1909, McBain (5) proposed the term *sorption* to embrace *ad*sorption on the surface, *ab*sorption by penetration into the lattice of the solid, and capillary condensation within the pores; perhaps for reasons of euphony the term has never enjoyed really wide usage and the designation *adsorption* is frequently employed to denote uptake whether by capillary condensation or by surface adsorption. The concept of capillary condensation followed from the theoretical demonstration by W. Thomson (6) (later Lord Kelvin) that the equilibrium pressure P over a concave liquid surface must be less than the saturated vapor pressure $P°$ at the same temperature. Thomson's original equation was not very suitable for application to adsorption isotherms; and as modified by later workers (7–9) it becomes

$$\ln\left(\frac{P}{P°}\right) = -\frac{\gamma v}{RT}\left(\frac{1}{r_1} + \frac{1}{r_2}\right) \tag{1}$$

where γ and v are the surface tension and molar volume, respectively, of the liquid adsorptive, and r_1 and r_2 are the principal radii of curvature of the surface, R and T having their usual meanings.

If the surface is hemispherical so that $r_1 = r_2 = r$, say, equation 1 simplifies to

$$\ln\left(\frac{P}{P°}\right) = -\frac{2\gamma V}{rRT} \tag{2}$$

the form usually referred to as the Kelvin equation.

Zsigmondy (10) in 1911 made use of Thomson's equation to interpret the earlier results of van Bemmelen (8) for the adsorption of water vapor on silica gel; Zsigmondy postulated that, as the pressure of the vapor over the gel was progressively increased, an adsorbed layer first formed on the walls of the pores and this was followed by capillary condensation in each pore of radius r and as soon as the relative pressure reached the value given by the Kelvin equation.

The subject was taken further by Foster (11) who used the Kelvin equation to calculate the pore size distribution of a number of xerogels from the adsorption isotherms of various organic vapors including benzene. Foster clearly recognized the necessity of making allowance for the thickness of the adsorbed layer on the walls, already referred to, which preceded capillary condensation; but he was handicapped at the time by the absence of any really reliable method for determination of the surface area of highly disperse solids, which was an essential prerequisite to the calculation of the film thickness corresponding to a given equilibrium pressure. The BET method (12), advanced in 1938 and developed in the years immediately following,

TABLE 1

Classification of pores (15) according to their width (w)

	$w/\text{Å}$
Micropores	Less than $\sim$ 20 Å
Mesopores[a]	Between $\sim$ 20 and $\sim$ 500 Å
Macropores	Above $\sim$ 500 Å

[a]This replaces the earlier terms "intermediate pores" and "transitional pores."

laid the foundations of a solution to this problem; from that time onwards a number of methods for the evaluation of pore size distribution, based essentially on the combined application of the Kelvin equation and the BET method, have been put forward.

In the early 1950s Dubinin, and independently Pierce (14), drew attention to the likelihood that in very fine pores the mechanism of adsorption would be modified because of the increased force acting on the molecules of adsorbate owing to the proximity of the walls; the result would be complete filling of the pores at very low relative pressures.

From the days of Zsigmondy onwards, the occurrence of capillary condensation has been associated with the presence of a hysteresis loop in the adsorption isotherm, and, in typical cases, application of the Kelvin equation shows that the region of the isotherm covered by the loop corresponds to pore widths in excess of $\sim$ 15 Å (1.5 nm); but that when the width exceeds $\sim$ 500 Å (50 nm) the lowering of relative pressure will be too small to measure by conventional means. The classification of pores quoted in Table 1, originally proposed by Dubinin and now officially adopted by the International Union of Pure and Applied Chemistry (15), is accordingly a very useful one.

1. The Origin of Pores

From general knowledge of the nature of solids, one would expect the presence of pores in solids, whether natural or artificial, to be the rule rather than the exception. Igneous rocks, having been formed by the solidification of a melt and therefore subjected to thermal stresses, almost inevitably contain actual cracks or planes of weakness which will become pores through attack by terrestial liquors; and sedimentary rocks, being essentially agglomerates of individual particles, will contain pores as the result of the imperfect packing together of these particles. All such pores may become partially or completely filled by deposition from ambient liquors, but will often be emptied again and perhaps enlarged by subsequent physicochemical attack.

Artificial solids, whether produced in the laboratory or in industry are even more prone to contain pores. Modes of treatment that will induce or intensify porosity include: the expulsion of a volatile product by heating (e.g., evolution of carbon dioxide in lime burning); removal of one constituent from the matrix of a second (e.g., production of Raney nickel by removal of aluminium from an alloy with nickel by treatment with alkali); formation of gel-like precipitates (precipitation from solution of numerous hydroxides or hydrous oxides, such as silica, alumina, chromia); formation of sublimates (evaporated films of metals); compaction of powders (as in pharmacy, powder metallurgy, the ceramic industry); and grinding, since the fine particles first produced reaggregate to form secondary particles which are porous.

2. Practical Importance

The practical importance of the adsorption of gases by porous materials is now widely recognized. Such adsorption finds a direct application in the removal of water vapor or other gaseous impurities from a gas stream by means of an adsorbent such as silica gel or charcoal; a variant of this use is in gas chromatography. Frequently, the adsorption is unwanted as in the dampening of materials of construction or clothing in the atmosphere.

A less direct, but very widespread, application of adsorption is for elucidation of the pore structure itself: analysis of the adsorption isotherm provides the most widely used—and often the only available—means of obtaining quantitative information about pore size distribution, and in suitable cases, as to pore shape also. Information of this kind is needed in numerous fields where porous solids find application; in heterogeneous catalysis, for example, chemical reaction occurs on the walls of the pores, which have to be reached by diffusion through the body of the pore. The rate of reaction of a porous solid with a gaseous or liquid reactant, or its rate of dissolution in a solvent, likewise depends on the pore structure of the solid. The manner in which a filler behaves in a composite, exemplified by carbon black in vulcanized rubber, will again be influenced by the structure of its pore system.

The case in which the pores are of molecular dimensions, that is, are "micropores" (Table 1), is of special interest. Micropores increase the tenacity with which vapors are held within the solid, and therefore affect the catalytic and chromatographic behavior, the performance of a drying agent or even the mechanical strength of the solid. The unsuspected presence of microporosity leads to erroneous values of the specific surface when this is determined by gas adsorption. Information as to the presence and nature of microporosity is to be gained by suitable studies of the adsorption of gases.

3. Adsorption Isotherms

The amount n_a of vapor taken up by a mass m of solid is in general proportional to m, and it depends also on the temperature T, the relative pressure $P/P°$ of the vapor, and the nature of both the solid and the vapor:

$$\frac{n_a}{m} = n = f\left(\frac{P}{P°}, T, \text{vapor, solid}\right)$$

In the experimental study of the subject, the approach most frequently followed is to determine the adsorption isotherm: The uptake of a particular vapor on a given solid is measured as a function of the relative pressure, the temperature being kept constant. It is usual to refer the uptake to unit mass of the solid:

$$n = f\left(\frac{P}{P°}\right)_{T, \text{ vapour, solid}}$$

The material actually adsorbed by the solid (*the adsorbent*) is termed the *adsorbate*, in contradistinction to the *adsorptive*, which is the general term for the material in the vapor phase which is capable of being adsorbed.

The graphical representation of the adsorption isotherm may assume one of a number of shapes according to the nature of the system, and particularly of the adsorbent; but the majority of isotherms in the literature are embraced within the five types of the classification of Brunauer, Deming, Deming and Teller (16), illustrated diagrammatically in Figure 1. Within recent years an

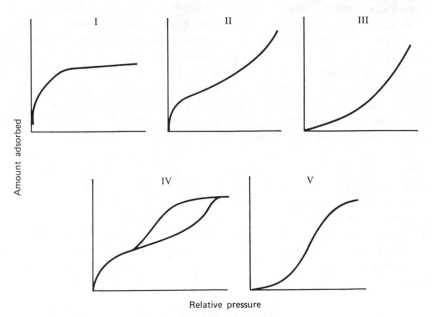

Fig. 1. The five types of isotherm classification by Brunauer, Deming, Deming, and Teller (16) (BDDT classification—alternatively the "BET" classification).

increasing number of stepped isotherms, which are not embraced within the Brunauer classification, have appeared in the literature.

The isotherm type is determined by the nature of the adsorbate and adsorbent considered jointly. A type II isotherm is characteristic of adsorption on a nonporous solid unless the adsorbent–adsorbate interaction is unusually weak, in which case a type III isotherm will result. A type IV isotherm is to be expected if the adsorbent is mesoporous and a type I if it is microporous; if however the adsorbent–adsorbate interaction is unusually feeble a porous solid will tend to yield a type V isotherm.

The *stepped* isotherm—a not very common form—seems to be characteristic of nonporous solids having very uniform surfaces.

The adsorption of vapor by a nonporous solid, which gives rise to a type II isotherm, serves as a useful starting point for the study of the more complicated process of uptake by a porous solid, where capillary condensation (in mesopores) or pore filling (of micropores) has to be considered. Section II will accordingly be devoted to a discussion of relevant features of adsorption on nonporous solids, as a preliminary to Sections III and IV, which deal with mesoporous and microporous solids, respectively. The remainder of the present section is concerned with the nature of the forces responsible for adsorption on the surface of a nonporous solid, and by implication on the walls of a mesoporous solid also.

B. Surface Forces (17)

For the present limited purpose of providing background material, a quite elementary treatment of surface forces will suffice. The impressive advances made in recent years in this field are in the nature of refinements rather than fundamental revisions of the basic ideas; since perforce they refer to idealized systems, they are still far from the state where the detailed course of an isotherm can be calculated a priori from molecular parameters of the vapor and the solid. An elementary treatment makes clear the kind of factors involved, and thus makes it possible to predict at least the direction in which adsorption will be affected, for example, by an increase in polarizability, or in polarity, of the vapor, or in the size of the adsorbate molecule.

The forces that bring about physical adsorption always include "dispersion forces" which are attractive, and short-range repulsive forces. If either the solid or the vapor, or both, is/are polar in nature, classical electrostatic forces will be present in addition.

Dispersion forces, whose existence was first recognized by London (18), derive their name from the connection between their origin and the cause of optical dispersion; they arise from the rapid fluctuations in electron density within a given atom, which induce an electrical moment in a near neighbor and thus lead to attraction between the two atoms. London, making use of quantum-mechanical perturbation theory, worked out an expression for the potential energy ε_D (r) of two isolated atoms separated by a distance r; developed (19) by later workers it reads:

$$\varepsilon_D(r) = - A_1 r^{-6} - A_2 r^{-8} - A_3 r^{-10} \tag{3}$$

and is valid provided the atoms are not too far apart. The negative sign, of course, implies attraction. In this expression A_1, A_2, and A_3 are the dispersion constants associated with instantaneous dipole–dipole, dipole–quadrupole, and quadrupole–quadrupole interactions, respectively.

In view of the unavoidable uncertainties and approximations inherent in the application of the expression to actual numerical calculations, the terms in r^{-8} and r^{-10}, which are of secondary importance only, are usually omitted and equation 3 simplifies to:

$$\varepsilon_D(r) = - A_1 r^{-6} \tag{4}$$

In addition there is a force of repulsion (arising from the interpenetration of the electronic clouds of the constituent atoms), which may be represented by the empirical equation (20):

$$\varepsilon_R(r) = B r^{-j} \tag{5}$$

where B and j are empirical constants, the latter usually being assigned the value $j = 12$.

The total potential between the two molecules thus becomes (21):

$$\varepsilon(r) = - A r^{-6} + B r^{-12} \tag{6}$$

often designated as the Lennard–Jones 6–12 potential (22). The general from of the curve of ε (r) against r is indicated in Figure 2. Various expressions have been advanced from time to time for the evaluation of the constant A. One of the best known is that of Kirkwood and Müller (23, 24), viz:

$$A = 6mc^2 \cdot \frac{\alpha_1 \alpha_2}{(\alpha_1/\chi_1 + \alpha_2/\chi_2)} \tag{7}$$

where α_1 and α_2 are the polarizabilities and χ_1 and χ_2 the diamagnetic susceptibilities of the constituent atoms; m being the mass of the electron and c the velocity of light. Both α and χ increase in value with the size of the molecule; but the dependence of α on size is greater than that of χ (in the simplest case α is approximately proportional to σ^3 and χ to σ^2, where σ is the diameter of the molecule considered as a spherical conductor). Thus, for a given molecule, characterized by α_1 and χ_1, the interaction energy with another molecule (α_2, χ_2) will increase as the polarizability α_2 of the latter increases.

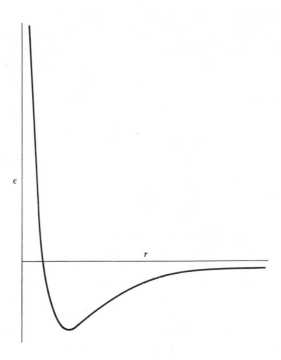

Fig. 2. The potential energy ε of two isolated atoms as a function of their distance apart. The curve for potential energy ϕ of a molecule as a function of its distance from a surface of a solid, is similar in general shape to this curve.

Constant B of equation 5 could in principle be evaluated by noting that at the equilibrium distance of separation r^* the net force acting between the atoms would be zero, that is, $\varepsilon_D(r^*) = \varepsilon_R(r^*)$, but unfortunately this procedure demands an accurate knowledge of r^*, which in general is difficult or even impossible to attain. Until recently (25) the general practice has been to assess $\varepsilon_R(r)$ empirically as 40% of the total potential $\varepsilon(r)$.

To apply these various equations to the adsorption of a gas on a solid, it is necessary to consider (26, 27) the interaction of the surface layers of a solid composed of atoms (or ions) of a substance Y, with an isolated molecule of a gas X. The individual interactions of each atom of X with each atom of Y have to be added up to obtain the potential of a single molecule of the gas with reference to the solid:

$$\phi(z) = \sum \varepsilon_{ij}(r_{ij}) \tag{8}$$

or

$$\phi(z) = -A_{ij}\sum_j r_{ij}^{-6} + B_{ij}\sum_j r_{ij}^{-12} \tag{9}$$

Here r_{ij} is the distance between the center of the molecule i in the gas phase (or, for a complex molecule, its atom or group i) and the center of an atom j in the solid. If an adsorbent having a known crystal lattice with a given location is considered, the various values of r_{ij} can be expressed in terms of a single quantity z: here z is the distance between the center of the gas molecule (or its atoms or groups) and the plane through the centers of the atoms in the outermost layer of the solid.

Various approximations have been introduced to lighten the mathematical task of carrying out the summation implied in equation 8. Thus, London (28) replaced the summation by volume integration, the solid being thus treated as a continuum, so that:

$$\phi = \rho_V \iiint \varepsilon_{ij} \, dV \tag{10}$$

where ρ_V is the number of atoms of adsorbent per unit volume, the integration being carried out over a semiinfinite volume.

With the readily availability of computers, however, it is now possible actually to sum up all the individual interactions over, say, 100 to 200 of the nearest pairs, and to confine the integration porcedure to the remainder (27).

If the solid is polar—if it consists of ions, or contains polar groups such as OH, or π electrons—it will give rise to an electric field (29), which will induce a dipole in the molecule X. The resulting interaction energy ϕ_p will be:

$$\phi_P = \frac{-1}{2} \alpha^2 F \tag{11}$$

where α is the polarizability of the molecule X, and F is the field strength at the center of the molecule. If in addition the molecule possesses a permanent dipole, this will interact with the field to give a further contribution $\phi_{F\mu}$ given by:

$$\phi_{F\mu} = -F\mu \cdot \cos\theta \tag{12}$$

where μ is the dipole moment of the molecule and θ the angle between the field and the axis of the dipole.

Finally, if the molecule has a quadrupole moment—examples are CO, CO_2, and N_2—it will interact strongly with the field gradient $\dot{F}$ to produce a further contribution, $\phi_{\dot{F}Q}$, to the energy (31).

Thus the overall interaction energy $\phi(z)$ of a molecule with the surface may in general be represented (32) by the expression:

$$\phi(z) = \phi_D + \phi_p + \phi_{F\mu} + \phi_{\dot{F}Q} + \phi_R \tag{13}$$

Here ϕ_D and ϕ_R are written for the terms in r^{-6} and r^{-12}, respectively, in equation 9; these contributions are universally present, whereas the electrostatic energies ϕ_P, $\phi_{F\mu}$ and ϕ_{FQ} may or may not be present according to the nature of the adsorbent and the adsorbate.

In principle equation 13 could be used to calculate the numerical value of interaction as a function of the distance z of any given molecule from the surface of a chosen solid. In practice however, the scope has to be limited to simple types of molecule on an idealized surface of a solid having a relatively simple lattice, such as an inert gas on potassium chloride or magnesium oxide. Even so, the inevitable uncertainties are such that the final result can be no more than an approximation to the true situation; not only are the expressions such as 4, 5, and 6 used in the calculation themselves approximations, but the distance z can only be defined by use of arbitrary assumptions as to the exact location of the "surface" of the solid. The general form of $\phi(z)$ against (z) however is not in doubt, and is similar to that for the two isolated atoms already given in Figure 2.

The potential energy depends on the distance z of the atom from the surface, and on the position of the atom in the xy plane relative to the lattice of the solid; for any such position the adsorption energy ϕ_0 will be equal to the value of $\phi(z)$ at the minimum in the potential curve (cf. Figure 2). Calculations of the value of ϕ_0, usually for an inert gas molecule, on various typical positions on the surface of an ionic solid, have been carried out by a number of workers (26, 33–37).

In the past few years Ricca and his co-workers (21, 25) have turned their attention to systems (until recently inaccessible to experimental investigation) that have the advantage of theoretical simplicity; they have carried out a number of calculations of ϕ_0 for crystalline xenon as the adsorbent and helium, neon, and argon as adsorptives (Table 2). Points of particular interest

TABLE 2

Potential energy ϕ_0 for He, Ne, and Ar atoms adsorbed on the (100) and (111) faces of crystalline xenon

	Values of $-\phi_0/10^{-23}$ J molecule[1] (rounded to the nearest unit)					
Adsorbed Atom	Position on (100) Face			Position on (111) Face		
	a	b	c	a	b	c
He	340	209	135	271	234	147
Ne	641	404	268	518	453	291
Ar	1251	855	608	1072	971	677

After Ricca, Pisani, and Garrone (25).
a. The center of an array of nearest neighbors on the surface atom of Xe.
b. Between adjacent surface atoms of Xe ("saddle position").
c. Directly over a surface atom of Xe.

which emerge from Table 2 are (1) the most favorable site (maximum interaction) on each face is at the center of an array (position (*a*)), being consistently higher for the fourfold coordination of the (100) face than for the three-fold coordination of the (111) face; (2) the interaction energy increases steadily with atomic number of the adsorbate molecule; and (3) for a given molecule there is considerable variation in ϕ_0 from position to position; there must accordingly be an energy barrier to translational movements in the *xy* plane, the barrier being higher the higher the atomic number of the adsorbate. For an argon atom (molecule) on the (100) face, for example, the easiest path from one preferred site to another is over the saddle (*b*) so that the energy barrier for movement is (1251–855), that is, 396×10^{-23} J/molecule; since the mean thermal energy $\sim kT$ at 87°K is 104×10^{-23} J/molecule, the argon molecule will have only limited mobility and will spend much more time in the region of site (*a*) than elsewhere—the adsorption will be *localized*. For helium on the same face, however, the energy barrier is only (340–209), that is, 131×10^{-23} J/molecule, implying a very low degree of localizsation.*

The difference in ϕ values as between the (100) and (111) faces is of significance in relation to the surfaces of real solids, which will rarely consist of an extensive area of a single type of face.

1. "Specific" and "Nonspecific" Adsorption

The part played by polarity in enhancing the force of adsorption has been discussed by Kiselev (38, 39) and his associates; they propose the designation "nonspecific" adsorption for those cases which involve only dispersion and repulsion forces ($\phi_D + \phi_R$), and "specific" adsorption for those cases where the electronic distribution in the adsorbate molecule is sufficiently asymmetric to produce a coulombic contribution as well (some or all of ϕ_P, $\phi_{F\mu}$, ϕ_{FQ}).

Adsorbents are divided into three types, containing:

(I)   No ions or positive groups (e.g., graphitized carbon);

(II)   Concentrated positive charges (e.g., OH groups on hydroxylated oxides);

(III)   Concentrated negative charges (e.g., $=O$, $=CO$).

Adsorbates are divided into four groups, having:

(a)   Spherically symmetrical shells or σ-bonds (e.g., noble gases, saturated hydrocarbons);

(b) π-Bonds (e.g., unsaturated, or aromatic, hydrocarbons) or lone pairs of electrons (e.g., ethers, tertiary amines);

---

*For this very light molecule, the calculation needs refinement through the use of quantum-mechanical calculation of probability density.

TABLE 3

Specific and nonspecific adsorption (38)

Adsorbate Group	Type of Adsorbent		
	(I)	(II)	(III)
(a)	n	n	n
(b)	n	n + s	n + s
(c)	n	n + s	n + s
(d)	n	n + s	n + s

s = specific adsorption
n = nonspecific adsorption

(c) Positive charges concentrated on peripheries of molecules;
(d) Functional groups with both electron density and positive charges concentrated as above (e.g., molecules with $-OH$ or $=NH$ groups).

The interactions resulting from combinations of the various types of adsorbent and adsorbate are summarized in Table 3.

To put the subject on a quantitative basis Kiselev estimated the energy of specific interaction ΔH_{sp} as the difference in heats of adsorption between a suitable pair of adsorbates, one of them belonging to group (a) and therefore showing only nonspecific interaction.

Thus, on dehydroxylated silica (type I) the heats of adsorption of ethane (group (a)) and ethylene (group (b)) were 4.2 and 3.8 kcal/mole, respectively; but on hydroxylated silica (type II) the corresponding values were 4.4 and 5.2 kcal/mole, so that the energy of specific interaction of ethylene with the hydroxylated silica was $\Delta H_{sp} \sim (5.2 - 3.8)$, that is, ~ 1.4 kcal/mole. Another example was n-pentane (group a) and diethyl ether (group (b)), which on graphitized carbon black (type I) gave heats of adsorption of 9.2 and 8.9 kcal/mole, respectively, whereas on hydroxylated silica the values were 7.3 and 15.0 giving an energy of specific interaction of the ether with hydroxylated silica of $\Delta H_{sp} \sim 8$ kcal/mole.

A slightly different approach has been followed by Barrer (40) in his attempt to assess the specific contribution to the heat of adsorption of a number of polar adsorbates on certain zeolites. He first estimated the nonspecific contribution $(\phi_D + \phi_P + \phi_R)$ (equation 13) for each of the polar gases, by interpolation on a reference curve of the heat of adsorption against polarizability for a series of simple nonpolar gases (H_2, Kr, C_2H_6, etc.); and by subtraction from the total heat of adsorption a figure for the specific contribution $(\phi_{F\mu} + \phi_{FQ})$ for the polar gas was arrived at (N_2, N_2O, NH_3, etc). In general it was found to be relatively large. With chabasite and nitrous oxide, for example, the contribution $(\phi_{F\mu} + \phi_{FQ})$ accounted for 6.2 kcal/mole out of a total heat of adsorption of 15.3 kcal/mole, respectively. Zeolites provide rather extreme examples of specific adsorption because of the in-

TABLE 4

Isosteric heat of adsorption (at half-coverage) q_{st} of nitrogen and argon[a]

Adsorbent	q_{st}(kcal / mole)[b] for		Specific Contribution for Nitrogen
	Argon	Nitrogen	
Graph. carbon black	2.7; 2.8	2.6; 2.7; 2.8	~0
Bone mineral	2.6	3.7	~1.1
Gamma-alumina	2.0	2.7	~0.7
Hydroxylated silica	2.1; 2.4	2.8; 3.2	0.8
Dehydroxylated silica	2.1	2.2	0
Polypropylene	1.6	1.7	0

[a]Reduced from the Table of Sing and Ramakrishna (41).
[b]1 kcal = 4.19 kJ.

tense fields present in the vicinity of the exchangeable cations; for the general run of adsorbents the specific contribution would be smaller:

Argon and nitrogen constitute a particularly interesting pair of adsorbates in the present context since they are so similar in polarizability and molecular size, and therefore in their non-specific adsorption behaviour. From Table 4 it is seen that the heat of adsorption for nitrogen is almost the same as for argon on nonpolar adsorbents such as polypropylene and dehydroxylated silica, but is significantly higher on adsorbents such as alumina or hydroxylated silica which can interact with the quadrupole of nitrogen (41).

2. Application to the Adsorption Isotherm

The lowering of potential energy of a molecule of vapor in the vicinity of a solid surface implies that the concentration of the vapor will be higher in the neighborhood of the solid surface than in the free space beyond, that is, that adsorption will occur. Despite the fact that the minimum of the potential curve (Figure 2) lies so close to the surface, the adsorbed film will not be limited to a single molecular layer in thickness, because an adsorbed molecule itself interacts (albeit more weakly) with molecules in the vapor phase; in effect it becomes part of a composite adsorbent. Thus, as the pressure of vapor is progressively increased, a multilayer will build up on the surface, to be succeeded if mesopores are present by capillary condensation. Unfortunately the theoretical treatment of adsorption forces touched on in earlier pages—valuable though it is for deepening our insight into the nature of adsorption itself—has not yet reached the point where it can predict the detailed course of an actual isotherm. For monolayer–multilayer formation a semiempirical approach is unavoidable, while for coping with capillary condensation the Kelvin equation—thermodynamic rather than mechanistic in nature—is still our best available tool. Sections II and III deal with these respective themes, and the special features of adsorption in microporous solids forms the subject matter of Section IV.

II. ADSORTPION ON NONPOROUS SOLIDS

A. Introduction

As was indicated in the previous section, the Kelvin equation has played a major role in the systematic study of the adsorption of vapors by porous solids. In recent years it has been found profitable to approach the problem also from a different direction, by examining the way in which the adsorption isotherm of a vapor on a nonporous solid is modified when the solid becomes porous. In the majority of cases the nonporous solid gives rise to a type II isotherm, and in this section some essential characteristics of this type will be described.

1. Experimental Findings

The ideal nonporous substance for reference measurements would be a sheet of a laminar solid, such as mica, having a surface free of imperfections and of readily measurable area. Unfortunately however the mass adsorbed on a convenient area of such a solid would be so small (around $0.05 \ \mu g/cm^2$, of benzene at a relative pressure of 0.5, for example) that its accurate measurement would be extremely difficult. In general, therefore, determinations have to be made on bodies of relatively large specific surface area, that is, on powders; and in highly disperse solids of this kind, the surfaces of individual particles are bound to be imperfect on an atomic scale owing to the presence of steps, point defects, and dislocations; moreover the powder as a whole cannot be truly nonporous because of the inevitable presence of interstices between the particles. Even so, experience has shown that it is possible to obtain powders which in their adsorptive behavior approximate to nonporous solids, inasmuch as the majority of interstices are so large as to fall in the macropore range of size; with sufficient care in the selection and handling one can ensure that the nature and concentration of surface imperfections is sufficiently reproducible for practical purposes. Then, if the particle size is reasonably uniform, the surface area can be estimated by electron and optical microscopy, and it then becomes possible to refer the amount adsorbed to unit area of the adsorbent.

Over the last three decades a considerable number of measurements of this kind have been made, mostly with nitrogen at 77°K as the adsorptive. In the vast majority of cases the resultant isotherms have the sigmoid form exemplified in Figure 3a, belonging to type II of the classification of Brunauer, Deming, Deming, and Teller (16) (cf. Figure 1). In general the type II isotherm displays a straight portion, commencing at "point B." It is now well established that the part AB of the type II isotherm corresponds to the formation of a monolayer, and the part BCD to the gradual building-up of a

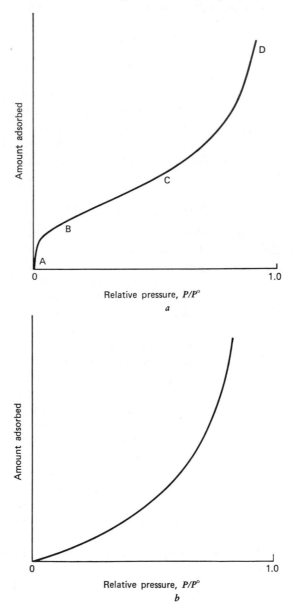

Fig. 3.   (*a*) A type II, and (*b*) a type III isotherm, of the BDDT classification.

multilayer on the surface of the solid, the adsorption at or near point B being equivalent to the monolayer capacity n_m, that is, the amount contained in a completed monolayer. From n_m, provided the area a_m occupied per molecule in the completed monolayer is known, the average area occupied per gram of solid—the specific surface—is readily calculated, since

$$S = a_m \cdot N \cdot n_m \tag{14}$$

Here N is the Avogadro constant and n_m is expressed in mole per gram of adsorbent.

Occasionally, a type III isotherm, convex to the pressure axis (Figure 3b) is obtained; but isotherms of this kind are comparatively rare; they arise only when the interaction between the adsorbent and the adsorbate molecule is unusually feeble, and until recently have attracted but little attention (cf. p. 255).

B. Some Mathematic Expressions for the Type II Isotherm

Various attempts have been made to express the course of the type II isotherm in mathematical terms, through the use of a suitable model of the adsorption process. From the nature of the problem any such model has to be based on simplifying assumptions, and it is not surprising therefore that no theoretical equation so far put forward has succeeded in representing the experimental results over the whole range of the isotherm from zero up to saturation pressure. In the present context any detailed consideration of such equations is unnecessary, and attention will be confined to such aspects of the theory of adsorption on nonporous solids as seem relevant to the discussion of adsorption on porous solids. The development of the subject of adsorption on "open" surfaces has been much influenced by two different types of treatment of the problem: that associated with the names of Brunauer, Emmett and Teller (12), and of Frenkel (42), Hill (43), and Halsey (44). A rather different approach has been advocated recently by Kiselev (45).

1. The BET Equation

The BET model represents an extension of the Langmiur model for monolayer to adsorption a multimolecular layer. The rate of condensation of molecules from the gas phase on to sites in any given layer i, is equated with the rate of evaporation from that layer. By summing up the adsorption in all layers from $i = 1$ to $i \to \infty$ an expression for the amount n adsorbed at any given relative pressure $P/P°$—an equation for the isotherm—is arrived at:

$$\frac{n}{n_m} = \frac{c(P/P°)}{(1 - P/P°)(1 + c - 1\ P/P°)} \tag{15}$$

(c is a constant, the significance of which will be discussed shortly). The summation is made with the aid of the simplifying assumptions that the surface of the solid is uniform, and that the adsorbate in all layers above the first is identical in properties with the liquid.

A somewhat more elegant means of deriving equation 15 from the BET model is afforded by statistical mechanics (46). The surface is assumed to possess B identical sites, each capable of adsorbing one molecule. Of the total N of molecules present, N_1 are distributed among the B sites i the first layer, the remainder being adsorbed "on top" of them. The partition functions of molecules in the first layer may be written as $j_1(1)$ and that of molecules in all the higher layers as $j_L(T)$, identical with the partition function of molecules in the bulk liquid. By following the usual procedures of statistical mechanics an equation identical with 15 is arrived at, the constant c being defined by the expression:

$$c = \frac{j_1(T)}{j_L(T)} \cdot \exp \frac{(E_1 - L)}{RT} \tag{16}$$

where E_1 is the heat of adsorption for first layer, and L the heat of condensation of the vapor to bulk liquid (equals heat of adsorption for all layers above the first).

The partition functions $j_1(T)$ and $j_L(T)$ are not easy to evaluate a priori, though the arguments of Kemball and Schreiner (47) indicate that the ratio $J_1(T)/j_L(T)$ may vary widely in value, by as much as 10^{-5} to 10, according to the particular system. For simplicity however it is usual to put

$$c = \exp \frac{(E_1 - L)}{RT} \tag{17}$$

and thereby to make the tacit, and scarcely justified, assumption that $j_1(T) = j_L(T)$, that is, that the partition function of the adsorbed molecule in the first layer is identical with that of the bulk liquid.

For convenience of plotting, equation 15 is best put into the alternative form:

$$\frac{P/P^\circ}{n(1 - P/P^\circ)} = \frac{1}{n_m c} + \frac{c-1}{n_m c} \cdot \frac{P}{P^\circ} \tag{18}$$

often referred to as the BET equation. The plot of $(P/P^\circ)/n(1 - P/P^\circ)$ against P/P° should, therefore, yield a straight line. In practice the linearity is usually found to be restricted to a relatively narrow range of relative pressures around point B. The range of linearity is often quoted as $0.05 < P/P^\circ < 0.30$, but in fact it frequently differs from this; for nitrogen adsorbed on Graphon at 77°K, for example, the plot was linear between $0.01P^\circ$ and $0.15P^\circ$, and for argon on rutile at 77°K, between $0.20P^\circ$ and $0.25P^\circ$. The

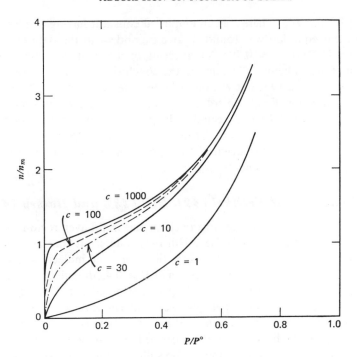

Fig. 4. Ideal "BET" isotherms for different values of c, which are marked on each isotherm.

role of parameter c in relation to the shape of the isotherm is brought out by writing equation 17 in the dimensionless form:

$$\frac{n}{n_m} = \frac{1}{1 - P/P^\circ} - \frac{1}{1 + \overline{c - 1}(P/P^\circ)} \tag{19}$$

and then constructing a set of isotherms in the form of curves of n/n_m against P/P°, for different values of c. Figure 4 shows such a set, where c takes in succession the values 1, 10, 30, 100, and 1000; and as is seen the knee of the isotherm becomes progressively sharper as c increases (48, 49). For value $c < 2$ the isotherm is actually convex to the pressure axis, and so is of type III rather than type II.

From time to time attempts have been made to extend the range of applicability of the BET equation, but they nearly always involve a third constant, which like n_m and c, is incapable of evaluation a priori and is, therefore, essentially empirical.

Brunauer, Skalny, and Bodor (220), for example, have put forward a modified equation in which the quantity (P/P°) of equation 19 is replaced throughout by $k(P/P^\circ)$, where k is an empirical constant having a value less than unity; this equation was found to give a good fit to the data of Shull (65) (Section II.D) if $k = 0.79$. As a result of a statistical—mechanical examination, Guggenheim (221) came to the conclusion that the appropriate parameter in equation 19 is not P°, but a more general parameter P^*, which unlike P° is characteristic of the solid and not merely of the vapor and the temperature. In general P^* is expected to be greater than P°; with nitrogen on an alumina-ferric oxide catalyst, for instance, $P^* = 1.20P^\circ$.

In the present context the original BET equation is to be preferred, on grounds of simplicity and general usefulness.

2. The Equations of Frenkel (42), Hill (43), and Halsey (44)

A different kind of theoretical approach to the multilayer region of the isotherm is made by Hill and Halsey. Hill (43) points out that as soon as the adsorbed layer reaches a thickness of three or four molecules, corresponding to a relative pressure in excess of 0.3 or 0.4, the effect of surface heterogeneity on the isotherm will be largely smoothed out. Consequently a theoretical treatment of this part of the isotherm can be based on an analysis of surface forces (cf. Section I). If the adsorbate is assumed to have the same density as the bulk liquid, the adsorbed layer may be regarded as a slab of liquid of thickness t adhering to the solid, and an expression for the potential energy of a molecule M on the surface of this slab (Figure 5) can be worked out with the aid of the 6–12 relation (cf. equation 6). The additional interaction of molecule M with the slab of liquid can then be handled by use of chemical potentials, that is, by a thermodynamic rather than a molecular force approach. The isotherm thus arrived at is

$$\ln \frac{P}{P^\circ} = -\frac{a}{\Gamma^3} \tag{20}$$

where Γ is the amount adsorbed per unit area of solid; constant a is a composite quantity, calculable in principle from properties of the adsorbent and adsorbate molecules, but in practice empirical. The index 3 arises from the triple integration implied by equation 10 of the r^{-6} term of the potential equation 6, the r^{-12} for repulsion being negligible for the distances in question.

The validity of equation 20 may be tested by plotting $\ln P/P^\circ$ against $\ln n$ (since $n \propto \Gamma$) when a straight line of slope 3.0 should be obtained for the multilayer region of the isotherm. In practice such a straight line is indeed often obtained, but the slope is rarely equal to 3.0. For nitrogen at 77°K, Zettlemoyer (50) and Young and Crowell (51) report values ranging from 2.12 to 3.0.

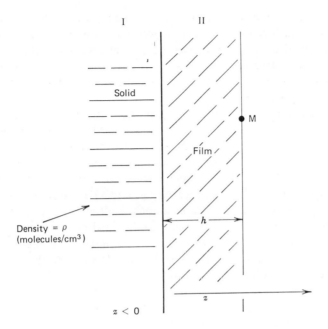

Fig. 5.   A molecule M on the surface of a slablike film of adsorbate.

Halsey (44) considered the case when the surface of the solid was energetically heterogeneous, and arrived at the equation

$$\ln\left(\frac{P}{P^\circ}\right) = -b\left(\frac{n}{n_m}\right)^{-r} \qquad (21)$$

where b is a constant, and the value of r may differ from 3; for nitrogen on a number of solids, for example, Halsey found $r = 2\cdot67$. The value of r may be taken as a guide to the strength of interaction between the adsorbate and the solid (50): a large value betokens short-range, specific forces and a small value, longer-range forces of the dispersion type.

The Frenkel–Hill–Halsey model helps one to visualize the way in which the effect of the solid on a given molecule of adsorbate M is enhanced by the presence of the underlying adsorbed layer—itself a resultant of the adsorption field of the solid. The effective range of the adsorption field is thereby increased, so that, in Halsey's phrase, a transmission of van der Waals forces across the adsorbed layer takes place.

To estimate the range of the transmission effect Halsey studied isotherms of argon adsorbed on layers of xenon which in turn had been pre-adsorbed on graphitized carbon black (52). In the absence of the xenon the argon isotherm had a steplike character, which remained until the thickness of the xenon film had reached six molecular layers. In a similar experiment with water pre-adsorbed on anatase, it was found that no further change in the isotherms of nitrogen or of argon occurred once the water film was more than three molecules thick. These findings have been supported by the recent work of Prenzlow (222).

A quantitative prediction of the range and magnitude of the transmission effect is a far more difficult matter. The Kirkwood–Müller (23, 24) and other expressions for the interaction energy of isolated molecules are themselves only approximations, and as pointed out by Freeman (53) the pairwise addition of potentials introduces a measurable error even into the calculation of the third virial coefficient of a free gas, and a correspondingly greater error when applied to solid–gas interactions. Indeed, according to Pitzer (54), a third-order perturbation—the effect of a third molecule on the interaction of two other molecules—can increase the interaction energy of two other molecules by as much as 10 to 20%, and he suggests that this effect may contribute to multilayer formation.

Transmission of van der Waals forces plays a particularly important part in a very narrow pores (micropores) where the width is less than five or six molecular diameters (cf. Section IV.).

3. The Virial Approach

As an alternative to the use of models, which necessarily have to be over-simplified and can only rarely correspond to actual systems, the adsorption isotherm may be expressed in a quite general mathematical form. Advantages of this approach have been emphasised by Kiselev (55), Barrer (56), and Pierotti (57): A "virial" equation is used, analogous to those employed in the treatment of imperfect gases and nonideal solutions. Kiselev has proposed the equation:

$$P = n \cdot \exp(C_1 + C_2 n + C_3 n^2 + \cdots) \tag{22}$$

where the coefficients C_1, C_2, C_3, etc. are constants at constant temperature.

By the use of sufficient constants (usually not more than four) it is possible to reproduce the course of an experimental isotherm over most of the mono-layer region.

The device is particularly useful in estimating the degree of similarity in the shapes of isotherms for the monolayer region (particularly at low and intermediate coverages), by comparison of the values of corresponding coefficients. No mechanism of adsorption needs to be assumed, a particularly valuable feature in relation to the treatment of micropore filling (Section IV).

C. The Packing of Molecules in the Monolayer. The Molecular Area a_m

The average area a_m occupied per molecule of adsorbate on the surface of an adsorbent will clearly depend on the degree of localization. In conditions of free mobility a_m will be determined by the size of the adsorbate molecules and their mode of packing in the surface, which might be expected to approximate that in the bulk liquid adsorptive. Assuming an arrangement of twelve nearest neighbors in the bulk liquid—equivalent to six on the (planar) surface of the adsorbent—one arrives at the relation (59).

$$a_m = 1 \cdot 09 \left(\frac{M}{\rho_L N} \right)^{2/3} \tag{23}$$

where M is the molecular weight of the adsorptive and ρ_L its density in liquid form, the coefficient 1.09 being the packing factor appropriate to the arrangement in question. Nitrogen when adsorbed on most adsorbents appears to exhibit high mobility at 78°K; and application of equation 23 leads to the familar value $a_m = 16.2 \times 10^{-20}$ m²/molecule'(16.2 Å²/molecule) for this adsorbate. If however the adsorption is completely localized the molecules reside wholly in the potential wells, whose positions are determined by the lattice parameters of the *solid*; the value a_m will now be determined by the size and arrangement of the surface atoms of the *adsorbent* and will therefore, be independent of the molecular size of the adsorbate (except when a molecule of adsorbate is large enough to cover two or more lattice sites.) Thus, in localized adsorption, different vapours on a given *solid* will exhibit the same value of a_m, whereas in completely mobile adsorption it is the same *vapor* that will show a constant value of a_m, on different adsorbents.

Complete localization is rare in physical adsorption, but is the general (though not universal) rule in chemisorption, since this involves some displacement of electrons. The molecular area a_m will now depend on both the molecular size of the adsorbate and the lattice parameter(s) of the adsorbent. The situation is too complicated to enable the value of a_m for a particular vapor to be predicted from the properties of the solid and the vapor; but experience indicates that the value of $a_m = 16.2$ Å² for nitrogen, already referred to, can provide a starting point to an empirical approach to the problem: this value, when used to convert monolayer capacity into specific surface (equation 14) leads to reasonable agreement ($\pm 10\%$) with the geometrical area determined by microscopy, for a number of well-characterized nonporous solids.

With other adsorbates, however, arbitrary adjustment of the value of a_m is usually necessary to obtain agreement between the value of S calculated from the monolayer capacity with the specific surface based on the ni-

TABLE 5

The average area a_m occupied per molecule of adsorbate in a completed monolayer[a]

Adsorptive	T (°K)	a_m (Å²)	
		Range in the Literature	Customary Value
Nitrogen	78	13–20	16.2
Argon	78	13–17	14
Krypton	78	17–22	20
Xenon	78	18–27	25
Oxygen	90	14–18	14
Ethane	78	20–24	21
Benzene	298	30–50	40

[a] Gregg and Sing (60).

trogen isotherm. The adjusted value of a_m which may differ by 50% or more from that calculated with the equation 23, varies according to the nature of the solid, reflecting the variation in the degree of localization of the adsorbate molecules. Consequently, it is rarely possible to assign a fixed value to the molecular area of a given adsorptive. From their comprehensive review based on 188 references, McClellan and Harnsberger (59) are able to arrive at only five "recommended" vaules of $a_m/Å^2$, viz., N_2(77°K), 16.2; Ar(77°K), 13.8; Kr(77°K), 20.2; n-C_4H_{10} (273°K), 44.4; C_6H_6 (293°K), 43.0. Gregg and Sing (60) prefer to quote *ranges* of value for each adsorptive (Table 5). Even for nitrogen a range can be found in the literature, but cases where a value other than $a_m = 16.2Å^2$ needs to be used are fortunately not very common.

1. The BET Method for Determination of Specific Surface. The Role of Parameter c

Critical examination shows that the BET equation can be used for the determination of the monolayer capacity n_m, only if the knee of the isotherm is sufficiently sharp to give rise to a clearly defined point B. (cf. Figure 3a). In practice this means that the value of c shall be sufficiently large (> 50 say, for nitrogen). The value of n_m calculated from the BET plot will then agree within a few percent (60, 61) with that obtained from point B, always provided that the BET plot extends to pressures on both sides of point B. On the other hand, if the value of c is too high, (> 200, say, for nitrogen), the BET method will lead to erroneous results for the specific surface, since the high value of c implies either that the degree of localization is considerable (in consequence of the high value of the heat of adsorption) or that the adsorbent is microporous; in the former case the value of a_m will be in doubt for reasons clear in the preceding section, and in the latter, the adsorption at point B will include a contribution for micropore filling as well as for monolayer coverage, so that the "BET" value for specific surface will be too high. The role of microporosity is dealt with in Section IV.

D. The Concept of a Standard Isotherm

From general considerations it is clear that one of the main factors determining the adsorption (per unit area of adsorbent) at a given relative pressure is the heat of adsorption ΔH. The greater the heat of adsorption the greater is the amount adsorbed at a given relative pressure in the monolayer region; that is, the sharper is the knee of the isotherm. This is illustrated by the ideal BET isotherms of Figure 4., where the progressive increase in c [$\simeq \exp (\Delta H - L/RT)$] is associated with an increasingly sharp knee. The heat of adsorption of a given adsorptive such as nitrogen must vary from one adsorbent to another, since the interaction energy depends on properties such as diamangetic magnetic susceptibility (cf. equation 7) which are characteristic of the particular solid. Thus the exact shape of the isotherm must vary from solid to solid; nevertheless, provided the variation in the heat of adsorption (and therefore in c) is relatively slight, the change in isotherm shape will be small. To a first approximation the different isotherms should, therefore, be capable of superposition by mere adjustment of the scale of ordinates as long as the value of c remains substantially the same on the various adsorbents. Such "normalization" can be brought about, following Kiselev (62), by plotting the isotherms with adsorption per unit area, n/S, or with the statistical number of monolayers, as proposed by Pierce (63), Sing (64), and others. In either case the points for all isotherms would be expected to fall close to a common curve.

A pioneer in the search for a standard isotherm was Shull (65), who showed that the isotherms of nitrogen at 77°K on a number of different solids could be represented by a common curve, the scatter in n/n_m being around $\pm 10\%$ except at low coverage. Cranston and Inkley (66), made a similar finding in 1957. In 1959 Conway Pierce (63) referred to a "compositve isotherm" for nitrogen, again based on numerous isotherms on solids such as carbon, oxides, and ionic crystals; he concluded that, on a relative basis at least, the effects due to the particular solid could be ignored; the composite isotherm could be represented over the multilayer range by the Halsey equation with index $r = 2.75$ and $b = 2.99$ (cf. equation 21). More recently Pierce (67) has compared the values of n/n_m for the various "universal" isotherms in the literature; over the range $0.20 < P/P° < 0.95$ he finds moderately good agreement ($\pm 5\%$) among the isotherms of Shull (65), Cranston and Inkley (66), and Pierce (63); the isotherm of Lippens, Linsen, and de Boer (68) diverged from the others more widely (Figure 6).

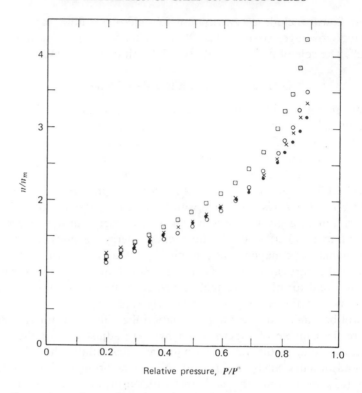

Fig. 6. Comparison of a number of standard isotherms of nitrogen at 77° K, plotted as n/n_m against $P/P°$. ○, Shull (65); ×, Pierce (67); □, Linsen and de Boer (68); ●, Cranston and Inkley (66).

It is hardly necessary to emphasize that if a solid is to give rise to a standard isotherm it must be free of pores. Enhanced adsorption will be brought about at low relative pressures by microporosity and at intermediate pressures (say $0.4 < P/P° < 0.95$) by mesoporosity. Unfortunately it is not easy to establish the absence of porosity in the solids used for adsorption experiments; the unsuspected presence of pores may well account for some at least of the discrepancies between different published versions of the "standard isotherm."

A second reason for the discrepancies has already been indicated: the strength of interaction of a given adsorbate with the adsorbent—as reflected in the value of $(\Delta H \sim L)$ or of parameter c—is not absolutely constant but varies somewhat from one adsorbent to another. If first-order precision is desired, therefore, one must abandon the search for a truly universal isotherm valid for a given adsorptive with all adsorbents, and accept the necessity for building up a *series* of standard isotherms, one for each solid substance.

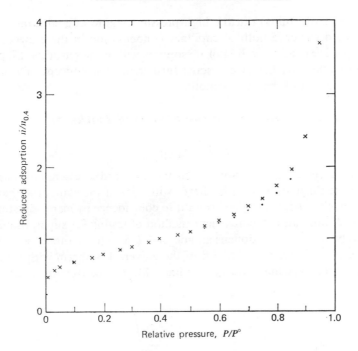

Fig. 7. Standard isotherms of nitrogen at 77°K on alumina and silica, according to Sing (69, 70). ●, silica; ×, alumina.

This is essentially the approach taken by Sing and his co-workers, dealing first with silica as adsorbent and nitrogen as adsorptive (69). These workers have collected the results for a wide variety of silicas, having surface areas ranging from ~1 to ~200 m²/g and including ground quartz as well as various samples of nonporous but amorphous silica. They have found that the experimental points for all these samples fit on to a common curve very closely (Figure 7). More recently, Sing (70) has put forward a corresponding curve, though based on fewer samples, for γ-alumina. These two standard curves are fairly similar but they do show differences that exceed those between the isotherms for different samples of the same substance. A standard isotherm for argon (at 77°K) on silica has also been put forward by Sing (71) and by Nicolaon and Teichner (72).

A number of attempts have been made within recent years to arrive at a standard isotherm for adsorption of nitrogen on graphitized carbon black (73–76). Standard isotherms for oxygen at 77 and 90°K, respectively, have been put forward by Brunauer and his co-worders (77). The case of water as adsorbate is referred to later (Section IV.E).

In his 1959 paper (63) Pierce advanced the then novel idea that deviation from the standard isotherm in the high-pressure region offered a means of detecting the occurrence both of capillary condensation in the crevices between the particles and the filling of mesopores within the particles. Lippens and de Boer (78) took the idea a stage further in their concept of *t*-plots, which is the subject of the next section.

1. Deviations from the Standard Isotherm

a. The t-plot

The task of detecting deviations from the standard is essentially one of comparing the shape of a given isotherm with that of the standard, that is, of finding whether the two can be brought to coincidence by mere adjustment of the scale of ordinates. A convenient method of testing for superposability is provided by the *t*-plot of Lippens and de Boer (78). The method is based on the *t-curve*, which is essentially a plot of the universal isotherm with *t* rather than n/n_m as the dependent variable (Figure 8). The conversion is accom-

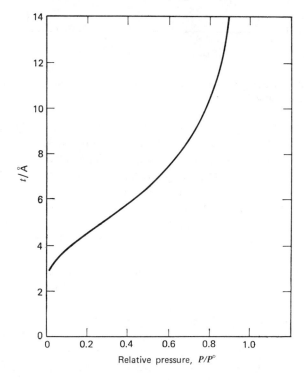

Fig. 8.   The *t*-curve (graph of $t/\text{Å}$ (= 3.54 n/n_m) against $P/P°$) corresponding to the standard isotherm of Fig. 6.

plished by taking n/n_m to be equal to the number of statistical molecular layers in the adsorbed film, and multiplying by the thickness σ of a single molecular layer, so that $t = (n/n_m)\,\sigma$. (Clearly t represents only a nominal, not the actual thickness, which must in general vary from place to place along the surface). For nitrogen at 77°K, Lippens and de Boer (78) put $\sigma = 3.54$ Å, a value based on the assumption that the arrangement of molecules in the film is hexagonal close-packing.

The isotherm under test is then redrawn as a *t-plot*, that is, a curve of amount adsorbed, n, against t rather than against $P/P°$ (the change of variable being effected by means of the *t*-curve). If the isotherm being tested is identical in shape with the standard, the *t*-plot must be a straight line passing through the origin; and its slope b_t is equal to n_m/σ, since:

$$n = \left(n\,\frac{\sigma}{n_m}\right)\cdot\left(\frac{n_m}{\sigma}\right)$$

$$= t\left(\frac{n_m}{\sigma}\right) \tag{24}$$

or
$$n = b_t t \tag{25}$$

Again, since $\qquad S = a_m \cdot N \cdot n_m \qquad$ (cf. equation 14)

we have $\qquad S = a_m \cdot N \cdot b_t \cdot \sigma \tag{26}$

By insertion of $\sigma = 3.54$ Å (t also being expressed in angstroms) and $a_m = 16.2 \times 10^{-20}$ m²/molecule for nitrogen, with $N = 6.02 \times 10^{23}$ molecules/mole, we obtain:

$$S = 3.45 \times 10^5 \, b_t \text{ m}^2/\text{g} \tag{27}$$

so that the specific surface can be evaluated directly from the slope of the *t*-plot.

b. The α_s-plot

The concept of the *t*-plot arose in the context of the necessity to make allowance for the thickness of the adsorbed layer on the walls of the pores, when calculating the pore size distribution from the (type IV) isotherm of nitrogen (III.C) For the rather different purpose of detecting deviations from the standard isotherm, however, the actual thickness of the film is irrelevant; in its essentials, the task now is merely one of comparing the shapes of the "experimental" and the standard isotherm, and it is not even necessary to involve the monolayer capacity or the statistical number of monolayers n/n_m in the treatment. It is simpler, and from the purely mathematical point of view sounder, to replace n_m by n_s, the adsorption at some fixed relative pressure, $P/P° = s$. This is essentially the device of Sing (80, 81), who plots $n/n_s\,(= \alpha_s)$

instead of n/n_m against $P/P°$, to obtain a standard α_s-curve instead of a t-curve. The α_s-curve is used in a manner closely analogous to that for the t-curve: an α_s-plot rather than a t-plot is constructed from the experimental isotherm, and if a straight line results one can infer that the experimental isotherm is identical in shape with the standard.

The equation to the line may be written

$$n = b_s\alpha_s \tag{28}$$

and the slope b_s will be equal to n_s, just as the slope b_t of the t-plot was equal to n_m/σ (cf. equation 24). Since the area of the reference sample will be known, usually from the BET–nitrogen method, the specific surface of the material under test can be obtained by direct comparison of its b_s value with that of the reference isotherm, since S is proportional to b_s with the same proportionality factor in both cases.

In practice Sing selects $s = 0.4$ and thus plots $\alpha_{0.4} = n/n_{0.4}$ against $P/P°$, $n_{0.4}$ being of course the adsorption at a relative pressure of 0.4. The reason for this particular choice was merely that at the relative pressure of 0.4 for nitrogen at 77°K the process of micropore filling, if present, is complete, and capillary condensation has not as yet begun.

An outstanding advantage of the α_s method is that it can still be used even when the standard isotherm does not possess a sharp knee, that is, when the value of c is small, whereas all methods based on the monolayer capacity as normalizing factor require c to be relatively large ($50 < c < 200$) so that the point corresponding to monolayer completion can be found with reasonable certainty. Thus, in principle, the α_s method is applicable to adsorptives other than nitrogen, even when—as with carbon tetrachloride on silica at 298°K— the standard isotherm is of type III rather than type II.

c. The Effect of Mesoporosity

If the adsorbent is mesoporous, capillary condensation will occur in each pore when an appropriate relative pressure is reached, and a type IV isotherm will result (Figure 9a). A fuller discussion of capillary condensation is deferred to the Section III; meanwhile we may note that when such condensation takes place, the uptake at any relative pressure will be increased by the amount of capillary condensate present in the pores. The t-plot or α_s-plot will, therefore, show an upward deviation commencing at that relative pressure at which the smallest pores are being filled (Figure 9b). If pores of suitable shape are present, capillary condensation can occur reversibly at pressures below the lower closure point of the hysteresis loop, and the t- or α_s-plot will enable its occurrence to be detected. Adsorption in mesoporous solids is the subject matter of Section III.

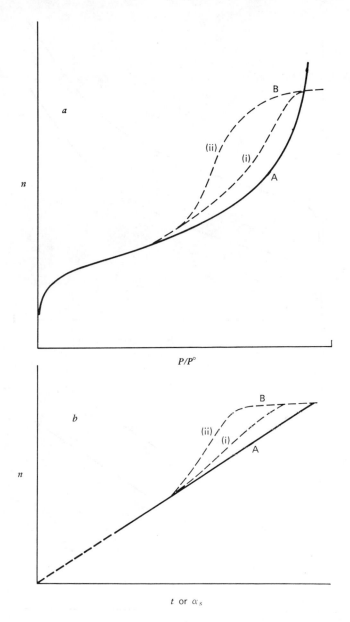

Fig. 9.   Effect of mesoporosity on the adsorption isotherm and the t-(or α_s-) plot. (a) (A) is the isotherm on a nonporous sample of the adsorbent; (B) is the isotherm on the same solid when mesopores have been introduced into it, (i) being the adsorption, and (ii) the desorption branch. (b) t-(or α_s-) plots corresponding to the isotherms of Fig. 9a. (Diagrammatic only.)

261

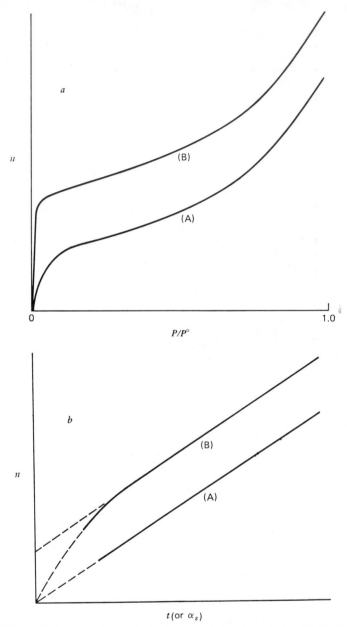

Fig. 10.   Effect of microporosity on the isotherm and the t-(or α_s-) plot. (a) (A) is the isotherm on a nonporous sample of the adsorbent; (B) is the isotherm of the same solid when micropores have been introduced into it. (b) t-(or α_s-) plots corresponding to the isotherms of Fig. 10(a). (Diagrammatic only.)

262

d. The Effect of Microporosity

If micropores are introduced into a solid that originally gave rise to a type II isotherm, the uptake at a given relative pressure is increased, particularly in the low pressure region (Figure 10a). The effect on the t-(or α_s-) plot is indicated in Figure 10b. The high pressure branch of the plot is still linear (provided mesopores are absent) but when extrapolated to the adsorption axis it gives a positive intercept that is equivalent to the micropore volume. The slope of the linear branch is now proportional to the *external* surface area of the solid. Microporosity is dealt with in Section IV.

III. ADSORPTION ON MESOPOROUS SOLIDS

A. Porosity and the Type IV Isotherm

In Section I it has already been indicated how application of the Kelvin equation to the type IV isotherm leads naturally to the hypothesis of capillary condensation in mesopores of a porous solid. From time to time various attempts have been made to demonstrate more directly the relationship between mesoporosity and the type IV isotherm, by comparing the isotherm of a vapor on a nonporous powder before and after the powder has been formed into a compact. The process of compaction produces pores in the form of the interstices between the particles of the original powder; such pores will tend to have dimensions of the same order as those of the constituent particles, and it can be arranged that these will fall within the mesopore range of size.

An early example is provided by the results of Carman and Raal (82) who found that the isotherm of CF_2Cl_2, on a silica powder changed from a well-defined type II before compaction to an equally well-defined type IV after compaction. Essentially similar results were obtained by Kiselev (83) for hexane on carbon black and by Zwietering (84) for nitrogen at 77°K on silica spherules. In these and similar studies (85–88) the reversible part of the isotherm was scarcely affected by the compaction; but, very significantly, both branches of the hysteresis loop were situated *above* the isotherm for the uncompacted material (cf. Figure 11); that is, the hysteresis was associated with an enhanced adsorption. This finding strongly indicates that the presence of pores has brought about a change in the nature of the adsorption process at the relative pressures corresponding to the hysteresis loop; if the adsorption process had remained one of monolayer–multilayer formation as it was on the open surface of the loose powder, (cf. Section II), the amount adsorbed would, if anything, have diminished owing to the slight loss of available area at interparticulate contacts.

A rather more extensive study, using a series of compacting pressures, was undertaken by Langford (85). With a powder composed of spherical particles

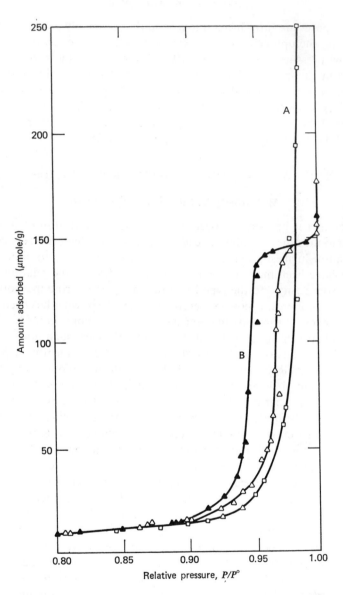

Fig. 11. Adsorption isotherm of *n*-hexane on P-33 carbon black. (A) loose powder, (B) compacted powder. (Courtesy Kiselev (83)).

264

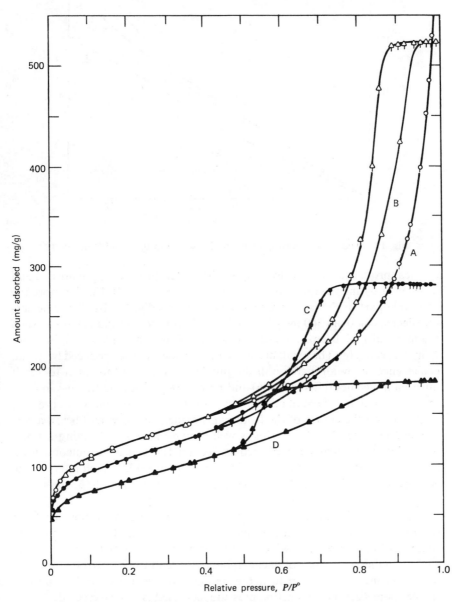

Fig. 12.   Adsorption isotherms of nitrogen at 77 °K on a nonporous silica (85). (A) Before compaction; (B), (C), (D), after compaction at 16, 64, and 130 tons in², respectively.

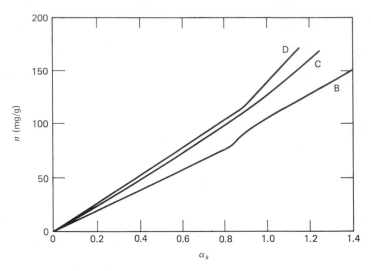

Fig. 13.   The α_s-plots corresponding to the isotherms B, C, and D of Fig. 12.

of silica (produced by flame hydrolysis) the hysteresis loop was depressed and broadened by increase in compacting pressure (Figure 12). Furthermore the reversible part of the isotherm was lowered markedly, the BET specific surface decreasing from 360 m²/g for the loose powder to 300 and 250 m²/g when the material had been compacted (64 and 130 ton/in²) at 9.9 and 20 × 10⁸ N/m² respectively. The available surface had evidently been reduced by actual coalescence or by an increase in the number of crevices too small to admit nitrogen molecules (cf. Wade (86) and Kiselev (87)). Despite the loss in area, the relative enhancement of adsorption in the hysteresis region is still present; In this region the course of the isotherm is higher than it would be on a corresponding "open" surface. This is well shown up by constructing a t-plot or α_s-plot in the manner explained in Section II, taking the umcompacted powder as reference material. As is seen from Figure 13, (which refers to the adsorption branch of the loop), the α_s-plot deviated upwards in the region corresponding to the hysteresis loop.

In these experiments the average particle size of the original powder (from the BET–nitrogen area) was around 75 Å; the size of the interparticulate spaces will of course vary widely from region to region of the compact according to the exact mode of packing, but it will tend to be distributed around this value, and will therefore lie in the mesopore range of sizes; a shift towards smaller pore sizes as compacting pressure increased is to be expected.

The results of Avery and Ramsay 88 are particularly interesting in that their nitrogen isotherms (on very fine silica and zirconia powders), which were of type II on the umcompacted materials, changed on progressive compaction first to type IV and then actually to type I (Figure 14) (see Section IV).

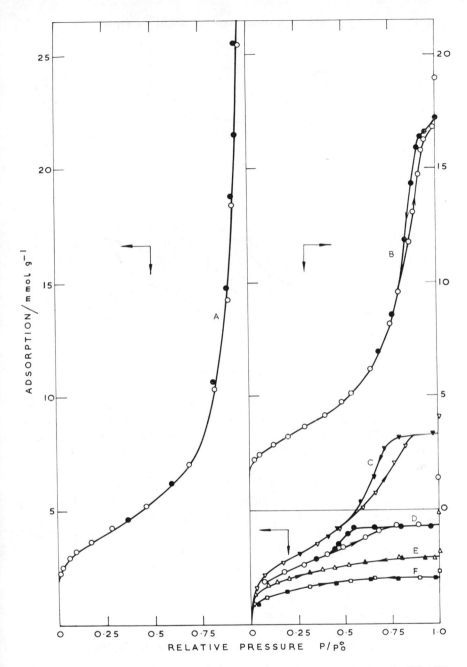

Fig. 14. Adsorption isotherms of nitrogen at 77°K on a nonporous zirconia (88). (A) Before compaction. (B), (C), (D), (E), after compaction at. 10, 40, 50, and 100 ton/in², respectively.

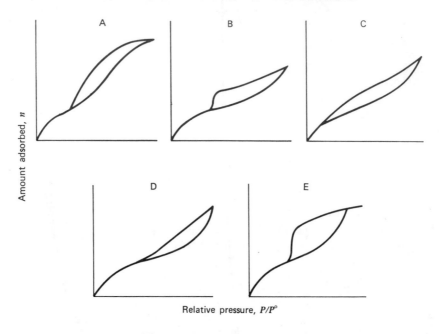

Fig. 15.   Types of hysteresis loop.

The results of all these and similar experiments clearly demonstrate that increased adsorption results from the presence of pores in the mesopore range; and the hypothesis of capillary condensation offers the most reasonable explanation for this enhancement.

1. Types of Hysteresis Loop

The hysteresis loops to be found in the literature are of various shapes. The classification put forward originally by de Boer (89) in 1958 has proved valuable, but it may be usefully modified in the light of subsequent experience: de Boer's type C is uncommon and is better replaced by that proposed by Everett (90), moreover in types B and D the closure of the loop at the high pressure end is rarely characterised by the vertical branch shown in the de Boer diagrams. These features have been taken into account in the revised classification presented in Figure 15.

2. The Surface Area of Mesoporous Solids

In several of the experiments just cited the isotherm was scarcely altered in the region up to, and somewhat beyond, point B, despite the conversion of the isotherm from type II to type IV. Clearly the calculated value of the monolayer capacity—whether obtained from point B or by the BET equation

—is the same before and after compaction; any reduction in uptake at point B which does occur is plausibly explained partly by loss of area by actual deformation and coalescence of flat or soft particles, and partly by loss of accessibility of the narrow crevices between a particle and its neighbors as the number of immediate neighbors increases with increasing comapction.

Now, since the calculation of specific surface by the standard procedures from point B or by the BET equation is soundly based for type II isotherms, it follows that the specific surface of mesoporous solids may be determined in like manner from the type IV isotherm of nitrogen, and, by implication of other suitable adsorptives. These procedures lead to a result for *total* surface area, whether internal (walls of mesopores) or esternal (those parts of the sample freely exposed to the exterior and visible in principle in the optical microscope).

As was indicated in Section II, if the t- or α_s-plot is straight, its slope is proportional to the specific surface area of the sample. Similarly, the branch at high relative pressures (where the isotherm has flattened out) is frequently linear, and may reasonably be interpreted as corresponding to the thickening of a multilayer on the external surface; its slope should, accordingly, be proportional to the area of the external surface. Difficulties in applying this concept are, first, that the standard isotherm is itself much less certain in this region than elsewhere (owing to its steep slope), and second that the slope of the branch is often so low (because of the smallness of the external surface) that its accurate determination demands many closely spaced and accurate experimental points in this difficult region.

B. Derivation of the Kelvin Equation (91)

The Kelvin equation may be derived in various ways. Since the phenomenon of capillary condensation is intimately bound up with the curvature of a liquid meniscus, it is instructive to follow an approach in which this connection is emphasised.

Picture an element of liquid surface having principal radii of curvature r_1 and r_2 which are defined by taking two planes at right angles to one another, both passing through a normal erected from a point in the surface. If P^β is the pressure on the concave and P^α that on the convex side of the meniscus, then by the equation of Young (92) and Laplace (93) the difference in pressure on the two sides of the meniscus is

$$P^\beta - P^\alpha = \gamma \left(\frac{1}{r_1} + \frac{1}{r_2} \right) \tag{29}$$

If we now write

$$\frac{1}{r_1} + \frac{1}{r_2} = \frac{2}{r_m} \tag{30}$$

so that r_m is the mean radius of curvature, equation 29 assumes the form

$$P^\beta - P^\alpha = \frac{2\gamma}{r_m} \tag{31}$$

or

$$\Delta P = \frac{2\gamma}{r_m} \tag{32}$$

We now turn to the process of capillary condensation. For the pure liquid (α) in equilibrium with its vapor (β), the condition for mechanical equilibrium is given by equation (31) and that for physicochemical equilibrium by the equation

$$\mu^\beta = \mu^\alpha = \mu^\sigma$$

where σ denotes the surface and μ chemical potential. If we now pass from one equilibrium state to another (an "equilibrium desplacement") then

$$dP^\beta - dP^\alpha = d\left(\frac{2\gamma}{r_m}\right) \tag{33}$$

and

$$d\mu^\beta = d\mu^\alpha \tag{34}$$

Each of the coexisting phases will be governed by a Gibbs–Duhem euqation so that

$$s^\alpha dT + V^\alpha dP^\alpha + d\mu^\alpha = 0 \tag{35}$$

$$s^\beta dT + V^\beta dP^\beta + d\mu^\beta = 0 \tag{36}$$

where s^α, s^β, and V^α and V^β are the molar entropies and molar volumes, respectively, of the two phases.

At constant temperature, equation 34 together with 35 and 36 leads to the simple relationship

$$V^\alpha dP^\alpha = V^\beta dP^\beta \tag{37}$$

so that equation 33 can be written either as

$$d\left(\frac{2\gamma}{r_m}\right) = \frac{V^\alpha - V^\beta}{V^\beta} dP^\alpha \tag{38}$$

or as

$$d\left(\frac{2\gamma}{r_m}\right) = \frac{V^\alpha - V^\beta}{V^\alpha} dP^\beta \tag{39}$$

If we now neglect V^α compared with V^β, and if further the vapor behaves as a perfect gas, then from (39)

$$d\left(\frac{2\gamma}{r_m}\right) = -\frac{RT}{V^\alpha P^\beta} dP^\beta \tag{40}$$

$$d\left(\frac{2\gamma}{r_m}\right) = -\frac{RT}{V^\alpha} d \ln P^\beta \tag{41}$$

In integral form this becomes

$$\frac{2\gamma}{r_m} = -\frac{RT}{V}\ln\left(\frac{P^\circ}{P}\right) \qquad (42)$$

or

$$\ln\left(\frac{P}{P^\circ}\right) = -\frac{2\gamma V}{RT}\cdot\frac{1}{r_m} \qquad (43)$$

where P° is the saturation vapor pressure, which of course corresponds to $r_m = \infty$. V is the molar volume of the liquid.

Equation 43 is conventionally termed *the* Kelvin equation. The tacit assumption is made, at the integration stage, that the liquid is incompressible (that V^α is independent of pressure).

From equation 43 it follows that the vapor pressure P over a concave meniscus (r_m a positive quantity) is lower than that over a plane surface of the liquid at the same temperature. In other words "capillary condensation" of a vapor within a pore should occur at some pressure below the saturation vapor pressure, provided the condensed liquid forms a concave meniscus in the pore (angle of contact < 90°).

It should be emphasised that when capillary condensation occurs the walls of the pore already carry a film of adsorbed vapor, the thickness t of which

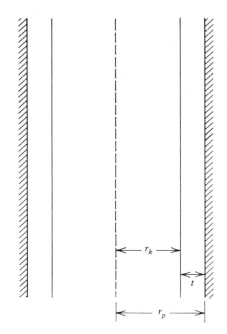

Fig. 16.   Cross-section, parallel to the axis of a cylindrical pore of radius r_p, with adsorbed film of thickness t. The "inner core" has the radius r_k.

will be determined by the value of $P/P°$ (cf. Section II). Thus capillary condensation takes place within the "inner core" of the pore, rather than directly in the pore itself (Figure 16). Phrased differently, the filling of a capillary with liquid adsorptive is a two-stage process: monolayer–multilayer adsorption followed by capillary condensation.

If the experimental temperature is below the triple point of the adsorptive, the question arises as to the appropriate reference state, that is, as to whether to calculate the relative pressure from the saturation vapor pressure of the supercooled liquid ($P°,^l$) or of the solid ($P°,^s$) at the experimental temperature. If the Kelvin equation is applied to the calculation of pore size distribution (Section III.C) the tacit assumption is made that the capillary condensate is in the liquid state, since the method depends on the existence of a meniscus. Direct evidence of the correctness of this assumption in the case of water as adsorbate is provided by Sidebottom and Litvan (94) when they show that the adsorption isotherms and the expansion isotherms (plots of fractional linear expansion of the adsorbent against $P/P°$), respectively, at a series of temperatures straddling $0°C$ ($+1.5$ down to $-35.3°C$) could be superposed within experimental limits if plotted against $P/P°,^l$, but were widely separated if plotted against $P/P°,^s$.

1. The Radius of Curvature

In order to apply equation 43 to the calculation of pore sizes from an adsorption isotherm, it is necessary to adopt a model of pore shape. On grounds of simplicity Zsigmondy, and his co-worker Anderson, assumed that the pores are cylindrical and the menisci therefore hemispherical, and later workers have usually followed this practice unless there is strong evidence of an alternative shape; the two principal radii of curvature of the meniscus are

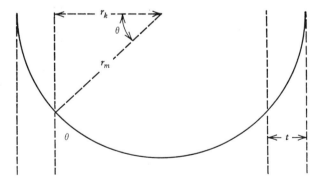

Fig. 17.   Relation between r_m of the Kelvin equation and the core radius r_k for a cylindrical pore with a hemispherical meniscus. θ is the angle of contact.

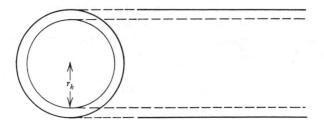

Fig. 18. A cyclindrical meniscus in a cylindrical pore.

then equal to each other and therefore to $r_m/2$ (cf. equation 30). Since, as already emphasized, the walls of a pore are covered with an adsorbed film of thickness t when capillary condensation takes place in it, the quantity r_m of equation 43 will be equal to the radius r_k of the *inner core* of the cylinder (Fugure 16), provided the angle of contact θ between the liquid and the film is zero. The radius r_p of the pore itself will, therefore, be given by $r_p = r_k + t$.

If the angle of contact is greater than 0° (but less than 90°), then by geometry we have $r_k = r_m \cos \theta$ (cf. Figure 17) so that equation 43 becomes

$$\ln \frac{P}{P^\circ} = \frac{-2\gamma V \cdot \cos \theta}{RT \cdot r_k} \qquad (44)$$

The alternative possibility that, in principle, the meniscus in a cylindrical capillary could assume a cylindrical form (Figure 18) was pointed out by Cohan (95) in 1938: The adsorbed layer would in effect constitute a layer of liquid with thickness t lining the walls of the pore. One radius of curvature would now be infinite, and the other would (if $\theta = 0$) be equal to r_k, the radius of the core; thus from equation 30, $r_k = r_m/2$, and the radius r_p of the pore itself would be given $r_p = r_m/2 + t$.

A second model, that of parallel-sided slits, has come into prominence in recent years, as electron microscopy has shown the prevalence of platelike particles in solids. The meniscus in these pores will be hemicylindrical in

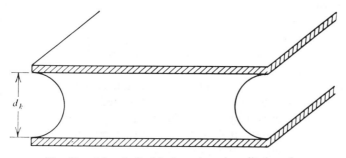

Fig. 19. A hemicylindrical meniscus in a slit-shaped pore.

shape (Figure 19); one radius of curvature will again be infinite and the other (if $\theta = O$) will be equal to one-half the distance d_k between the films on the walls; consequently by equation 30, $r_m = d_k$, and the distance between the actual walls will be given by $d_p = r_m + 2t$.

2. The Corrected Kelvin Equation

Implicit in the conventional application of the Kelvin equation to capillary condensation there is a contradiction: the very existence of an adsorbed film on the walls shows that the chemical potential of the adsorbate is being influenced by the adsorption field produced by the walls—an effect expressed through, for example, the t-curve (II.D.1); yet when the Kelvin equation is applied to the meniscus, it is assumed that the reduction in chemical potential is produced solely by the curvature of the meniscus, and expresses itself through a difference in hydrostatic pressure on the two sides of the meniscus. A question of fundamental significance, therefore, is the state of affairs at the junction of the meniscus and the adsorbed film: in this region the chemical potential of the adsorbate must be the resultant of the effect of the wall and the curvature of the meniscus, acting jointly. As Derjaguin pointed out some years ago, however, the conventional treatment involved the tacit assumption that the curvature falls jumpwise from $1/r_m$ to zero at the surface of the adsorbed layer, whereas the change must actually be a continuous one.

Derjaguin (96) arrived at a "corrected" Kelvin equation:

$$-\frac{dv}{dS} = \frac{V}{RT \ln(P^\circ/P)}\left[\gamma \cos \theta + RT \int_P^{P^\circ} \frac{\Gamma}{P} \, dP\right] \qquad (45)$$

where Γ is the adsorption per unit area on the walls of the pore; dv/dS is the volume : surface ratio of an elementary pore and for a cylinder is equal to one-half of the radius of the core.

Recently, the problem has been taken up again by Everett and Haynes (97), who emphasize that the condition of diffusional equilibrium throughout the adsorbed phase requires that the chemical potential shall be the same at all points within the phase; and since, in the vicinity of the wall, the interaction energy (adsorption potential) varies with distance from the wall, the internal pressure of the condensate must also vary by way of compensation.

Figure 20 is a plot of potential against distance from the wall for a liquid in a capillary of sufficient width for its middle A to be outside the range of forces from the wall. Since the capillary condensate is in equilibrium with the vapor, its chemical potential $\mu^a(= \mu^g)$, represented by the horizontal line EF, will be lower than that of the free liquid $\mu^{\circ,l}$; the difference in chemical potential of the condensate at A, represented by the vertical distance AF, is brought about entirely by the pressure drop, $\Delta P^\infty = 2\gamma/r_m$, across the

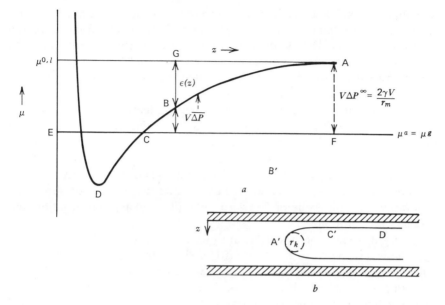

Fig. 20. (a) Chemical potential in a liquid in contact with a solid surface, as a function of distance from the surface. The liquid is in equilibrium with vapor at pressure P. (b) Modification to the profile of a meniscus in the vicinity of the walls: the curvature at A' is equal to the Kelvin radius r_m as measured by the reduction in vapor pressure. (Courtesy Everett & Haynes (97).)

meniscus (cf. equation 32); but at some point B, say, nearer the wall, the chemical potential receives a contribution represented by the vertical line GB, from the adsorption potential. Consequently, the reduction ΔP in pressure across the meniscus must be less at B than at A, and, again by equation 32, the radius of curvature of the meniscus must be larger than at A.

At the middle of the capillary where the effect of the walls on surface potential is negligible, the radius of curvature will be equal to r_m as calculated by the Kelvin equation (43); but it will become progressively larger than this figure as the wall is approached.*

3. Theories of Hysteresis (99, 100)

The origin of adsorption hysteresis has been the subject of discussion ever since the existence of the phenomenon was first recognized. The earliest

---

*De Boer and Broekhoff (98) have recently proposed a method for taking account of the effect of the wall on chemical potential of the adsorbate, based on an expression relating the chemical potential of the film to its thickness t.

attempt at an explanation appears to have been that of Zsigmondy (10), who related it to the familiar observation that the radius of curvature of the meniscus is greater when a liquid is rising in a capillary tube (often due to the presence of air as an impurity) than when the liquid is retreating from the already wetted surface, that is, to the fact that the advancing contact angle is in general higher than the receding angle.

Zsigmondy's explanation has failed to gain general currency, largely on account of the difficulty of measuring the relevant contact angles, and particularly of ensuring that the walls of the fine pores of an adsorbent are identical in surface properties with the material, almost invariably in slab form, which is used in conventional determinations of the contact angle. In the face of difficulties of this kind, the usual practice is to assume, somewhat arbitrarily, that the value of θ is zero for both capillary condensation and capillary evaporation.

Present day views on hysteresis are much influenced by the "ink bottle" hypothesis, attributed to Kraemer (101) and to McBain (102), in which the effect of narrow constrictions on the ease of meniscus formation is considered. In essence, the problem is one of the nucleation of the liquid phase; and in this connection Cohan (95) contrasted the properties of the cylindrical and the hemispherical meniscus. (Section III.B.1). With the cylindrical meniscus (Figure 18), as already pointed out in Section III.B.1, the core radius is equal to $r_m/2$, so that the Kelvin equation becomes

$$\ln\left(\frac{P}{P^\circ}\right) = -\frac{\gamma V}{RT} \cdot \frac{1}{r_k} \tag{46}$$

as compared with the form (cf. equation 43)

$$\ln\left(\frac{P}{P^\circ}\right) = -\frac{2\gamma V}{RT} \cdot \frac{1}{r_k} \tag{47}$$

for the hemispherical meniscus.

Let us consider (89) first a cylindrical pore of core radius r_k, open at both ends; as the relative pressure is gradually increased from zero, a *hemispherical* meniscus is unable to form because there is no means of nucleating it; the only possibility will be a *cylindrical* meniscus, so that capillary condensation will occur at a relative pressure of $P/P^\circ = \exp(-\gamma V/RT \cdot r_k)$. If however the pore is closed at one end a hemispherical meniscus can form there, so that capillary condensation will take place at a relative pressure of exp $(-2\gamma V/RT \cdot r_k)$. In both systems capillary evaporation will occur from a hemispherical meniscus, at the relative pressure exp $(-2\gamma V/RT \cdot r_k)$ so that hysteresis will be found in the first case, but not in the second.

The course of events in an ink-bottle type of pore as pressure increases or

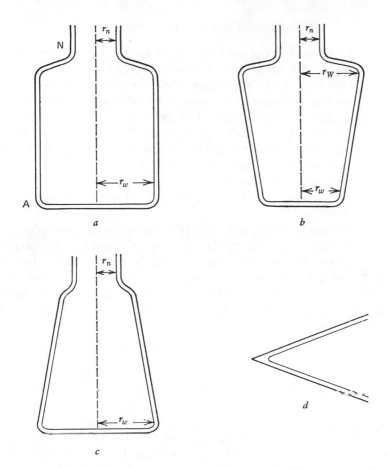

Fig. 21. (*a–c*) "Ink-bottle" pores; (*d*) cone-shaped pore. The filling and emptying of the "ink-bottles" is described below.

(*a*) (i) If $r_w < 2r_m$: Condensation commences at the base (hemispherical meniscus) at $P/P^\circ = \exp\left(-2\gamma V/RT\cdot r_w\right)$ and the pore fills up at this relative pressure. Evaporaton commences at the neck (hemispherical meniscus) at $P/P^\circ = \exp\left(-2\gamma V/RT\cdot r_n\right)$. (ii) If $r_w > 2r_n$: Condensation commences in the neck (cylindrical meniscus) at $P/P^\circ = \exp\left(-\gamma V/RT\cdot r_n\right)$; the pore fills up at $P/P^\circ = \exp\left(-2\gamma V/RT\cdot r_w\right)$. Evaporation from neck as in (i).

(*b*) (i) If $r_w < 2r_n$, and $r_w < 2r_n$: Condensation commences at the base at $P/P^\circ = \exp\left(-2\gamma V/RT\cdot r_w\right)$; the pore filling extends over the range of relative pressure up to $\exp\left(-2\gamma V/RT\cdot r_w\right)$. Evaporation commences from hemispherical meniscus in the neck at $P/P^\circ = \exp\left(-2\gamma V/RT\cdot r_n\right)$. (ii) If $r_w > 2r_n$: Condensation commences in the neck (cylindrical meniscus) at $P/P^\circ = \exp\left(-\gamma V/RT\cdot r_n\right)$; but filling of the body will not commence until $P/P^\circ = \exp\left(-2\gamma V/RT\cdot r_w\right)$. Evaporation as in (i).

(*c*) (i) If $r_w < 2r_n$: Pore filling and emptying as in case (i) of Fig. 21*a*. (ii) If $r_w \; 2r_n$ and $r_w > 2r_n$: Pore filling and emptying as in case (ii) of Fig. 21*a*.

(*d*) A cone-shaped pore. Filling and emptying occur without hysteresis.

decreases can be worked out with the aid of these concepts. The simplest case is a "bottle" with a narrow neck of radius r_n and a cylindrical body of radius r_w, such that* $r_w < 2r_n$ (Figure 21a); since nucleation of a hemispherical meniscus can occur at the base at the relative pressure $(P/P°)_I =$ exp $(-2\gamma V/r_wRT)$, whereas formation of a cylindrical meniscus in the neck would require the higher relative pressure $(P/P°)_{II} =$ exp $(-\gamma V/r_nRT)$, capillary condensation will commence at the base at the pressure P_I, and will continue at this pressure until the whole pore is filled. Evaporation, however, will take place from a hemispherical meniscus in the neck, at a relative pressure $(P/P°)_{III} =$ exp $(-2\gamma V/r_nRT)$, and since this is lower than $(P/P°)_{II}$, hysteresis will be found.

In the converse case where $r_w > 2r_n$, $(P/P°)_{II}$ will be less than $(P/P°)_I$, so that condensation will occur in the neck on a cylindrical meniscus; as pressure increases the condensation will extend into the body, which will become full when the relative pressure rises to $(P/P°)_I$. Capillary evaporation will commence from the neck (hemispherical meniscus) at the relative pressure $(P/P°)_{IV} =$ exp $(-2\gamma V/r_nRT)$, and since $(P/P°)_{IV}$ is less than the equilibrium pressure $(P/P°)_I$ for the body, the pore will empty completely at the pressure $(P/P°)_{VI}$. Since the pressure for evaporation is less than that for condensation, hysteresis will present.

If the body of the "bottle" is tapered, the exact way in which the pores are filled and emptied will depend on the values of the ratios $r_n : r_w$ and $r_n : r_W$, where r_w is the core radius of the narrowest part of the body and r_W that of the widest part. Some typical examples are briefly described in the legends to Figure 21. As Everett points out, however (103), the analogy of a pore as a narrow-necked bottle is overspecialized, and in practice a series of interconnected pore spaces rather than discrete bottles is more likely. The progress of capillary condensation and evaporation in pores of this kind has been discussed by de Boer (89) and, more recently, by Everett (103). In a pore of the shape indicated in Figure 22a, for example (the cross section being circular, and $r_w > 2r_n$) capillarly condensation will first occur at a cylindrical meniscus in the constrictions, at a relative pressure of $(P/P°)_V =$ exp$(-\gamma V/r_nRT)$; as relative pressure increases, the condensation will extend on the hemispherical meniscus thus formed, into the wider parts of the tube until a relative pressure of $(P/P°)_{VI} =$ exp $(-2\gamma V/r_wRT)$ is reached; the spherical bubble now remaining will immediately collapse so that the pore is completely filled. Capillary evaporation will commence at the two open ends at a relative pressure of $(P/P°)_{VII} =$ exp $(-2\gamma V/r_wRT)$ from hemispherical menisci, which will recede as the pressure decreases until, at $(P/P°)_{VIII} =$ exp

---

*Throughout this subsection, all the terms in r refer to the *core*, that is, to the pore diminished by the adsorbed layer (thickness t) on the walls.

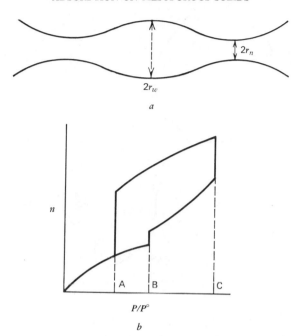

Fig. 22. Capillary condensation and evaporation. (a) A pore of circular cross-section with gradually varying radius. It is assumed that $2r_n < r_w$. (b) Adsorption isotherm corresponding to (a). (Diagrammatic only.)

$-2\gamma V/r_n RT$), each will have reached a constriction. Since further evaporation must lead to the formation of wider menisci with higher values of equilibrium pressure, the whole of the remaining liquid is unstable and will evaporate at once. The values of relative pressure at corresponding amounts adsorbed are higher during condensation than during evaporation, so that hysteresis will be present (Figure 22b). Other models of particular interest in relation to hysteresis are cone-shaped pores; parallel-sided slits; and the interstices between touching spheres. In a pore of cone shape (Figure 21d) there is no hindrance to nucleation, so that a meniscus will be formed first at the apex of the cone and will extend steadily as the relative pressure increases until the pore is full; since both evaporation and condensation will occur at the same hemispherical meniscus hysteresis will be absent.

Parallel-sided slits present an extreme case of hysteresis, since the mean radius of curvature of planar walls is infinite; thus, no condensation can occur until the saturation vapor pressure is reached, and the pore then fills up immediately; a hemicylindrical meniscus (Figure 19b) can now form with a radius of curvature of $d_k/2$ (d_k is the distance between the films on the walls); consequently the pore will empty completely at the relative pressure of

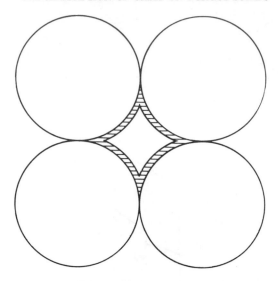

Fig. 23.   The interstice between spheres in contact.

$\exp\left(-2\gamma V/d_k RT\right)$. This simple picture will of course be modified if the sides of the slit are not truly parallel or are not perfectly planar.

Finally, in the case of touching spheres (Figure 23), condensation can commence in the crevice between contiguous spheres, to form a torus of liquid that will extend as relative pressure increases until adjacent tori coalesce; the spherical cavity thus produced will fill up completely without any further increase in pressure. When relative pressure is reduced, evaporation will first occur from a hemispherical meniscus formed in the foramen (window) leading into the cavity, when the relative pressure has fallen to $(P/P^\circ) = \exp\left(-2\gamma V/r_f RT\right)$, where r_f is the radius of curvature of the foramen. The value of r_f will, of course, differ from the mean radius of curvature of the meniscus present during the condensation process, since this will be given by the two principal radii of curvature (one positive, one negative) of the surface of the torus already referred to. The adsorption isotherm would thus have the general shape of a type A isotherm (Figure 15), but with steep sides to the hysteresis loop.

Table 6 from a paper by Karnaukhov (104) summarizes the position for a variety of models in a slightly different way: the values of r_m calculated by the simple Kelvin equation (43) from the two branches of the isotherm are compared with the actual radius of the pore. The line of thought was again to consider the location of the meniscus during capillary condensation and capillary evaporation, respectively; thus, in the case of the open cylinder (model 2) the meniscus during adsorption will be of cylindrical rather than

TABLE 6

Comparison of Kelvin radius r_m (equation 43) with actual radius for various model porous solids

		Ratio of r_m to Actual Radius of Pore[a]	
Model of Pore		Adsorption Branch	Desorption Branch
1	Closed cylinder	1	1
2	Open cylinder	2	1
3	Cone	1	1
4[b]	Slit with parallel walls	2	2
5[b]	Tapering slit	2	2
6[c]	Spaces between parallel rods	2	1
7[c]	Capillary of square section	2	1
8[d]	Throat of bottle-shaped pore	<1	1
9[e]	Constrictions between touching spheres with:		
	coordination number 12	3	1
	coordination number 6	5	1

[a] For simplicity, the thickness of the adsorbed film on the wall is neglected.

[b] In models 4 and 5 the "radius" of a pore is taken as half the distance between its opposite walls.

[c] In models 6, 7, and 9 the radius of the pore is taken as the radius of the circle inscribed in it.

[d] For model 8 the ratio for the adsorption branch may vary over a wide range, depending on the ratio of the sizes of the throat and the cavity.

hemispherical form, so that the factor 2 will appear in column 3. (cf. equations 46 and 47).

A more general theory of hysteresis has been advanced by Everett (105,106) in the attempt to explain the behavior of an adsorption system in relation to scanning, that is, the process of crossing between the two branches of the hysteresis loop. Despite the thermodynamic irreversibility of the process, all changes are found to be confined to the area of the loop, the branches of which therefore represent quasi-equilibrium states.

Everett divides the whole adsorption space into pore domains, accessible from neighboring domains through pore constrictions. In each such domain the adsorbate is able to exist in at least two physical states, and the whole system passes through different sets of microscopic states when following different scanning paths.

The "domain" theory has inspired a comprehensive program of experimental studies, mainly in Everett's laboratory, designed to test and develop the underlying ideas. Interpretation of scanning behavior should be of signal value in providing a means of testing the internal consistency of the various

procedures for calculation of pore size distribution from adsorption isotherms.

Other, quite different causes of hysteresis remain to be mentioned: mechanical–structural changes in the solid accompanying adsorption and desorption;
penetration into the body of the solid, either through very narrow orifices
leading into cavities (cf. Section IV) or along very fine cracks or cleavage
lines, which leads to low pressure hysteresis (p. 255); quasichemical changes
such as slow hydroxylation or hydration of the surface of an oxide. For discussion of effects such as these, the concepts of reaction kinetics are helpful.
These processes all involve an energy of activation, so that the extent of
change registered during any particular observation will depend both on the
actual duration of the observation, and on the total time of the experiment up
to that point. Consequently the recorded points will not in general represent
a state of equilibrium, and the amount adsorbed at any relative pressure is
likely to be greater if the point on the isotherm is being approached from
higher relative pressure, in the conventional type of experiment in which the
adsorption branch is measured first.

C. Use of the Kelvin Equation for Calculation of Pore Size Distribution

One of the main uses of the Kelvin equation is for the calculation of pore
size distribution from the adsorption isotherm: the pore radius is obtainable
from the relative pressure and the pore volume from the amount adsorbed
(when converted to a liquid volume). Early efforts in the field, such as that of
Foster already cited, perforce led to a distribution of *core* size, rather than of
pore size, owing to the lack of information as to the thickness t of the adsorbed film on the walls. Following the formulation of the BET method for
the determination of specific surface, and indirectly therefore of t, it has become possible to devise procedures for the calculation of pore size distribution that take into account the adsorbed film (107).

To appreciate the fundamental role of film thickness it is helpful to consider
the progressive emptying of pores, initially full, as the relative pressure is
lowered in steps from its starting value of unity. Let the pores be divided into
groups, with core radii r_1^k, r_2^k, etc., corresponding to pore radii r_1^p, r_2^p, etc., and
relative pressures $(P/P°)_1$, $(P/P°)_2$, etc.* When the relative pressure is reduced
to $(P/P°)_1$, the first group of cores lose their capillary condensed liquid, but
still carry a film, of thickness t_1, on their walls. Since the amount of capillary
evaporated liquid is $(n_0 - n_1)$, the volume v_1^k of this group of cores is
$(n_0 - n_1)V$ (n_0 and n_1 are the amounts adsorbed at relative pressures of
unity and $(P/P°)_1$, respectively). The corresponding volumes of the *pores* is

---

*In this and Sections III.C.1 and III.C.3, the symbols r^p and r^k, etc. are for clarity
substituted for r_p and r_k, etc.

$Q_1 v_1^k$, where Q_1 is the factor for converting the core volume v_1^k into the pore volume v_1^p, and is a function both of pore shape and of the thickness of the film on the walls (cf.III.C.3).

When the relative pressure is reduced to $(P/P°)_2$, the second group of cores lose their capillary-condensed liquid, but in addition the pores of the first group yield up some of their remaining adsorbate owing to the decrease from t_1 to t_2 in the thickness of the film on their walls. Similarly, when the relative pressure is further reduced to $(P/P°)_3$ the decrement $(n_2 - n_3)$ in the amount adsorbed will include contributions from the walls of pore groups 1 and 2 (as the film thins down from t_2 to t_3) in addition to the contribution from the cores of group 3.

It is this composite nature of the amount given up at each step which complicates the calculation of the *pore* size distribution curve. The various devices proposed from time to time for dealing with it differ among themselves in mathematical procedure rather than basic physical principles. Typical examples will be outlined, not necessarily in chronological order of their appearance in the literature.

1. Methods Involving the Area of Pore Walls

In the method of Orr and Dalla Valle (108), modified from that of Pierce (109), the correction for thinning of the adsorbed film is accomplished through intervention of film area. The stepwise reduction of relative pressure starting at $P/P° = 1$, is again considered. For each step, the area ΔS^p of the walls of those pores (or average radius r^p) that have lost capillary condenstae during that step is calculated. The volume of adsorbate released from all the exposed walls during this step is, thereore, $\Delta v^f = \Delta t \sum(\Delta S^p)$, where Δt is the reduction in film thickness during the step, and $\sum(\Delta S^p)$ is the area of walls of all the pores that have lost their capillary condensate.

The change Δv^k in pore volume during the step is then given by $\Delta v^k = \Delta v - \Delta v^f$ where Δv is the total loss, expressed as a liquid volume, of adsorbate during the step.

If the pores are cylindrical, the corresponding change Δv^p in pore volume is given by

$$\Delta v^p = Q \cdot \Delta v^k \tag{48}$$

where $Q = (r^p/r^k)^2$; r^p in turn is given by

$$r^p = r^k + t, \tag{49}$$

and the surface area ΔS^p of the pores of radius r^p, by

$$\Delta S^p = \frac{2\Delta v^p}{r^p} \tag{50}$$

In practice the computation is somewhat intricate, requiring close attention to detail because of the cumulative nature of the area of film $\sum(\Delta S^p)$. Furthermore, each of the conversions such as that implied by equation 48 or 50 is an individual calculation, in contrast to the standard conversions of the Roberts method (Section III.C.3).

These and similar procedures rest on the tacit assumption that the surface area of the film in a given pore does not change as the relative pressure diminishes, which is true for parallel-sided slits but not for cylinders, where the film area must increase as the film thins down. The procedure of Dollimore and Heal (110,111), and of Roberts (112), described in the next two subsections, are free from this defect.

A detailed procedure for the calculation of pore size distribution for slit-shaped pores taking into account the t-correction, and based on the earlier work of Steggerda (113), has been described by Lippens, Linsen, and de Boer (79).

2. Methods Involving the Length of Pores

In several methods for the calculation of pore size distribution such as those of Shull (65), Wheeler (107), Barrett, Joyner, and Halenda (114), and Cranston and Inkley (66), the length of the pores (assumed cylindrical) appears as an intermediary. The calculations are somewhat involved, and a modified procedure has been put forward by Dollimore and Heal (110), who take pore radius, rather than relative pressure, as the independent variable; the pore system is divided into groups of pores ranging in average radius r from 95 Å down to 7.5 Å. Since only *pore* radii appear directly in the analysis, the symbol r rather than r^p will be used for simplicity.

Consider stage 1 in the desorption process where the thickness of the adsorbed film is t_i, and the pores of radius r_i have just lost their capillary condensate; the volume of multilayer lining the pores of any radius r (where $r > r_i$) will then be

$$\pi \left[r^2 - (r - t_i)^2 \right] L(r) \, dr$$

or

$$\pi \left[2rt_i - t_i^2 \right] L(r) \, dr \tag{51}$$

where $L(r)$ is the length of pores with radius between r and $r + dr$.

The volume v^f of the adsorbed film on the walls of all those pores already emptied of capillary condensate will be

$$v^f = \int_{r_i}^{\infty} \pi \left[2rt_i - t_i^2 \right] L(r) \, dr$$

$$= t_i \int_{r_i}^{\infty} 2\pi r L(r) \, dr - \pi t_i^2 \int_{r_i}^{\infty} L(r) \, dr \tag{52}$$

$$= t_i \sum (S_i) - \pi t_j^2 \sum (L_i) \tag{53}$$

Here $\sum(S_i)$ is the total area, and $\sum(L_i)$ the total length, of all pores of radius in excess of r_i

The diminution $\Delta v_i{}^f$ in film volume during stage i is thus

$$\Delta v_i{}^f = \Delta t_i \sum (S_i) - 2\pi t_i \Delta t_i (L_i) \tag{54}$$

and if Δv_i is the total amount (expressed as a volume of liquid) given up during stage i, then the volume $\Delta v_i{}^k$ of cores emptied during this stage must be

$$\Delta v_i{}^k = \Delta v_i - \Delta v_i{}^f$$

and the corresponding pore volume

$$\Delta v_i{}^p = Q_i(\Delta v_i - \Delta v_i{}^f) \tag{55}$$

where

$$Q_i = \left(\frac{r - t_i}{r}\right)^2 \tag{56}$$

The calculation implied in equation 55 is carried out for each stage commencing with stage 1, that is, saturation, where $S_i = 0$ and $L_i = 0$. For subsequent stages one uses the geometric relationships $S_i = 2v_i/r_i$ and $L_i = S_i/2\pi r_i$.

In their paper Dollimore and Heal give a table of r values (in suitable steps of 10 Å at the coarse and 1 Å at the fine end of the range) together with corresponding values of (P/P°), t, and Q. In a later publication they examine the manner in which the calculated pore size distribution (for a number of typical solids) depends on the particular t-curve chosen; they recommend the expression $t/\text{Å} = 3.54(-5/\ln P/P^\circ)^{1/3}$, as giving results showing the optimal internal consistency. (cf. equations 20 and 21).

3. A Method Involving neither Length nor Area of Pores (Roberts) (112)

In the Roberts procedure the progressive increase in core radius due to the thinning down of the adsorbed film on the walls, already referred to, is allowed for by a neat device.

The pore radius is again taken as the independent variable and the pore system is divided into groups of pores, all pores within a given group being assigned the same average radii r^p.

The range from 100 to 20 Å is divided into eight groups of 10 Å each, and that from 20 to 10 Å into two, giving ten groups in all, numbered 1 to 10. (In principle the range could be extended at the upper end, to say 250 Å with appropriately wider steps, say 33.3 or 50 Å; it should not extend at the lower end, for reasons which will be apprarent from Sections III.D and IV).

For each group i the pore volume $v_i{}^p$ is related to the corresponding core volume $v_i{}^k$ by a model, usually either cylinders or parallel-sided slits; but allowance of course needs to be made for the fact that film thickness may correspond to some lower relative pressure $(P/P°)_j$ rather than $(P/P°)_i$. Thus, for a particular group of cylindrical pores having pore radius $r_i{}^p$, film thickness t_j, and core volume $v_{ij}{}^k$ we have

$$\frac{v_i{}^p}{v_{ij}^k} = \left(\frac{r_i{}^p}{r_i{}^p - t_j}\right)^2 \tag{57}$$

that is

$$v_i{}^p = Q_{ij} \cdot v_{ij}{}^k \tag{58}$$

where

$$Q_{ij} = \left(\frac{r_i{}^p}{r_i^p - t_j}\right)^2 \tag{59}$$

It will be noted that the first suffix of Q_{ij} refers to the pore radius and the second to the film thickness.

Consider now the first three groups of pores, which are initially all full. When the relative pressure falls from unity to $(P/P°)_1$, an amount w_1 (expressed as a liquid volume) is given up; the volume $v_1{}^p$ of this group is, by equation 58,

$$v_1{}^p = Q_{11}v_1{}^k = Q_{11}w_1$$

When P falls to P_2, the cores of group 2 give up a volume $v_{22}{}^k$, but since the thickness t has diminished to t_2, the *total* loss from group 1 is now $v_{12}{}^k$; thus the combined loss w_2 from both groups is

$$w_2 = v_{22}{}^k + v_{12}{}^k$$

By equation 58 the pore volume of the second group is

$$\begin{aligned} v_2{}^p &= Q_{22}v_{22}{}^k \\ &= Q_{22}(w_2 - v_{12}{}^k) \\ &= Q_{22}\left(w_2 - \frac{v_1^p}{Q_{12}}\right) \end{aligned}$$

By an extension of the argument, the volume $v_3{}^p$ of the third group is

$$\begin{aligned} v_3{}^p &= Q_{33}v_{33}{}^k \\ &= Q_{33}\left(w_3 - \frac{v_2^p}{Q_{23}} - \frac{v_1{}^p}{Q_{13}}\right) \end{aligned} \tag{60}$$

The procedure is readily extended to the other groups up to group 10.

It is now necessary to draw up a table of values of Q_{ij} calculated with equation 59 for all values of i and j from 1 to 10. The $r_1{}^p$ is 95 Å, $r_2{}^p$ is 85 Å, and so on. First, the values of r^k and of t are calculated for a succession of values of $P/P°$, quantity r^k being found from the Kelvin equation and t from a standard

TABLE 7

The conversion factor Q_{ij} in the Roberts method for the calculation of pore size distribution.[a]

		Values of $1/Q_{ij}$ for the First Four Groups of Pores[b]			
j		4	3	2	1
$t_j/\text{Å}$		10.66	11.29	11.87	12.36
$r_j/\text{Å}$		60	70	80	90
$\bar{r}_j/\text{Å}$		65	75	85	95
i	$\bar{r}_i/\text{Å}$				
4	65				
3	75	0.736			
2	85	0.765	0.752		
1	95	0.788	0.776	0.766	
$1/Q_{ii}$		0.699	0.722	0.740	0.757

[a] Reduced from the Table of Roberts (112).
The values of t_j are those of Barrett, Joyner, and Halenda (114).
[b] Example: $1/Q_{24} = 0.765$

t-curve (II.D.1). Then, by use of the relation $r^p = r^k + t$ (if the pores are cylindrical) a curve of r^p against $P/P°$ can be constructed; it can then be used to find the values of $P/P°$ corresponding to the succession of pore radii $r_1{}^p$, $r_2{}^p$, etc., and through them to find the corresponding values t_1, t_2, etc. of t. It is thus possible to build up the required table of Q_{1j} values through equation (59). An extract from the Table of Roberts, for the first three groups, is given in Table 7 by way of illustration.

If the pores are parallel-sided slits, the conversion factor $Q_{ij}{}^s$ will be $r_i{}^p/(r_i{}^p - 2t_j)$

4. The "Modelless" Method of Brunauer, Mikhail, and Bodor (115)

The methods discussed so far all depend on the assumption of a definite and simple model of pore shape, and this can rarely corresopnd to reality. Brunauer, Mikhail, and Bodor have attempted to circumvent this limitation in their "modelless" method, though as they themselves point out, the method in itself yields only a *core* size distribution; and that if a pore size, distribution is required, or if correction is to be made for thinning of the adsorbed film, it is still necessary to postulate a model and to have recourse to a t-curve. In their view, however, the core size distribution is sufficient for most practical purposes.

In place of the Kelvin radius r_m, Brunauer, Bodor, and Mikhail use the hydraulic radius r_H, which for a pore of uniform cross section is equal to the ratio of the volume of an element of pore to the surface area of its walls, that is,

$$r_H = \frac{dv}{ds} \tag{61}$$

The method involves the calculation of the surface area of the pores by means of the relation used for this purpose by Kiselev (116) in 1945:

$$\gamma dS = -(\mu - \mu°) \, dn \tag{62}$$

Here S is the surface area of the cores, μ the chemical potential of the adsorbate within the pores, and $\mu°$ that of the saturated vapor, n being the amount adsorbed.
Since $\mu - \mu° = RT \ln (P/P°)$ equation 62 becomes

$$\gamma \, dS = -RT \ln \left(\frac{P}{P°}\right) dn \tag{63}$$

This equation is then integrated for a succession of small steps in the isotherm, step 1 corresponding to the range of relative pressure 1.0 to 0.95; step 2 to 0.95 to 0.90, and so on. If n_1 is the amount still adsorbed at the end of step 1, viz. when $P/P° = 0.95$, n_2 the corresponding amount at the end of step 2, viz. when $P/P° = 0.90$, and so on, then the area ΔS_1 of the cores of group 1 is

$$\Delta S_1 = -\frac{RT}{\gamma} \int_{n_0}^{n_1} \ln \left(\frac{P}{P°}\right) dn \tag{64}$$

where n_0 was the amount adsorbed at saturation.
The corresponding hydraulic radius $r_{H,1}$ is then immediately given by

$$r_{H,1} = \frac{\Delta v_1}{\Delta S_1} = \frac{\Delta n_1}{\Delta S_1} \cdot V$$

where $\Delta n_1 = (n_0 - n_1)$, is the amount given up in this step and Δv_1 is its liquid volume (V as usual is the molar volume of the liquid adsorptive).
The calculation is extended to the remaining steps, the limits of integration being in succession n_2 and n_1, n_3 and n_2, and so on. A curve of cumulative core volume $(n_0 - n)V$ against r_H can then be constructed and from it a core size distribution curve, that is, a plot of dv/dr_H against r_H. At this stage the method of analysis is completely "modelless."
In the "corrected modelless" method allowance is made for the thinning of the adsorbed film. To make this correction for stage 2 it is necessary to calculate n_2', the amount evaporated from the walls of group 1 during this stage. The amount of liquid evaporated from the *cores* of the second group is then

$(n_2 - n_2')$ and the core volume of the group is $(n_2 - n_2')V$, so that the corrected surface of the second group is

$$\Delta S_2 \text{ (corr)} = -\frac{RT}{\gamma} \int_{n_1}^{(n2-n'2)} \ln\left(\frac{P}{P^\circ}\right) dn \tag{65}$$

and the corrected hydraulic radius

$$r_{H,2} \text{ (corr)} = \frac{(n_2 - n_2') V}{\Delta S_2 \text{ (corr)}}$$

The evaluation of n_2', n_3', etc. necessitates the use of a model and a t-curve. For cylindrical pores, for example

$$n_2'V = \left\{\Delta S_1 (t_1 - t_2) + \frac{\Delta S_1}{4r_{H,1}} \cdot (t_1 - t_2)^2\right\}$$

where t_1 and t_2 are the values of t at the ends of step 1 and step 2, respectively.

Finally, the corrected core size can be converted into a pore size essentially by use of a Q-factor as in other methods, and therefore by reference to a t-curve and a pore model. The *pore* size distribution curve can then be constructed.

Recently, the validity of the local hydraulic radius as a means of defining the capillary properties of a pore has been examined by Everett and Haynes (97), who refer to the work of Gauss (118) of 1830. Gauss had considered the perturbation of an interface having constant curvature defined by solid boundaries such that an increase dv in the volume of liquid is accompanied by a change dS^{sl} in the area of solid–liquid interface. On geometrical grounds, it is shown that these quantities are related by the equation

$$\frac{d(S^{lg} - S^{sl} \cos \theta)}{dv} = \frac{2}{r_m} \tag{66}$$

where r_m is the mean radius of curvature and θ the angle of contact between the liquid and the solid.

For the special case of a cylinder of radius r completely wetted by the liquid, we have $\theta = 0$, and since the cross section is constant we also have $dS^{lg} = 0$, consequently

$$-\frac{dS^{sl}}{dv} = \frac{2}{r_m} \tag{67}$$

and since, for a hemispherical meniscus (with $\theta = 0$) $r = r_m$, this leads to

$$-\frac{dS^{sl}}{dv} = \frac{2}{r} \tag{68}$$

The hydraulic radius is given by

$$\frac{1}{r_H} = \frac{2\pi r dl}{\pi r^2 dl} \text{ , that is, } \frac{1}{r_H} = \frac{2}{r} \tag{69}$$

(where dl is the additional length of capillary wetted), and from 67, 68, and 69 it therefore follows that

$$r_H = \frac{r_m}{2}$$

thus the hydraulic radius is directly related to the core parameter.

If the cross section of the (uniform) capillary is not circular, the line of contact between solid, liquid, and vapor no longer lies in a plane; but if the hydraulic radius is calculated as $r_H = A/P$, where A is the area of the projection of the boundary of contact on to a plane normal to the axis of the cylinder, and p is the perimeter of this projection, then r_H should, as before, be equal to $r_m/2$. When, however, the pores are no longer uniform in cross section, then the change dS^{lg} which accompanies the change dv in volume of liquid, may be positive or negative, depending on the direction of taper of the pore; it is then essential to use equation 66. The problem of the movement of a meniscus through the window spaces between spheres is a particularly difficult one, which so far has defied solution either with or without the concept of hydraulic radius.

5. Use of the Adsorption or the Desorption Branch for Pore Size Distribution Calculations

The question whether to use the adsorption or the desorption branch of the hysteresis loop for calculations of pore size distribution is a long-standing one. The making of a sound decision clearly involves a judgment as to the origin of hysteresis, and as yet there is no completely general agreement on this subject though opinion in general leans towards the views of de Boer (89) and Everett (103, 105, 106).

The difficulty is that the Kelvin equation, by virtue of its thermodynamic origin, is strictly applicable only to truly reversible processes, whereas the very existence of adsorption hysteresis shows that this condition cannot hold along at least one of the branches of the loop.

The adsorption branch of the loop was chosen for pore size calculations by Rao (119) and also by Cranston and Inkley (66). On the other hand, Kington and Smith (120) from determinations of the heat of adsorption and of desorption for argon on porous glass concluded that the *adsorption* branch corresponded to the equilibrium state, the desorption branch being connected with spontaneous evaporation of liquid from the pores. Karnaukhov (104) has argued that though spontaneous processes are to be expected in very different structures during both filling and emptying of pores, they are nevertheless always preceded by equilibration of the vapor with the liquid meniscus; the final stages of equilibrium may, therefore, be used for the calculation of the critical size of the pores as well as the size of constrictions responsible for sudden condensation and evaporation of adsorbate; consequently the presence of spontaneous processes should not prevent the use of the Kelvin equation for calculation of pore sizes. Karnaukhov concludes that it is the desorption branch of the isotherm which is to be preferred for this purpose; he supports this view by reference to results of Zhdanov (121), who measured the isotherms of methanol and ethanol on artificially prepared mixtures of plates and powders of porous glasses with known, different, pore sizes. The desorption branch of the isotherms led to a pore size distribution curves in good agreement with the curve predicted from the size distributions of the separate components, whereas the adsorption branch gave no indication of the bidisperse nature of the mixture.

6. Adsorbates other than Nitrogen

Hitherto the vast majority of calculations of pore size distribution have referred to nitrogen as adsorbate. Karnaukhov (104) has urged the desirability of extending the method to other adsorbates, in particular those that can be used at or near room temperatures.

From the Kelvin equation

$$\ln\left(\frac{P}{P^\circ}\right) = -\frac{2\gamma V}{RT}\cdot\frac{1}{r_m} \tag{43}$$

it is seen that the higher the value of $\gamma V/T$ (see Table 8) the lower will be the relative pressure at which capillary condensation occurs in a pore of given size r_m. Since the thickness of the adsorbed film on the walls increases with increasing P/P°, it would seen at first sight that the t-correction is automatically reduced if the adsorptive has a high value of $\gamma V/T$, but since $t = N\sigma$

TABLE 8

Values of $\gamma V/T$ for typical adsorptives[a]

Adsorptive	$T\,(°K)$	γ(dyn/cm)	V (cm³/mole)	$\gamma V/T$ (dyn cm²/mole T)
Nitrogen	78	8.88	34.7	3.95
Argon	87.5	13.20	28.53	4.30
Methanol	293	22.60	40.42	3.12
Carbon tetrachloride	293	26.75	96.54	8.72
Benzene	293	28.88	88.56	8.76

[a] After Karnaukhov (104).

(where N is the number of statistical molecular layers and σ is the molecular diameter of the adsorbate) the value of t immediately prior to capillary condensation in a given pore will depend not only on $P/P°$ but also on N and σ. Now the value of N at a given value of $P/P°$ will be greater the greater the net heat of adsorption, as expressed, for example, through the constant c of the BET equation. It may therefore happen that the beneficial effect of a high value of $\gamma V/T$ in reducing the value of $P/P°$ for capillary condensation in a given pore may be offset by the deleterious effect of large values of c and σ, and therefore of t.

The point is illustrated in Table 9 which compares the effect of the factor $\gamma V/T$ on the "t" correction for nitrogen and benzene, respectively. In the absence of a standard isotherm for benzene, it has been assumed for simplicity that the standard isotherms of benzene and of nitrogen are identical; and σ for benzene has been estimated from the expression $\sigma = V/Na_m$ with (122) $a_m = 40.3$ Å^2, to give $\sigma = 3.6$ Å, a value fortuitously close to that for nitrogen, viz. 3.54 Å. As will be seen, though the t-correction is consistently smaller for benzene, the difference is much less than might have been expected from the respective values of $\gamma V/T$; at present, the small advantage gained in this respect is more than offset by the general uncertainty as to the value of a_m and the course of the standard isotherm for benzene. It is to be hoped that, in time, these disavdantages attending organic adsorptives will be overcome, so that the overdependence of the adsorption method for pore size calculation on the properties of one adsorptive, nitrogen, will be removed.

Karnaukhov has pointed to a further substantial advantage that would be gained by the use of adsorptives having reasonably large vapor pressure at room temperature; if the adsorbent is coarsely porous the part of the isotherm of major interest will lie close to saturation, and at low temperatures such as 77°K it is difficult to obtain accurate results owing to the temperature instability of the refrigerant. With adsorptives such as benzene, on the other hand, the isotherm can be measured at a temperature slightly above ambient,

TABLE 9

Values of t for nitrogen and benzene at different values of r_m.

r_m(Å)	$P/P°$		t (Å)	
	N$_2$	C$_6$H$_6$	N$_2$	C$_6$H$_6$
20		0.50		5.65
30	0.72	−0.50	8.6	6.8
50	0.83	0.86	10.5	7.9
300	0.97	0.935	17.5	14.5
500	0.98	0.96	20.6	16.7

so that a conventional thermostat can be used; furthermore, the temperature difference between the sample and the bath, which at 77°K is large enough to produce major errors into the isotherm, is negligible when the bath is close to room temperature (123, 124).

7. Pore Shape

For the simple pore models referred to hitherto, the relationship between the quantity r_m of the Kelvin equation and the parameter or parameters describing pore size, is readily worked out. For more complex shapes of pore one may use the relationship implied by equation 67

$$\frac{dS}{dv} = \frac{2}{r_m} \tag{70}$$

which remains valid so long as the pore is of constant cross section and the angle of contact is zero; dS is the additional area wetted when the volume of capillary condensate increases by dv. For a particular pore shape, provided it is not too complicated, the relationship between r_m and the pore parameter(s) can be arrived at by working out an expression for the quantity dS/dv through application of elementary geometrical principles. Results for some typical systems, obtained by Everett (125), are given in Table 10, whence it is seen that the relationship between r_m and the pore size parameter(s) may differ considerably from unity.

TABLE 10

Comparison of Kelvin radius r_m and pore size parameters for some idealized pore systems.[a,b]

System	Size Parameter	Kelvin Radius r_m
Nonintersecting cylindrical pores	Radius of cylinder = R	$r_m = R$
Parallel-sided fissures	Distance apart of walls = d	$r_m = d$
Interstices between packed spheres	Radius of spheres = R	
Cubic packing		$r_m = 0.613R$
Rhombohedral packing		$r_m = 0.229R$
Intersecting square capillaries	Width of capillaries = 2R Separation of capillaries = 2δ	$r_m = R\{1 + R/\delta\}$

[a] Reduced, with modified symbols from the paper by Everett (125).

[b] The width of pore refers in all cases to the *core*, since the walls are covered with a film of thickness t.

TABLE 11

Capillary condensation and evaporation. Form of pores and menisci in different porous solids[a]

Solid	Form of Pores or Form of Particles	Form of Meniscus	
		Adsorption	Desorption
I. *Etched structures*			
Porous glasses	Cylindrical, or close to cylindrical, closed and open pores, bottle-shaped pores	Cylindrical and spherical	Spherical
Activated charcoals	Spherical pores, connected by narrow channels	Cylindrical (throats) spherical cavities	Spherical
Anode aluminum oxide film	Cylindrical closed pores	Spherical	Spherical
II. *Corpuscular structures*			
Silica, alumina, and titania gels, aluminosilicate catalysts, carbon blacks, aerosils	Spaces between contacting globules	Saddle-shaped	Spherical or saddle-shaped
Montmorillonite, oxidized graphite	Spaces between parallel and inclined planes (slitlike pores)	Cylindrical	Cylindrical

[a] Reduced from the paper by Karnaukhov (104).

More recently the subject has been taken up again by Karnaukhov (104), who has drawn on the considerable body of information which has become available—notably from electron microscopy—as to the forms of the pores in actual solids; some typical examples are listed in Table 11.

D. The Lower Limit of the Hysteresis Loop—The Tensile Strength Hypothesis

In 1965 Harris (126) drew attention to the fact that the hysteresis Loops of nitrogen isotherms at 77°K frequently close at a relative pressure around 0.42, but never below (except in cases where there is "low pressure" hysteresis: Section IV.G). It had earlier been noted by Dubinin (127) that the isotherms of nitrogen at 77°K of five different active carbons likewise had a closure point at $P/P° = 0.42$; but Harris pointed out that of more than one hundred isotherms in the literature having hysteresis loops, over one-half gave a sharp drop in adsorption in the relative pressure range 0.42 to 0.50, the remainder showing a lower closure point at pressures in excess of $0.42P°$.

Interpreted by means of a Kelvin-type analysis, these observations would imply that a large proportion of adsorbents possess an extensive pore system in the very narrow range $17 < r_p < 20$ Å with a sudden cut-off at exactly the same radius $r_p = 17$ Å corresponding to $P/P° = 0.42$ (cf. Section III.C.2). The improbability of this state of affairs led Harris to suggest that a change in the mechanism of adsorption occurred at this point, but he gave no details.

Further evidence as to the reality of the effect is provided by the results of Langford (85), who measured isotherms of nitrogen at 77°K on a number of different substances of various particle shapes, both before and after they had been formed into compacts under high pressure in a die (cf. Section III.A); in some cases the compacted material was further modified by ball-milling or by heating. From the results it was clear that there was a minimum value for loop closure in the range of relative pressure around 0.42 to 0.45, despite variation both in pore shape and in pore size distribution among the samples. Ramsay and Avery's results (88) (III.A), on compacts of silica and zirconia and also for isotherms of nitrogen, give further evidence of a minimum closure point at $P/P° = 0.42$.

For adsorbates other than nitrogen, suitable experimental evidence is less plentiful; but with benzene at 293°K Dubinin (127) and his co-workers found loop closure at $P/P° = 0.17$ on five different carbons. For benzene at 298°K, examination of a number of isotherms in the literature suggests a minimum value of relative pressure for loop closure around 0.20 (e.g., on alumina (128) xerogel, 0.20; on titania (129, 130) xerogel, 0.20, 0.22; on alumina, active carbon and ammonium silicomolybdate (131), 0.17–0.20).

Similarly carbon tetrachloride at 298°K gives indication of a lower limit for loop closure around (fortuitously) the same relative pressure, 0.20 (e.g., on dehydrated gibbsite (128), 2.20; on titania xerogel (129, 130), 0.20–0.25; on partially dehydrated (133) gypsum, 0.20 to 0.23; on stannic oxide xerogel (134), 0.21; on ferric oxide (135), 0.22; on calcined vermiculite (136), 0.20 to 0.25).

The results obtained by Hickman (137, 138) offer particularly convincing evidence that the lower closure point is not affected by the pore structure of the adsorbent. Hickman measured the isotherms of butane at 0°C on samples of an artificial graphite withdrawn at intervals from a batch which had been ball-milled for a period of 1044 hr. The specific surface area, as measured by nitrogen adsorption, increased from 9 to 590 m²/g during this period, implying a 63-fold diminution in an average particle size, and other evidence indicated that the particle shape had also changed. In all cases a hysteresis loop of type B was obtained, which showed a sharp rise at F (Figure 24), and each isotherm showed low pressure hysteresis (cf. EF of Figure 24), but this is an extraneous effect almost certainly due to swelling (p. 257), and in its absence

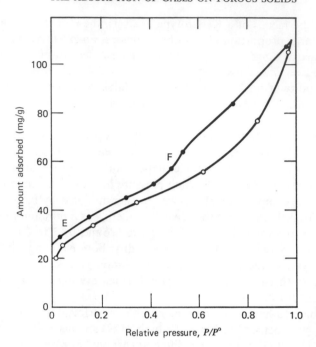

Fig. 24.   Adsorption isotherm of butane at 273° K on a sample of artificial graphite ball-milled for 192hr (137, 138). The shoulder F appeared at a relative pressure which was the same within experimental limits for all five samples in one milling experiment, all six in a second such experiment, and also two of the milled samples that had been compacted. The milling time varied between 0 and 1044 hr.

the loop would close at F. With all nine samples, including compacts made from the 192 and 684 hr samples, the step F was located at almost exactly the same relative pressure, 0.50. Yet the pore structure (the pores being the interstices between the particles, which themselves were almost nonporous) must have varied widely throughout the series, both in absolute size and in distribution of sizes.

An explanation for the existence of a lower closure point of this kind had hinted at, as early as 1948, by Schofield (139), who suggested that the absence of hysteresis in the isotherms of gels having very small pores, might be attributed to the tensile strength of the liquid adsorbate. This idea has been elaborated by Flood, (140), Everett (141), Dubinin (142), and Melrose (143).

According to the Young–LaPlace equation, when a liquid condensed in a capillary tube is in equilibrium with its vapor at a relative pressure of $P/P°$ the presence of the meniscus will produce a negative pressure, or tension τ, given by

$$\tau = \frac{2\gamma}{r_m} \tag{71}$$

so that by the Kelvin equation

$$\tau = -\left(\frac{RT}{V}\right) \ln\left(\frac{P}{P^\circ}\right) \tag{72}$$

Now the maximum tension that a liquid can withstand is of course equal to its tensile strength τ_0; consequently there will be a minimum value, $(P/P^\circ)_h$, say, of relative pressure compatible with the continued existence of capillary-condensed liquid, given by

$$\ln\left(\frac{P}{P^\circ}\right)_h = \left(\frac{V}{RT}\right)\tau_0 \tag{73}$$

This minumum value will be constant for a given adsorptive at a given temperature, no matter what the adsorbent. Liquid present in any finer pores than those given by $r_{m,h}$ of the expression

$$r_{m,h} = -\frac{2\gamma V}{RT} \ln\left(\frac{P}{P^\circ}\right)_h \tag{74}$$

will therefore be unstable and evaporate as soon as the relative pressure is reduced to the critical value $(P/P^\circ)_h$.

This will mean that for each adsorptive at a given temperature the closure of the hysteresis loop may occur above, but never below, a critical value of relative pressure, $(P/P^\circ)_h$, which is characteristic of the adsorptive but independent of the nature of the adsorbent. The experimental facts thus receive qualitative explanation.

The most straightforward quantitative test of the "tensile strength" hypothesis would be to compare the value of τ_0 calculated from $(P/P^\circ)_h$ by equation 73 with the value of the tensile strength of the liquid adsorptive obtained directly. Unfortunately however the direct measurement of the tensile strength of liquids is experimentally very difficult and fraught with uncertainty. Of the few values published (144, 145) the following (given as τ_0/bar) are relevant; carbon tetrachloride 276; benzene, 157; chloroform, 317; water, 275. They may be compared with values calculated from $(P/P^\circ)_h$: carbon tetrachloride (0.22), 410: n-butane (0.50), 174; benzene (0.20), 191; n-hexane, (0.30), 232: cyclohexane (0.20), 370; (all at 298°K except butane, 273°K; figures in brackets denote $(P/P^\circ)_h$).

Though the agreement between the two sets of values, is not close it may nevertheless be reported as satisfactory—in view of the uncertainty in the literature values—that the difference between them is less than twofold.

Because of the inadequacy of the direct test, it is necessary to resort to indirect ones. Thus, Kadlec and Dubinin (142) make use of an expression for theoretical tensile strength,

$$\tau_0 = \frac{2.06\gamma}{d_0} \tag{75}$$

which is based on the 6–12 relation (cf. equation 6) for molecular forces. (d_0 is the distance between neighboring molecules in the bulk liquid, and can be calculated from the liquid density; cf. Gardon (46)). Since $\tau_0 = 2\gamma / r_{m,h}$ (cf. equation (71)), we obtain the further relation

$$\frac{d_0}{r_{m,h}} = 1.03 \tag{76}$$

which should for hold for all adsorptives. Results for five different adsorptives are quoted in Table 12 from Kadlec and Dubinin's paper, and as is seen the values of the ratio $d_0/r_{m,h}$ do not (apart from that for water) differ widely among themselves, though they do deviate appreciably from the theoretical value 1.03. Dubinin and Kadlec consider the agreement between expectation and experiment to be satisfactory, however, in view of the crudity of the model of the liquid state. The question as to how far the bulk properties of the liquid, such as tensile strength, retain their macroscopic values in these very fine pores, is also relevant (p. 238).

Everett and Burgess (141) contend that it is preferable to calculate τ_0 directly from $(P/P°)_h$, essentially by use of equation 73 rather than by the intermediate calculation of $r_{m,h}$, so as to eliminate the surface tension from the analysis. They point out that the temperature dependence of τ_0 should correspond with that expected for the tensile strength of a bulk liquid obeying an equation of the van der Waals type. In their paper they give values of τ_0 calculated from the closure point $(P/P°)_h$ of the isotherms of eight different adsorptives (including nitrogen), most of them at several temperatures. Evidence as to whether the values quoted do actually represent the *minimum* point of closure for the particular adsorptive at the given temperature is not

TABLE 12

Test of "tensile strength" hypothesis[a]

	$d_0/\text{Å}$	$r_{m,h}/\text{Å}$ (eq. 76)	$d_0/r_{m,h}$
Argon	3.87	10.9–11.9	0.32–0.36
Benzene	5.60	13.0–15.4	0.36–0.43
n-Hexane	0.40	16.8	0.38
Dimethyl formamide	5.37	15.1	0.36
Water	3.30	11.0–15.5	0.21–0.30

[a] After Kadlec and Dubinin (142).

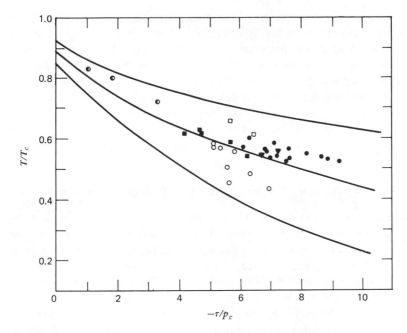

Fig. 25.   Test of the 'tensile strength' hypothesis (Everett and Burgess (141)). T/T_c is plotted against $-\tau_0/p_c$, where T_c is the critical temperature and p_c the critical pressure of the bulk adsorptive; τ_0 is the tensile strength calculated from the lower closure point of the hysteresis loop. ◐, benzene; ○, xenon; □, 2–2 dimethyl benzene; ■, nitrogen; ▼, 2,2,4-trimethylpentane; ●, carbon dioxide; ◆, n-hexane. The lowest line was calculated from the van der Waals equation, the middle line from the van der Waals equation as modified by Guggenheim, and the upper line from the Berthelot equation.

adduced: the values given for benzene at 298°K, for example, vary from 0.21 to 0.34, some of this variation possibly arising from difficulty of deciding on the exact location of the closure point.

Everett and Burgess plot their results in the form of T/T_c against τ_0/P_c (T_c and P_c are the critical temperature and critical pressure, respectively, of the bulk adsorptive), along with lines calculated from the van der Waals equation, the Berthelot equation, and the van der Waals equation as modified by Guggenheim. As is seen (Figure 25), the points show a considerable spread around the "Guggenheim" line, but they fall well within the limits set by the other two equations.

All in all, the circumstantial evidence in favor of the hypothesis is strong, but further work is required before it can be regarded as fully substantiated. In particular, the existence of a minimum value of relative pressure for loop closure, characteristic of the adsorptive at a given temperature, still awaits clear demonstration for a representative range of adsorptives other than nitrogen. Once the hypothesis is accepted, the implication for pore size distribution calculations is highly important: the Kelvin equation can give no information as to the presence or absence of pores having an r_m value below about 10 to 15 Å (the exact value depending on the particular adsorbate); with cylindrical pores, or parallel-sided slits, this figure would correspond to a pore diameter, or pore width respectively of 25 to 40 Å. Isotherms exhibiting type B or type E hysteresis loops will, therefore, inevitably give a peak in the pore size distribution curve around this value and a sudden cutoff immediately below it, irrespective of the actual pore size distribution.

1. The Range of Validity of the Kelvin Equation

The Kelvin equation is based on thermodynamic arguments, and by implication its applicability is limited to macroscopic systems. The question inevitably arises therefore of how far the equation is valid for systems in which one or more dimensions are of a molecular order of magnitude, as in the micropores and the narrowest mesopores of a solid. The problem is thrown into perspective by visualization in terms of a molecular model. Figure 26 represents molecules of nitrogen (taken as spherical, rather than ellipsoidal, for simplicity) adsorbed in pores only a few molecular diameters wide. Fugure 26a represents the state of affairs in a cylindrical pore at a pressure $P/P_0 = 0.5$, where the adsorbed film has a statistical thickness close to 2.0 layers (Section II.D.1). Thermal agitation will, of course, blur the positions of the molecules, but the "meniscus", that is, the surface of an envelope touching all the molecules when in their mean positions, could not be a perfect hemisphere. In Fugure 26b, which corresponds to $P/P_0 = 0.3$, the position is even less simple. The value of t is now 5.6 Å so that the mean thickness is around 1.5 molecules; the actual thickness at any given site naturally varies with time.

It would seem clear therefore that—quite apart from the limitations arising from the tensile strength of the liquid adsorbate, already noted—the Kelvin equation can no longer apply in very narrow pores. The quatitative question as to the actual width of pore at which the equation begins to fail may be formulated in terms of the effect of curvature on the surface tension of a liquid. Some time ago Guggenheim(147) came to the conclusion that surface tension becomes independent of capillary radious only when this exceeds ~ 500 Å. A direct experimental test of the Kelvin equation—for droplets rather than menisci, however—carried out by Blackman, Lisgarten, and Skinner (148), indicates that the equation may hold reasonably well with r values as low as 100 Å. These workers measured the rate of evaporation of spherical particles of lead (liquid), bismuth (liquid), and silver (solid) as a

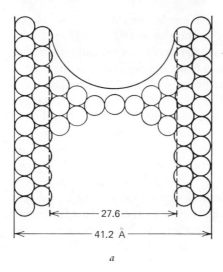

a

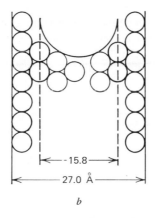

b

Fig. 26.   Model of nitrogen molecules (regarded as spheres) in narrow cylindrical pores. (*a*) At a relative pressure of 0.5. (*b*) At a relative pressure of 0.3.

function of particle radius r, the particles being photographed at intervals on the stage of an electron microscope. For all three materials the rate of evaporation $(dr/dt)_r$ was found to agree within experimental limits with that predicted from the Kelvin equation, and the values of γ calculated from the results agreed satisfactorily with those determined independently. For bismuth at 677°K, for example, the value of $\gamma = 374$ mJ/m² was obtained, as compared with the literature value of 374 to 380 mJ/m².

The problem has been taken up recently from a theoretical angle by Melrose (149), who extends the original treatment of Gibbs, paying particular attention to the position of the "dividing surface." Melrose derives the equation $\gamma = \gamma_\infty (1 - \Delta\lambda \cdot C)$ where $\Delta\lambda$ is the distance between the dividing surfaces defined in two different ways ($\Gamma = 0$ and $c = 0$, respectively) and γ_∞ is the surface tension for a plane surface of liquid; $C(=1/r_m)$ is the curvature of the interface. Assuming the interfacial region to be 4 to 6 molecular diameters thick, Melrose arrives at the result $2 \text{ Å} < \Delta\lambda < 4 \text{ Å}$.

The plot of γ/γ_∞ (for $\Delta\lambda = 3 \text{ Å}$ and $\Delta\lambda = 5 \text{ Å}$ respectively) against the radius r of a hemispherical meniscus (so that $r_m = r/2$) is given in Figure 27, from which it is seen that when r is equal to a few molecular diameters only, the value of r/r_∞ deviates appreciably from unity; when $r_m = 20 \text{ Å}$, for example, the value of γ/γ_∞ is 1.3 if $\Delta\lambda = 3 \text{ Å}$. Thus, when the Kelvin equation (43) is applied in the usual way to obtain pore sizes, by insertion of the normal value γ_∞ of surface tension, the calculated values of r may be too high by as much as 1.3-fold at the lower end of the mesopore range.

The use of partition functions to calculate the change in surface tension as a function of curvature has been described by Chang (150) and his co-workers, who studied five adsorptives, nitrogen, argon, cyclohexane, benzene, and water. A marked dependence was found, the value of γ being lower for a convex, and higher for a concave, meniscus than for a plane surface.

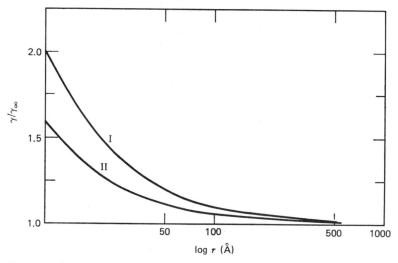

Fig. 27.   Plot of γ/γ_∞ [calculated by the equation of Melrose (143) $\gamma = \gamma_\infty (1 - \Delta\lambda/r_m)$] against $\log_{10} r_m$. (I) $\Delta\lambda = 5\text{Å}$. (II) $\Delta\lambda = 3\text{Å}$.

For nitrogen at 77°K with $r = 20$ Å, for example, the calculated value of γ was 13.55 mJ/m², as compared with the value for a plane surface of $\gamma_\infty = 9.59$ mJ/m², so that $\gamma/\gamma_\infty = 1.41$. In terms of the Melrose relationship this would imply that $\Delta\lambda = 4.1$ Å, a not unreasonable figure.

Treatments of this kind suffer from the disadvantage—as Everett and Haynes have emphasized—of resting on the tacit assumption that thermodynamic and statistical mechanical theory may be applied to systems containing a relatively small number of molecules, where the properties must exhibit wide fluctuations from their mean values; and unfortunately the thermodynamics of small systems has not yet been developed to the point where it can cope quantitatively with the behavior of matter in fine capillaries. A rather more experimental way of finding the point at which the Kelvin equation beings to fail has been described by Dubinin (151, 152).

The Kelvin equation (43) may be written in the form, say,

$$\frac{2\gamma V}{r_m} = RT \ln \frac{P^\circ}{P} = A \tag{77}$$

where $A (= -\Delta G)$ is the affinity of adsorption.

Now at a given degree of pore filling θ ($=n/n^\circ$, where n° is the adsorption when $P = P^\circ$), r_m and P/P° both have definite values. Differentiation of equation 77 gives

$$\left(\frac{\partial A}{\partial T}\right)_\theta = \frac{2}{r_m}\left(\frac{\partial(\gamma V)}{\partial T}\right)_\theta = \frac{A}{\gamma V}\cdot\left(\frac{\partial(\gamma V)}{\partial T}\right)_\theta$$

or

$$\left(\frac{\partial A}{\partial T}\right)_\theta = A\left(\frac{\partial \ln (\gamma V)}{\partial T}\right) \tag{78}$$

It happens that over short ranges of temperature $\partial \ln (\gamma V)/\partial T$ is almost constant so that $(\partial A/\partial T)_\theta$ is almost proportional to A ; a plot of $(\partial A/\partial T)_\theta$ against A should, therefore, be a straight line of slope $\partial \ln (\gamma V)/\partial T$.

Dubinin has tested this prediction by reference to the isotherms, at neighboring temperatures, of benzene on silica and on ferric oxide, taken from the literature. For the temperature range in question a linear decrease of $\ln (\gamma V)$ with temperature was confirmed, the value of $\partial \ln (\gamma V)\partial T$ being $-0.0041/K$.

In Figure 28 the derivative $(dA/dT)_\theta$ is plotted against A, the line DE being drawn with the slope $0.0041/K$ just referred to. As is seen the experimental points fit excellently on to the theoretical line down to point E which corresponds to $r_m = 15$ Å, but then deviate strongly from it, the slope actually changing sign. A closely similar plot was obtained for benzene on ferric oxide, the deviation commencing at ~ 16 Å

The deviation is so sudden and so large as to suggest that there is a com-

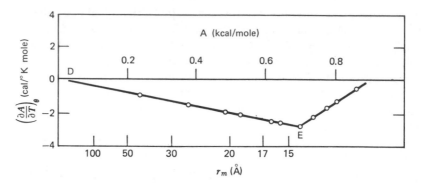

Fig. 28.   Test of the validity of the Kelvin equation for fine pores (Dubinin (151, 152)). Plot of dA/dT against A ($= RT \ln P/P°$).

plete change of mechanism at this point. Adsorption in pores finer than those corresponding to point E is no longer susceptible to treatment in terms of capillary condensation, and requires a quite different approach. This theme will be developed in the following section.

IV. ADSORPTION IN MICROPORES

A. The Force Field in Fine Pores.

Reference has been made earlier (Section II.B.2) to the overlap of the fields of force from neighboring walls in very narrow pores, which will enhance the energy of interaction of an adsorbate molecule with the surrounding solid. The possibility of an enhancement of this kind was recognized as long ago as 1932 by Polanyi (153), and shortly afterwards by de Boer and Custers (154), and it has frequently been discussed since in a semiquantitative manner by various workers (155–159). Some attempts have been made at the difficult task of making detailed calculations of the potential energy within micropores (34, 160, 161); they indicate that in pores of molecular dimensions the interaction energy is considerably increased, to an extent depending, naturally, on the ratio of the pore width to the molecular diameter. Such calculations tend, if anything, to underestimate the magnitude of the effect, since they have not been able to take into account the "transmission" of van der Waals forces by molecules already adsorbed, but even so they show that in very narrow pores the enhancement may rise by several fold.

The calculations of Anderson and Horlock (34) for argon adsorbed in slit-shaped pores of magnesium oxide provide an illustration. The maximum enhancement of potential (which included allowance for induced polariza-

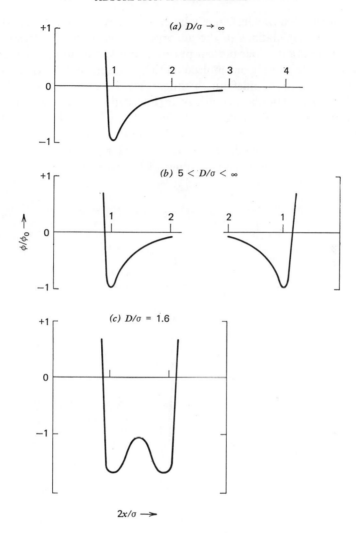

Fig. 29.   Interaction potential ϕ of a molecule of diameter σ with an adsorbent in the form of a cylinder of diameter D, having walls only one molecule thick. In (a), (b), and (c) the values of the ratio D/σ are marked on the diagram. ϕ_0 is the value of interaction potential for an open surface of the solid. (After Gurfein et al. (161).)

tion (Section I.B) occurred with a pore width of 4.4 Å, the corresponding interaction potential being -3.2 kcal/mole, as compared with values of -1.12, -1.0, and -1.07 kcal/mole for positions over a cation, an anion, and the center of a lattice cell, respectively, in an open, plane (100) surface of magnesium oxide.

The contribution of Gurfein, Dobychin, and Koplienko (161) who have made detailed calculations of the adsorption energy in cylindrical pores is particularly relevant to our present purpose. For simplicity the cylinder walls were supposed to be only one molecule thick, but the method can in principle be extended to walls of indefinite thickness. The Lennard–Jones 6–12 relation (22) was used to describe the energy of interaction of a single molecule of adsorbate with a single molecule of the wall, and then by integration the interaction energy ϕ of the adsorbate molecule with the whole of the wall was calculated in terms of the distance z of the center of the molecule from the nearest part of the wall. Different values of the cylinder diameter were taken in succession, and the results plotted as ϕ/ϕ_0 against $2z/\sigma$, where σ is the dia-

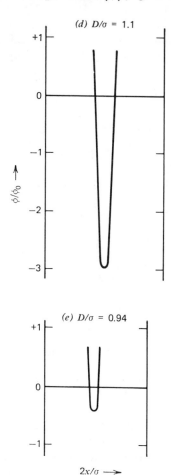

Fig. 30.   See legend of Fig. 29. In (d) and (e) the values of the ratio D/σ are marked.

meter of the (spherical) molecule and ϕ_0 the energy of interaction with a plane surface. Figures 29 and 30 indicate how the curves vary according to the relative diameter D/σ of the cylinder.

When D/σ is in excess of 1.5 there are two minima in ϕ/ϕ_0 across an axial plane, situated at a distance of $\sigma/2$ from either wall (Figure 29b and c), However, when $D/\sigma < 2$ the pore can of course accommodate only one molecule in its cross section, and in the range $1.5 < D/\sigma < 2.0$ the molecule will be located in a double minimum in the cross section, that is, in a ring minimum in the pore itself (Figure 29c). The apparent density of the adsorbate must then be lower than that of the adsorptive in bulk form, and Russian workers suggest that the increase in apparent pore volume often observed when probe molecules of a progressively increasing size are used may arise from this change in the packing density of the adsorbate in small pores, rather than from a simple molecular-sieve effect. Values of pore volume or apparent density based on molecular probe experiments must accordingly be treated with caution.

For very narrow pores where $D/\sigma < 1.5$ the two minima merge into a single minimum (the ring shrinks to a single depression) whch progressively increases in depth as D/σ diminishes to ~ 1.1 (Figure 30d); at this point the adsorption potential has been enchanced by a factor of 3.37 over that for a plane surface. (Steele and Halsey arrived at a figure of ~ 4.8, but they were

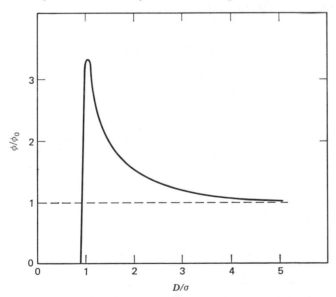

Fig. 31.   Variation of the ratio D/σ of the interaction potential ϕ of a molecule of diameter σ with a cylinder of diameter D (see legend to Fig. 29). ϕ_0 is the corresponding interaction potential for an open surface.

considering the case of a cylindrical pore in a solid matrix rather than the space inside a cylinder formed from a sheet.)

Since the repulsive energies referred to in equation 6 are not of the hard core type, the interesting possibility arises that the molecule may be present in a cylinder whose diameter is actually less than that of the molecule itself, that is, $D/\sigma < 1$. The interaction energy would thereby be diminished (cf. Figure 30e) until when $D/\sigma = 0.93$, it would have fallen to zero. The variation of relative adsorption energy ϕ/ϕ_0 with relative pore diameter is shown in Figure 31 whence it is seen that ϕ/ϕ_0 becomes unity (i.e., no enhancement) when $D/\sigma = 5$, that is, when the pore is about five molecular diameters wide.

The model is admittedly an idealized one; the walls are only one molecule thick, and by implication are composed of atoms (or ions) so small and tightly packed that the surface has no geometric or energy relief; and the adsorbate molecules are spherical. Even so, there can be little doubt as to the soundness of the general picture: The enhancement in energy of interaction within narrow pores may increase by as much as three- or fourfold, and will not disappear entirely until the pore is some five molecular diameters wide. In particular, the decisive factor is the diameter of the pore relative to the molecular diameter of the adsorbate, rather than the absolute diameter of the pore itself. Consequently the critical width of pore at which the "micropore effect" first appears will tend to be greater the diameter of the adsorbate molecule in question.

1. The Type I Isotherm

Adsorption within very narrow pores has long been associated with interpretation of the type I isotherm. The Langmuir mechanism when applied to a single layer of molecules gives rise to an isotherm of type I shape; consequently, it was argued, a type I isotherm must result if the pores of the solid are so narrow that they can accommodate only a single molecular layer of adsorbate on their walls. The plateau of the isotherm would then correspond to the completion of the monolayer, so that the uptake at a relative pressure of unity would be equal to the monolayer capacity, whence the specific surface could be immediately obtained (II.C.1). When applied to typical experimental isotherms, usually with active carbons as the adsorbent, this interpretation led to very large values of specific surface, often in the region of 2000 or 3000 m²/g. These figures are so high that they would require a very large proportion of the carbon atoms—around nine-tenths if $S = 3000$ m²/g—to be accessible to the gas molecules, that is, to be "surface" atoms. The tenuous structure thus implied is very difficult to reconcile with the mechanical properties of the solid. This, and other considerations, led to the alternative view proposed by Pierce, Wiley, and Smith (162), and independently by Dubinin (163), according to which the plateau of the isotherm represents not the completion of a monolayer but the filling up of the volume of the pores with adsorbate (164) in a liquidlike condition.

This alternative view is supported by a variety of experimental evidence. Thus, in the majority of cases, those systems which yield type I isotherms are found to conform to the Gurvitsch rule (165), the uptake at saturation $(P/P^\circ = 1)$ of different adsorptives on the same adsorbent being nearly equal ($\pm 5\%$) if expressed as volume of liquid. (Exact equality is not to be expected owing to steric effects, inevitable in very fine pores.)

Further support comes from the result obtained with samples of Saran char which had been progressively "burnt off" in oxygen (166); the isotherms were all of type I, whether the burn-off was 0, 33, or 70% by mass, the uptake at saturation increasing progressively as the degree of burn-off increased. In terms of the classical "surface coverage" hypothesis this would mean that burning off consistently produced channels of width between 2σ and 3σ no matter how far they extended (σ = molecular diameter). In fact, the value of S calculated by equation 14 increased from 1300 to 2450 m^2/g as burn-off increased from 0 to 70% and this would imply that the burning off extended the length and breadth of the channels without increasing their cross section. While a preferential oxidation of this kind is not impossible—it could stem from some special structural property of the char—it is more probable that the channels were widened as well as lengthened. The idea of widening is supported by the fact that the knee of the isotherm became more rounded, implying that the pore size distribution changed in favor of wider pores, as burn-off increases.

Evidence of a different kind is afforded by the results of Avery and Ramsay (88) already referred to, in which powders of silica and zirconia were compacted at a series of pressures and the isotherms of nitrogen at 77°K measured on the compacts. On the uncompacted powders the isotherms were of type II with no hysteresis, but on the compacts they first assumed a type IV character, the hysteresis loop moving progressively towards lower relative pressures as the compacting pressure increased; then at the highest compacting pressure the loop disappeared and the isotherm became type I in shape, further increase in pressure producing a lowering of the whole isotherm (Figure 14). The change in the type IV isotherms is readily explained in terms of a continuous reduction in average pore size and total pore volume as compacting pressure increased: the transition to the type I isotherm and the lowering of the isotherm on further compaction then fall naturally into place as a continuation of the same process. The older interpretation of the type I isotherm could not provide a plausible explanation of these changes.

B. The Evaluation of Microporosity

In the majority of cases, microporosity is associated with either mesoporosity or an external surface, or both. The effect of microporosity on the isotherm (of nitrogen, say,) is readily appreciated by consideration of Figure 32. In Figure 32a curve (i) is the isotherm for a solid which is exclusively microporous and curve (ii) that for a powder made up of nonporous particles. If the particles of powder are microporous (the total pore volume being given by the plateau of curve (i), the isotherm will assume the form of curve (iii), which represents a summation of curves (i) and (ii). As is seen, the composite isotherm (iii) is also of type II; it exhibits a steep initial portion arising from the contribution of the type I component, which will manifest itself in an increased value of the BET constant c as compared with that for a

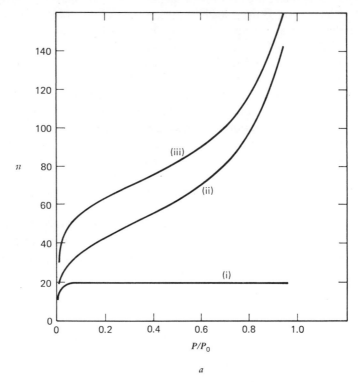

a

Fig. 32. (*a*) Adsorption isotherm for (i) a solid that is entirely microporous; (ii) a powder made up of nonporous particles; (iii) a powder with the same external surface as in (ii) but made up of microporous particles having a total pore volume given by the plateau of (i). n is the amount adsorbed in arbitrary units. The solid in (ii) and (iii) is assumed to have standard properties. Next page: (*b*) The t-plots corresponding to isotherms (ii) and (iii) respectively of Fig. 32a.

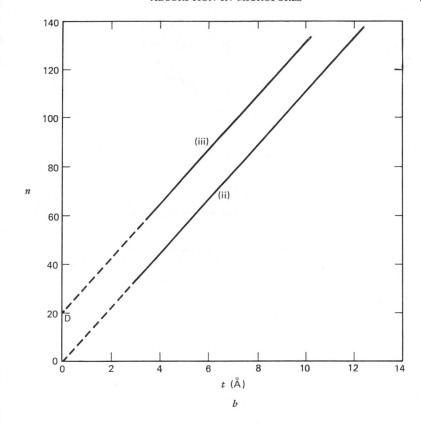

b

nonporous powder. In the multilayer region the composite isotherm (iii) is parallel to the isotherm (ii) for the nonporous solid.

Figure 33*a* refers to the case where the nonmicroporous adsorbent is composed of grains of mesoporous solid rather than fine particles of impervious powder; the isotherm on the purely mesoporous solid (curve (ii)) is now of type IV, and when micropores are also present, the isotherm (curve (iii)) will still be of type IV—again with an enhanced value of c.

It can be seen, therefore, that while an impervious powder will give rise to a type II isotherm, the converse is not true: The existence of a type II isotherm does not necessarily denote that the solid is free of micropores. Correspondingly, while a type IV isotherm signifies that the solid is mesoporous, it cannot be taken as evidence for the absence of microporosity. Clearly there is need for some way of analyzing a type II (or a type IV) isotherm which will reveal the presence of a micropore contribution and enable the micropore volume to be estimated. Various approaches have been suggested: (1) by the use of the t- or α_s-plot; (2) by filling the micropores with a pre-adsorbed

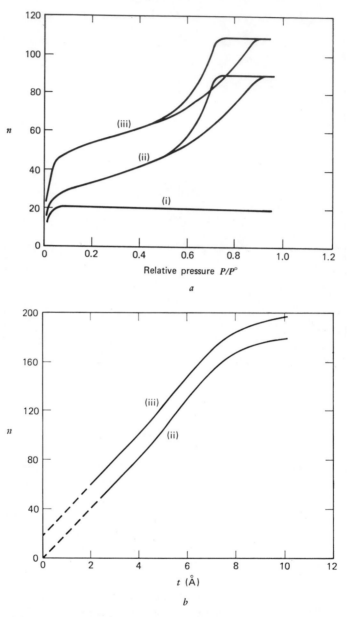

Fig. 33. (*a*) Adsorption isotherms for (i) a solid that is entirely microporous; (ii) a solid made up of mesoporous grains; (iii) a granular solid with the same mesoporous system as in (ii) but containing also micropores with a total volume given by the plateau of (i). *n* is the amount adsorbed in arbitrary units. (*b*) *t*-plots corresponding to isotherms (ii) and (iii).

312

material and thus in effect subtracting the contribution of isotherm (i) from the composite isotherm (iii) of Figure 32a or 33a; (3) by application of the Dubinin–Radushkevich plot.

1. The t- or α_s-Plot.

A method of assessing micropore volume using the t-plot of Lippens and de Boer was proposed by Sing (167) in 1967. If the surface of the solid has standard properties, the t-plot for the nonporous solid corresponding to isotherm (ii) of Figure 32a will be a straight line passing through the origin, with slope proportional to the specific surface of the solid (curve (ii) of Figure 32b). For the microporous sample that gives isotherm (iii) of Figure 32a, the t-plot will have the form of curve (iii) of Figure 32b; the linear branch of (iii) will be parallel to (ii), since it corresponds to the area of the outside of the particles, which is the same as that of the original powder. The vertical separation of plots (ii) and (iii), which is equal to the intercept OD, gives the micropore contribution to the isotherm which can be converted into the micropore volume by use of the liquid density. Because of the inevitable presence of steric effects, the value so obtained is subject to uncertainty, but with an adsorbate of small molecular size, the difference from the true volume of the micropores should in general not exceed a few percent. Figure 33b gives the t-plots corresponding to the mesoporous solid referred to in Figure 33a.

A direct test of the method is reported by Mieville (168), who measured the nitrogen isotherm on a number of mixtures of a mesoporous silica gel and a microporous zeolite, as well on as the separate components; the isotherm of the mixtures had the shape to be expected from those of the components, and the estimates of the micropore volumes from the various t-plot agreed well with the value calculated from the plateau of the type I isotherm of the microporous zeolite.

The entirely analogous use of the α_s-plot for micropore evaluation has been described by Sing (169).

It is perhaps hardly necessary to stress that both methods depend on the existence of a standard isotherm for the particular adsorbate and adsorbent in question, in order that the standard t-curve or α_s-curve can be constructed. This reinforces the desirability of building up a library of standard isotherms for the common adsorbent materials and the most frequently used adsorbates (170).

2. Pre-adsorption

The device of filling the finer pores in a solid with large molecules to isolate their effect on the adsorption isotherm has been adopted by Kiselev and his coworkers (171). The solid was soaked in a solution of polymer and the isotherm of nitrogen determined after the polymer-laden adsorbent had been dried and outgassed. With molecular sieve crystals, Barrer (172) was able to use water vapor as the pre-adsorbate since, once adsorbed, the water vapor escapes only very slowly on outgassing unless the temperature is elevated; it was therefore possible to determine the (small) external area of the crystals by nitrogen or krypton adsorption.

Actually, as early as 1949 Juhola and Wiig (173) had used the pre-adsorption of water vapor to study the pore structure, but since at that time the monolayer interpretation of the type I isotherm still held way, the significance of the results in terms of present day views on microporosity was not then apparent.

In 1969 Gregg and Langford (85) chose n-nonane as pre-adsorbate, since this substance was known from earlier work (131) to escape very slowly from micropores on pumping at room temperature. The method was applied to an adsorbent, Mogul carbon black, whose particles had been rendered microporous by exposure to oxygen at 500°C. This material had a well-defined external surface area, which could be evaluated by electron microscopy. The isotherm of nitrogen was determined at 77°K both before and after the micropores had been filled with n-nonane by adsorption from the vapor phase, and also after the micropores had been partially emptied by pumping at elevated temperatures (cf. Table 13). The *modus operandi* was to expose the outgassed solid at 77°K to the vapor of n-nonane, allow its temperature to rise to ambient, and then open up to the pumps. The nonane was thereby

TABLE 13

Adsorption of nitrogen at 77°K on a microporous carbon after pre-adsorption of n-nonane.[a]

	Isotherm				
	A	B	C	D	E
n_a (mg/g)	63	48	29	16	0
T (°K)	293	408	453	497	723
n_m (mole/g)	1.16	1.27	1.74	2.48	3.67
$S(N_2)$ (m²/g)	114	124	170	243	360
c(BET)	59	176	410	1200	1940

[a] Gregg and Langford (85).

n_a = amount of pre-adsorbed nonane;

T = temperature of outgassing of nonane-charged sample.

n_m = monolayer capacity calculated from BET plot (equation (18)).

$S(N_2)$ = specific surface calculated from n_m with $a_m(N_2)$ = 16.2 Å².

The specific surface calculated from particle size determined by electron microscopy was 110 m²/g.

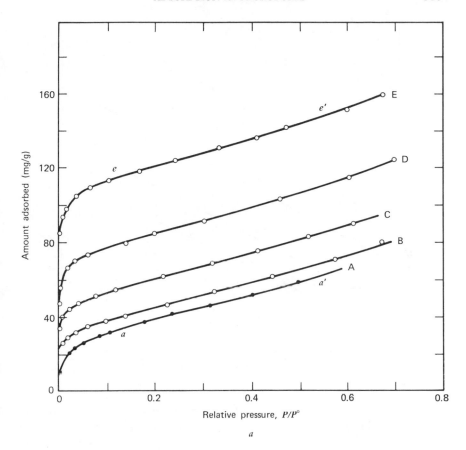

Fig. 34. (*a*) Adsorption isotherms at 77°K of nitrogen on a sample of Mogul I carbon black charged with different amounts n_a of pre-adsorbed nonane (85). Some points are omitted at low pressures for the sake of clarity. Values of n_a (mg/g: A, 63; B, 48; C, 29; D, 16; E, 0. (See Table 13.) Next page: (*b*) The α_s-plots corresponding to the isotherms B, C, D and E respectively of Fig. 34*a*. The isotherm of the fully charged sample (isotherm A) was taken as internal standard for construction of the α_s-curve.

removed from the external surface of the particles but not from the pores. The isotherms are shown in Figure 34*a* where curve A refers to the sample thus prepared, and curves B, C, and D, to the corresponding isotherms determined after removal of the nonane in stages by outgassing at progressively higher temperatures. Isotherm E was obtained with the fully outgassed sample. As is seen, the multilayer branch ee' of isotherm E is parallel to the corresponding portion aa' of isotherm A, and this is just as expected, since both aa' and ee' represent multilayer formation on the same external surface.

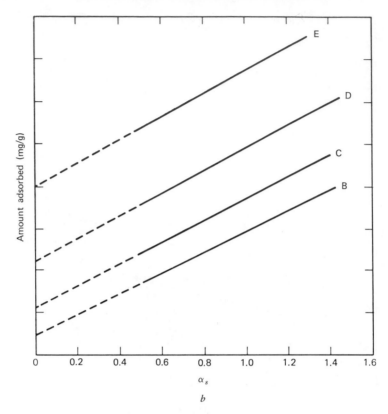

b

The vertical separation of aa' and ee' is a measure of the contribution of the micropores to the total adsorption, and the micropore volume itself can be obtained by expressing this separation as a volume of liquid.

Curves B, C, and D are also parallel to curves A and E in the multilayer region, and this, likewise, is only to be expected since the external surface is the same throughout, the differences between the isotherms merely reflecting the varying proportions of the micropore volume that have been emptied of nomane.

It is particularly instructive to examine the values of the apparent monolayer capacity n_m and the BET constant c calculated from the isotherms by the standard BET procedure. These appear in Table 13 along with the respective values of apparent specific surface calculated by use of $a_m = 16.2A^2$ for nitrogen.

Reference to Table 13 shows that the BET specific surface calculated from isotherm A, which corresponds to adsorption on the external surface only, agrees well with the value of the area calculated from the particle size as assessed by electron microscopy. The value of c is reasonable for an oxygenated surface of carbon; graphitized carbon black having 80% coverage with pre-adsorbed methanol (174), for example, gives $c = 40$. As more and more of the micropore volume is made available to the nitrogen, the value of c increases markedly, until in the case of the isotherm E, which corresponds to full availability of the micropores, it reaches the very high figure of $c = 1940$. This increase in c is of course a reflection of the fact that the heat of adsorption in micropores would be higher than on an open surface. The table demonstrates in a striking manner how the unsuspected presence of micropores could lead to an erroneous value of specific surface calculated from the isotherm by application of the usual BET procedure; the micropores in the present instance account for 246 ($= 360 - 114$) m²/g of the nominal surface area, but since the distribution of width within the micropores is unknown one cannot say what is the true area of the walls.

Figure 34b shows the α_s-plots corresponding to the isotherms of Figure 34a. In the absence of an agreed standard isotherm for carbon blacks, isotherm A has been taken as an internal standard for comparative purposes, as it corresponds to adsorption on the external surface. As is seen, the plots are (except near the origin) straight lines parallel to one another, the plot for isotherm A passing through the origin. Though these features follow necessarily from the choice of standard combined with the parallelism of the multilayer regions of the isotherm, they nevertheless provide a clear illustration of the usefulness of the α_s-plot in relation to micropore detection and evaluation.

As already emphasised in Section II, the α_s- or t-plot may be regarded as a means of assessing the degree of upward distortion of the isotherm from its standard course in the low pressure region; in principle such distortion could arise not only from microporosity but also from chemisorption or from the presence of active centers (impurities or other defects) on the surface where preferential adsorption could occur. Since, however, nonane would not be chemisorbed nor retained on active centers after outgassing at room temperature, any reduction in nitrogen uptake found after n-nonane has been pre-adsorbed, must be due to microporosity rather than to the other two causes.

It would clearly be of interest to discover how far the nonane pre-adsorption method can be used with adsorbates other than nitrogen.

A study along these lines with two organic adsorptives has been carried out by Tayyab (175), but a discussion of his rather unexpected results is best deferred until the role of fine constrictions has been considered (Section IV. D.1). Meanwhile it may be noted that the applicability of the method seems to be restricted to adsorptives such as nitrogen or argon which have negligible solubility in n-nonane.

3. The Dubinin–Radushkevich (DR) Plot.

An expression for the calculation of the micropore volume from the low pressure part of the isotherm has been put forward by Dubinin and Radushkevich (176, 177). It is based on the Polanyi theory of adsorption (178), in which the adsorption data are expressed in the form of a temperature invariant characteristic curve:

$$W = nV = f(\Delta G)$$

Here W is the volume of pores filled with adsorbate (assumed to have the density of the bulk liquid) when the relative pressure is $P/P°$, the corresponding molar free energy of adsorption being $\Delta G \ (= RT \ln (P/P°))$. By use of the assumption that W is a continuous function of ΔG (a gaussion distribution is proposed*) an expression containing the total pore volume W_0 is arrived at:

$$\log_{10} n = \log_{10}\left(\frac{W_0}{V}\right) - D\left(\log_{10}\frac{P}{P°}\right)^2 \tag{79}$$

Here D is a constant at constant temperature, characteristic of the particular system.

A plot of $\log_{10} n$ against $(\log_{10} (P/P°))^2$ should, therefore, be a straight line with intercept equal to $\log_{10} (W_0/V)$, which would lead immediately to the micropore volume W_0 itself. In principle, therefore, the DR equation

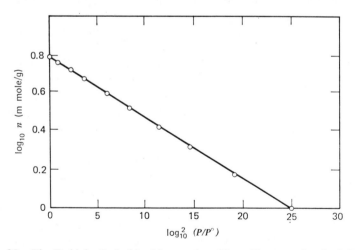

Fig. 35. The Dubinin–Radushkevich equation. Plot of log n against log^2 ($P/P°$) for the adsorption of benzene vapour at 293 °K on an activated carbon made from sucrose (180).

---

*Marsh and his co-workers (179) argue in favor of a Rayleigh distribution.

should provide a means for calculation of the micropore volume from adsorption measurements in the low pressure region of the isotherm—the part below the plateau of a type I isotherm.

With a substantial number of adsorption systems the DR plot is indeed a good straight line (180, 182) extending over a wide range of relative pressures; Figure 35, in which the plot is linear over a range of relative pressures from 10^{-5} to ~ 1, is not untypical. Frequently, however, the plot is found to deviate from linearity (181), in one of several ways: it may turn upwards at the high pressure end (Figure 36a); it may fall into two straight lines, the line at higher pressures being of lesser slope (Figure 36b); it may be a curve convex (Figure 36c), or concave (Figure 36d), to the $\log^2 (P/P°)$ axis.

The upward turn in Figure 36a could perhaps be explained in terms of capillary condensation in mesopores or multilayer adsorption on the external surface, and the value of W_0 calculated from the intercept obtained by extrapolation of the long linear portion would not then be obviously incorrect; but in Figure 36b extrapolation from low pressure measurements would lead to a value higher than the actual amount adsorbed at saturation; and the same would be true of Figure 36d, if one attempted extrapolation from a short, apparently straight, part of the curve for low pressure (high $\log^2$

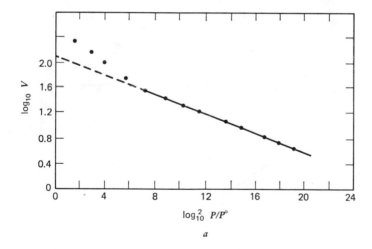

a

Fig. 36.   Dubinin–Radushkevich Plots. (a) Ethyl chloride adsorbed at 273°K on a poly-vinylidene carbon prepared at 1123°K. Page 320: (b) Carbon dioxide adsorbed at 293°K on Linde molecular sieves. $\bigcirc$, powder 5A; $\bullet$, powder 4A. (c) Carbon dioxide adsorbed at 195°K on activated (71.5% burn-off) carbon prepared from polyfurfuryl alcohol at 1123°K. Page 321: (d) Sulfur dioxide adsorbed at 273°K on an activated sugar charcoal (219). In Fig. 36a, b, c, the amount adsorbed (V) is expressed in cm³ (STP). These diagrams are reduced from the originals of Lamond and Marsh (181, 218).

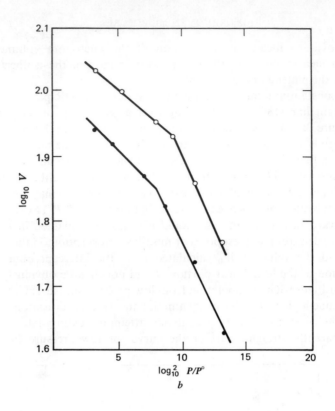

b

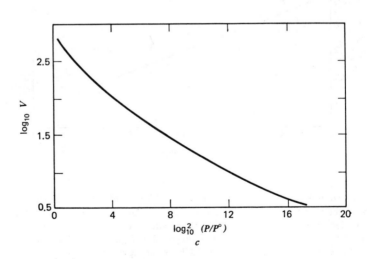

c

320

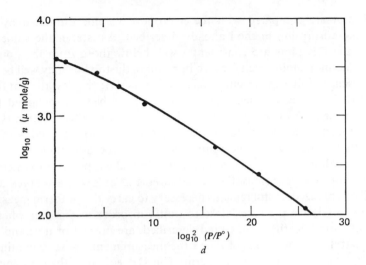

$(P/P°)$; conversely, for Figure 36c the extrapolation value of W_0 would be lower than the saturation adsorption.

An attempt to compare the extrapolated value of W_0 with an independent estimate of micropore volume has been made by Marsh and Rand (181), who measured isotherms of carbon dioxide at 273°K on two series of progressively activated carbons prepared from polyfurfuryl alcohol and polyvinylidene chloride, respectively. In Table 14 the values of W_0 obtained from

TABLE 14

Micropore volume W_0 in activated carbons (181)[a]

Carbon	Burn-off (%)	Micropore volume (cm³/g)	
		DR plot (carbon dioxide at 273°K)	Nitrogen Displaced by Nonane
Polyfurfuryl alcohol	0	0.15	0.0
	21	0.25	0.28
	51	0.31	0.42
	71	0.39	0.58
Polyvinylidene chloride	0	0.27	0.38
	21	0.40	0.94
	37	0.54	0.64
	56	0.54	0.72
	82	0.27	0.81

[a] Density of adsorbed phase/ g cm⁻³ : nonane, 0.72; nitrogen, 0.81; carbon dioxide, 1.10.

the DR plot are compared with the micropore volume estimated by the nonane pre-adsorption method already described. As is seen the values of W_0 from the DR plots are consistently well below those from the nonane method. The anomalous result for zero burn-off in the top line may well be due to the presence of fine constrictions (cf. Section IV.D); the deviation at 82% burn-off (last line) could mean that the pores have been so enlarged that they no longer act as micropores towards carbon dioxide (cf. Section II.A).

Taken together the results still leave the position as to the validity of the extrapolated values rather uncertain, and it must be admitted, therefore, that the early hope that the DR equation would make it possible to calculate the micropore volume from isotherm measurements at low relative pressures, has not been realized. It still remains necessary to carry the isotherm measurements up to relative pressures close to unity, in order to discover whether deviations from linearity, of the kind described, are present or not; and this implies that if the isotherm is of type I the measurements must be continued into the plateau region of the isotherm. The DR equation then becomes a convenient, but in practice empirical, means of estimating the saturation value of adsorption; it is perhaps more satisfying mathematically than finding the "Gurvitsch volume" by mere inspection of the isotherm, but in practice the results by the two methods cannot be very different. If the isotherm is of type II or type IV the DR method is no longer applicable, for it is not possible to estimate the relative contributions made by micropore filling and surface coverage to the measured adsorption in the low pressure region, that is, the portion lying below the "knee" of the isotherm (Section II.B).

4. The Dubinin–Radushkevich and Langmuir Plots Compared (183)

The validity of the value W_0 of pore volume obtained from the Dubinin–Radushkevich plot is sometimes tested by comparison with a supposedly independent estimate from a Langmuir plot. It is, therefore, of interest to examine the mathematical relationship between the two plots to find how far the resultant values of micropore volume are truly independent.

The Langmuir equation may be regarded as a special case of the BET equation for adsorption limited to a monolayer, and it may be written as

$$\frac{n}{n_m} = \frac{c(P/P^\circ)}{1 + c(P/P^\circ)} \tag{80}$$

Here n_m is the asymptotic value of n, but in practice (since the plateau of a type I isotherm will be nearly horizontal) n_m will be almost equal to the uptake n_s at saturation ($P/P^\circ = 1$) which in the absence of appreciable external surface area, is taken as equivalent to the micropore volume.

If then one puts $n/n_m = \theta \simeq n/n_s$, the Langmuir equation (80) may be written as

$$\theta = \frac{c(P/P^\circ)}{1 + c(P/P^\circ)} \tag{81}$$

and the DR equation (79) as

$$\log_{10} \theta = -D(\log_{10}(P/P^\circ))^2 \tag{82}$$

With equation 81 as basis, a series of pairs of values of P/P° and are calculated, with parameter c taking in succession the values 5, 10, 18, 50, and 100; insertion of these pairs into equation 82 gives the DR plots of Figure 37. When $c = 18$ the plot is seen to be linear (with slope $D = 0.18$) over a long range and it passes through the origin; extrapolation leads to the same value of θ, viz. $\theta = 1$ at $P/P^\circ = 1$, as that demanded by the Langmuir equation. For all other values of c however, the DR plot is not straight, being convex to the $\log^2 (P/P^\circ)$ axis for $c < 18$ and concave for $c > 18$, the deviation from linearity increasing as c diverges more and more widely from 18.

From the point of view of evaluating micropore volume, both the Langmuir and the DR equations must be regarded as empirical extrapolation

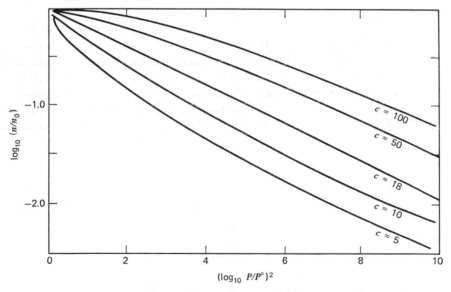

Fig. 37. Relationship between the Dubinin–Radushkevich and the Langmuir–BET equations. The value of n/n_s ($=\theta$) was calculated from the Langmuir–BET equation (81) for a series of values of c. The corresponding DR plots were then constructed with the aid of the DR equation (82). The values of c are marked on the respective curves. The line for $c=18$ is almost but not quite straight up to a point corresponding to $\log_{10}^2 (P/P^\circ) = 16$, or $P/P^\circ = 4$.

formulas. The analysis just given demonstrates that the same set of experimental data will be capable of representation—though not necessarily with a high, or the same, degree of accuracy—by either of the two equations, and that the extrapolated values of micropore volume will not be widely different. Consequently, values of micropore volume resulting from the use of the one equation cannot be used to "prove "the validity of the other.

C. The Heat of Adsorption in Micropores

From Section I it follows that the heat of adsorption in micropores should be higher than on an open surface. Striking confirmation of this view is to be found in the results obtained—notably by Barrer (32) and by Kiselev (184, 185)—in work on zeolites, bodies which contain cavities and windows of known dimensions lying well within the micropore range of size, and determined by the structure of the aluminosilicate framework and the disposition of the exchangeable cations. The heat of adsorption of vapors, even where purely physical adsorption is involved, is found to be much higher on typical zeolites than on, say, nonporous silcas (Table 15), especially with polar vapors such as nitrogen or benzene where coulombic interaction with the cation of the solid is to be anticipated.

With non-zeolitic adsorbents it is difficult to obtain evidence of microporosity that is truly independent of adsorption data. One must perforce remain content with establishing a correlation between a high heat of adsorption on the one hand, and features of the isotherm—such as a steep initial rise, or the form of the α_s-plot—indicative of microporosity, on the other. Several examples are to hand. Thus Sing and his co-workers (186) have calculated the isosteric heat of adsorption of nitrogen on four different silicas for which the type of porosity had been inferred from the shape of the iso-

TABLE 15

Heat of adsorption q_{st} of various vapors on silica and on a zeolite.[a]

Adsorbate	q_{st} Dehydroxylated Silica (kcal/mole)	q_{st} Hydroxylated Silica (kcal/mole)	q_{st} Cationated Zeolite NaX (kcal/mole)
Argon	2.1	2.1	3.1
Nitrogen	2.2	2.8	5.2
Ethane	4.2	4.4	6.2
Ethylene	3.8	5.2	9.2
n-hexane	8.8	8.8	14.7
Benzene	8.6	10.2	18.0

[a]After Kiselev (185).

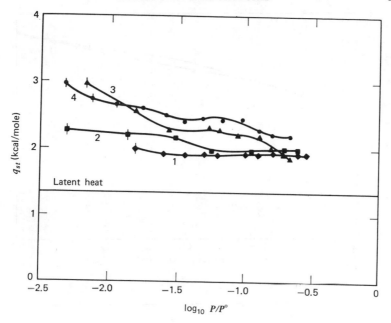

Fig. 38. Isosteric heat of adsorption q_{st} of nitrogen adsorbed at 77°K on nonporous and porous silicas (186). 1, ◆, "Fransil," nonporous; 2, ■, gel A, mesoporous; 3, ▲, gel B, mesoproous and microporous; 4, ● gel D, microporous.

therms and the form of the α_s-plots. Gels A and B (type IV isotherm in each case) were mesoporous, gel B having microporosity in addition; gel D (type I isotherm) was wholly microporous, while Fransil gave a type II isotherm very close to that of a standard nonporous silica. The results, plotted for more effective comparison against $P/P°$ rather than against n, are shown in Figure 38; and though the experimental accuracy was inevitably somewhat limited especially for the rather low-area Fransil, the general pattern is quite clear, the presence of microporosity in gel D and to a lesser extent in gel A, being reflected in significantly higher heats of adsorption.

A similar pattern emerged with carbon tetrachloride adsorbed on the same four adsorbents (81).

The carbon black used by Diano (187) in his experiments with n-hexane, was very similar to the one referred to in Section B, in which the presence of microporosity had been demonstrated by the nonane pre-adsorption method. The hexane isotherm (Figure 39a) displayed the expected steep rise AB from the origin characteristic of microporosity, followed by the normal multilayer region CD; the heat of adsorption (calculated from measurements of the heat of immersion in liquid hexane) was markedly higher at adsorptions corresponding to the micropore region AB of the isotherm (Figure 39b). At the

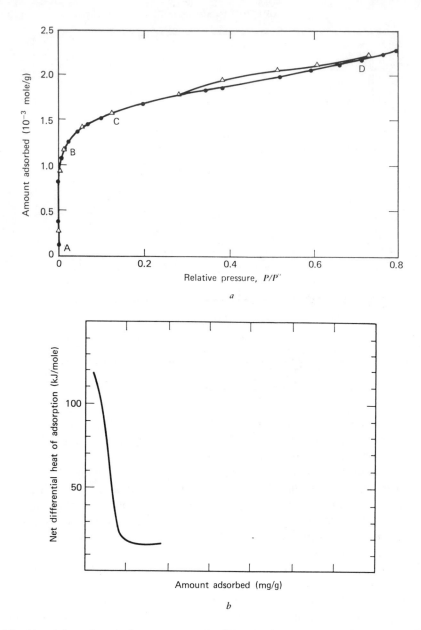

Fig. 39.   Adsorption of *n*-hexane vapor at 298 °K on a microporous sample of carbon black 86, similar to that referred to in Fig. 34. (*a*) The adsorption isotherm. (*b*) The net differential heat of adsorption (calculated from the heat of immersion in liquid hexane, of samples of the carbon black containing different amounts of adsorbed hexane) as a function of the amount adsorbed.

326

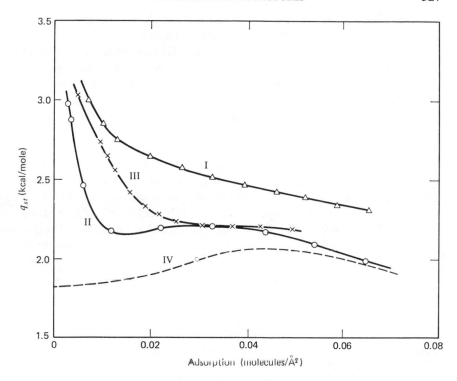

Fig. 40. The isosteric heat of adsorption q_{st} of argon on various samples of MgO plotted against coverage. (I) Mg(OH)$_2$ decomposed in vacuo at 300°C for more than 10 hr. (II) As in (I) + 1hr *in vacuo* at 1050°C. (III) As in (I) + 20 hr *in vacuo* at 1050°C. (IV) Theoretical curve for adsorption on plane (100) surface. (Redrawn from the original diagram of Anderson and Horlock (34).)

lowest uptake a value of 36 kcal/mole was obtained, some 3.5 times higher than the heat of adsorption on an open surface.

Anderson and Horlock's results (34), summarized in Figure 40, arose in the course of an investigation on active magnesium oxide (prepared by thermal decomposition of the hydroxide) which the *t*-plot showed to be almost wholly microporous. The isosteric heat of adsorption of argon, calculated from the isotherms at 77 and 90°K (curve I), was much higher than the theoretical value, calculated as in Section I, for adsorption on a freely exposed 100 plane of magnesium oxide (curve IV); the alternative possibility that the enhancement might be due to active sites on the surface was excluded by consideration of the effect of further heating at 1050°K for 1 hr (curve II) and 20 hr (curve III), respectively: Curve III lies *above* curve II, whereas the opposite behavior would be expected with active spots, since these would be progressively annealed out on prolonged heating. The higher position of curve III is plausibly accounted for by the formation of new pores, possibly by elimination of the elements of water.

In his pioneer work on the subject Kiselev (83) studied the adsorption of benzene vapor on a number of silica gels of various pore widths, plotting his isotherms on a unit area basis. For a specimen stated to have a pore width of 22 Å the isotherm was distorted towards higher adsorptions as compared with the isotherm for coarser-pored specimens (75 Å, 102 Å) and the heat of adsorption was correspondingly higher for this fine-pored specimen.

It should be emphasized (as already indicated in Section B.2), that while the presence of microporosity will lead to a high heat of adsorption, the converse is not true; a high heat of adsorption may arise not only from micropore adsorption, but also from chemisorption or from the presence of active centers produced by impurities or other defects. Furthermore heterogeneity of the surface originating in the exposure of different lattice planes may lead to a curve for heat of adsorption against amount adsorbed similar in shape to that associated with microporosity.

D. Constrictions in Micropores

As has already been stressed, adsorption is an exothermic process and from ordinary thermodynamic principles the amount adsorbed at a given relative pressure must, therefore, diminish as temperature increases, once equilibrium has been reached. In practice however the contrary is often found, in that the isotherm for T_2 lies above that for T_1 $(T_2 > T_1)$. Such behavior, characteristic of a system that is not in equilibrium, represents the combined effects of temperature on the rate of approach to equilibrium and on the position of equilibrium itself.

In the explanation proposed by Maggs (188) and independently by Zwietering and van Krevelin (189), diffusion of adsorbate molecules occurs through very narrow constrictions in micropores into cavities beyond. When the width of a constriction is very close to the molecular diameter of the adsorbate, the molecules will encounter an energy barrier to their progress through a constriction, so that entry into the cavity will be a rate process with a positive temperature coefficient; thus the number of molecules actually reaching the cavity during the period of a measurement—and therefore the amount adsorbed—will *increase* with increasing temperature. At sufficiently high temperatures, however, the rate will become fast enough for equilibrium to be set up within the time of the measurement, and the measured uptake will thence forward diminish with rising temperature in the usual way.

In the experiments of Figure 41 the adsorbate was butane (190), the adsorbent being a carbon rendered microporous by partial "burn-off" in oxygen. The increase in uptake with increase in temperature is very marked,

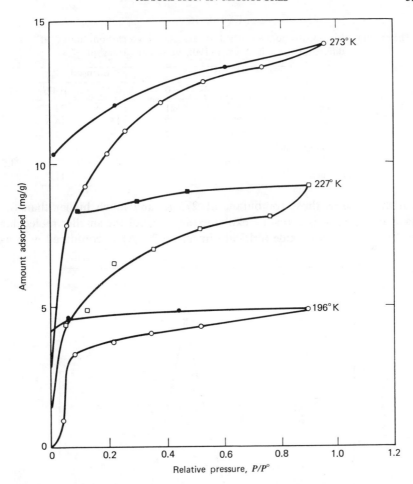

Fig. 41.   Adsorption isotherms of butane vapor at different temperatures (190) on a sample of carbon (prepared by heating a mixture of coke and pitch at 600 °C) burnt-off by 0.27 %.

and the extensive hysteresis provides supporting evidence for the presence of an activated process; in accordance with this view the slope of the desorption curve increases as temperature rises.

In Table 16, the results for other degrees of burn-off given, the adsorbates being carbon dioxide and nitrogen as well as butane. For butane, the amount v_s taken up at saturation (calculated as liquid volume) consistently increases with increase in temperature from 196 to 273°K; the proportionate increase is somewhat different for the three samples of carbon, pointing to a distribution of constriction size. The values of v_s for carbon dioxide at 196°K are

TABLE 16

The amount of various vapors adsorbed at saturation on a carbon "burnt-off" to different extents in oxygen (190). Values of v_s(mm³/g)

Solid	Vapor	Temp (°K)	Burn-off (%)			
			0	0.27	0.42	2.77
C(600)	Butane	273	21	24	24	17
		227	8	15	14.5	–
		196	2.4	8.5	7.2	–
	Carbon dioxide					
	dioxide	196	21	22	22.5	24.5
	Nitrogen	78	5.5	–	11.7	22.2

about the same as those for butane at 273°K, and much higher than for butane at 196°K; this is readily explained in terms of the smaller molecular size (191) of carbon dioxide (critical dimension 2.8 Å) as compared with *n*-

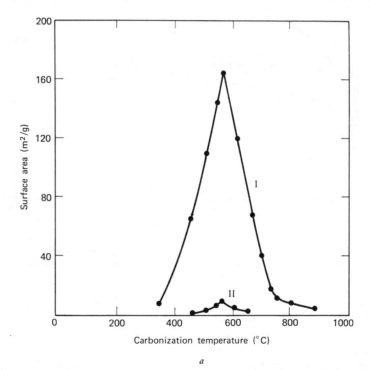

a

Fig. 42. (*a*) Apparent surface area of a carbon prepared by pyrolysis of dibenzanthrone, as calculated from adsorption isotherms of (I) carbon dioxide at 195°K, (II) nitrogen at 77°K. (Reduced from the original diagram of Marsh and Wynne-Jones (192). Page 331: (*b*) Apparent surface area of a carbon prepared by pyrolysis of polyvinylcyanide, as calculated from adsorption isotherms of (I') carbon dioxide at 195°K, (II') nitrogen at 77°K. (Reduced from the original diagram of Marsh and Wynne-Jones (192).)

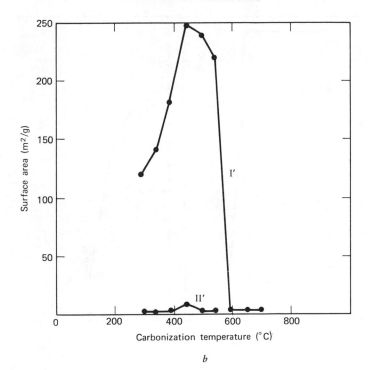

Surface area (m²/g)

Carbonization temperature (°C)

b

butane (critical dimension 4.9 Å). Nitrogen (critical dimension 3.0 Å) at 78°K also gives higher values than does butane at 196°K, a result explicable in a similar way. The concordance between the v_s values for carbon dioxide at 196°K and nitrogen at 77°K for the 2.77% burn-off sample, suggests that in this sample all constrictions finer than 2.8 Å have been burnt away.

Another example is provided by the results of Wynne-Jones and Marsh (192), obtained with carbon prepared from dibenzanthrone and from poly-vinylcyanide by pyrolysis at various temperatures (Figure 42*a* and *b*). Though quoted as "surface areas" the ordinates would be nearly proportional to the saturation uptakes, since the isotherms were of type I. The striking difference between the curves for carbon dioxide at 196°K (curves I and I′) and those for nitrogen at 78°K (curves II and II′) suggests that virtually the whole of the pore system is reached through constrictions whose critical dimension is between that for carbon dioxide and that for nitrogen.

Further information as to the size of the constrictions within micropores can be obtained from measurements of the heat of immersion of the solid in a range of liquids differing in molecular size. The method is well illustrated by the study of Barton, Beswick, and Harrison (193) who worked with microporous carbons prepared from polyvinylidene chloride (Saran and polymer A) in two ways, viz. by pyrolysis, and by prolonged treatment with

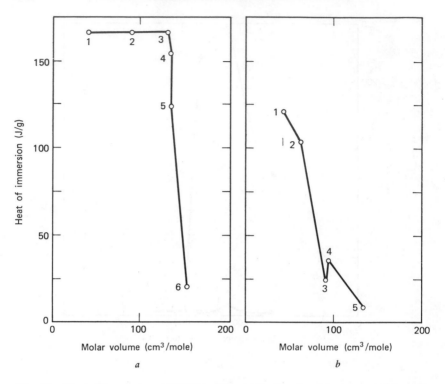

Fig. 43.   Heat of immersion at 300 °K in different liquids of a carbon prepared from Saran Polymer A, by (a) pyrolysis (Fig. 43a) and (b) treatment with KOH for 12 hr. The liquids for points 1–6 were: (1) methanol; (2) benzene; (3) n-hexane; (4) 3-methyl benzene; (5) 2–2 dimethyl-butane; (6) 2-2-4 trimethylpentane. The abscissae represent the molar volumes of the liquids. (Redrawn from the original diagram of Barton, Beswick and Harrison (193).

potassium hydroxide at room temperature. Figure 43a shows results for the pyrolyzed material. Points 3, 4, 5, and 6 correspond to adsorptives (immersing liquids) which contain respectively none, one, two, or three branch methyl groups; the heat of immersion falls in value steeply from point 3 to point 6, even though the molecular volume has increased only slightly. With the three methyl groups, indeed, the heat of immersion has fallen nearly to zero. These results clearly show that the critical factor is the magnitude of the smallest dimension of the adsorbate molecule relative to the size of the pore entrances: The majority of pores have to be reached through entrances narrower than the width of three methyl groups, which would probably be about the same as the minimum kinetic diameter of neopentane (194), 6.2 Å. The fact that the heat of immersion in benzene is almost the same as in methyl alcohol despite the much larger molecular volume of benzene suggests that the pores are slit-shaped, since benzene is a "flat" molecule with a thickness around 3.5 Å, approximately the same as the diameter of a methyl group.

The solid used for Figure 43b was the product of a 12-hr treatment with potassium hydroxide. The heat of immersion in hexane is now much smaller than before (Figure 43a) indicating that the pore entrances are narrower than in the pyrolyzed material. The fact that with the straight chain alcohols the heat diminishes markedly with increasing chain length can be explained if the constrictions are so narrow that the chains are forced to lie parallel to their walls; in this orientation the energy of (dispersion) interaction of a molecule with the walls is almost proportional to the chain length (195), and it is this energy which determines the energy of activation for passage of the molecule from the constriction into the cavity (Figure 44a); thus the area "wetted" by the adsorbate molecule in the time of a (short) experiment, and with it the heat of immersion, will diminish with increasing chain length.

A closer examination of the process of passage through a constriction suggests that the value of the ratio w/σ plays a crucial role in relation to the height of the energy barrier (w is the width of the constriction and σ is the minimum dimension of an adsorbate molecule). Let us visualize a pore system where the mesopore E leads through the constriction C into the microporous cavity V. When w/σ for the constriction is slightly in excess of unity (say $w/\sigma = 1.1$, of Figure 30d) the potential ϕ_0 in the constriction will be $\sim 3.5\ \phi_0$, where ϕ_0 is the minimum potential for the surface of the mesopore: and if the value of w/σ for the cavity V is such as to produce a potential $-2\phi_0$, then in passing from the constriction C into the cavity, a molecule will

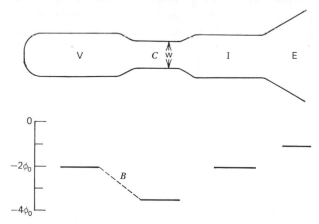

Fig. 44. The energy barrier brought about by a constriction of width w in a micropore. B = energy barrier; C = constriction; E = mesopore; I = micropore; V = cavity. w/σ is assumed to be slightly greater than unity ($w/\sigma \simeq 1.1$, say). The potential energy ϕ_0 relates to Fig. 30d.

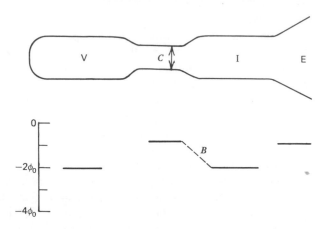

Fig. 45. The energy barrier brought about by a constriction of width w in a micropore. B = energy barrier; C = constriction; OE = mesopore; I = micropore; V = cavity. w/σ is assumed to be slightly less than unity (w/σ = 0.94, say). The potential energy ϕ_0 relates to Fig. 30e.

encounter an energy barrier B of 1.5 ϕ_0 (Figure 44). Suppose now that the value of w/σ for the constriction is slightly less than 0.94 in the model of Section IV.A, then the potential ϕ_0 within the constriction will actually become positive when the repulsion forces predominate over the attractive. Consequently the molecule now encounters an energy barrier on passing from the mesopore E into the constriction C (Figure 45).

In both cases, then, a molecule meets an energy barrier in its passage from the mesopore into the cavity, but in the case of the vary narrow constriction the barrier is situated in the step from the mesopore into the constriction, whereas in the case of the rather wider constriction it is located in the step from the constriction into the cavity. In between there must of course be a critical value of the ratio w/σ such that the energy barrier completely disappears so that movement through the constriction is no more hindred than in a mesopore.

1. Constrictions and the Nonane Pre-adsorption Technique

The role of constrictions in relation to the pre-adsorption method for the evaluation of microporosity has been touched on in Section IV.B. The work of Tayyab (175), there referred to, was carried out with two adsorbates other than nitrogen, viz. n-hexane and carbon tetrachloride. The adsorbents were ammonium salts of the three heteropolyacids, silicomolybdic, phosphomolybdic, and phosphotungstic; by application of the nonane-nitrogen

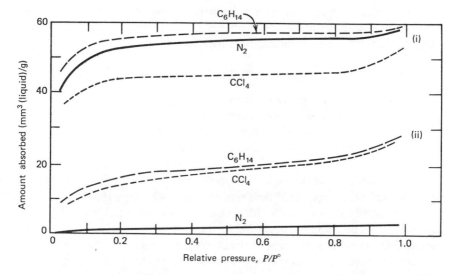

Fig. 46. Test of the pre-adsorption technique for evaluation of micropore volume, with two organic vapors, carbon tetrachloride and *n*-hexane (175). The adsorption isotherms of N_2 at 77°K, and of CCl_4 and *n*-hexane at 298°K, determined gravimetrically (i) before, and (ii) after, pre-adsorption of *n*-nonane, are plotted as amount adsorbed (expressed as a volume of liquid) against relative pressure.

technique all of them were shown to be highly microporous, and one of them, the phosphotungstate, to possess also a considerable external surface.

For ease of comparison the isotherms were all plotted with the uptake expressed as a volume of liquid. The volume adsorbed near saturation was consistently lower with the organic adsorptives than with nitrogen, except in the case of hexane on ammonium phosphomolybdate when it was almost the same as the nitrogen volume. These findings are readily explicable in terms of molecular sieve effect, the organic substances having larger molecule than nitrogen. After pre-adsorption of nonane, however, the isotherms of the organic adsorptives were always markedly *higher* than the corresponding isotherm of nitrogen (cf. Figure 46 which is typical). The results can be explained on the supposition that the adsorbed nonane is blocking the entries to cavities, rather like a stopper in a bottle; since the *n*-hexane and the carbon tetrachloride would be somewhat soluble in the (solid-like) nonane, they would be able to diffuse slowly through the "stopper" into the "bottle" beyond and thereby cause an increase in weight.

From these experiments it seems clear that the *n*-nonane method for the evaluation of microporosity is restricted to adsorptives such as nitrogen or argon which have negligible solubility in *n*-nonane. It is also likely that the temperature of measurement of the isotherm needs to be low, so that any diffusion of adsorbate through the nonane "stopper" which could result from a slight solubility, would be negligibly slow.

E. Size Distribution in Micropores

The use of molecular probes can in principle yield information as to the volumes of proes reached through entrances of a given size, but the task of evaluating the *distribution* of widths in a micropore system is much more difficult; such attempts as have been made hinge on the *t*-method of de Boer and his associates.

As explained in Section IV.B, if a powder is made up of microporous particles the *t*-plot will assume the form of Figure 47 where the linear branch QR corresponds to multilayer adsorption on the external surface of the particles, its slope being proportional to the area of this surface. The curved portion PQ, which corresponds to the steeply rising part of the isotherm, is the combined result of monolayer adsorption on the outer surface and filling of the micropores. Now de Boer and his collaborators (196) have argued that the slope of a tangent to the curve at any value of t, say t_1, is (if slit-shaped pores are assumed) proportional to the surface area of pores having a width in excess of $2t_1$: such pores would be incompletely filled, whereas those of widths less than $2t_1$ would already be full, since the adsorbed layers of thickness t_1 on opposite walls would have coalesced; to evaluate the pore size distribution therefore, the part PQ of the *t*-plot was divided into j steps, each corresponding to an increase $2\Delta t$ in the thickness of the multilayer. At the beginning and end of each step a tangent to the *t*-plot was drawn and the respective slopes converted into surface areas S_j and S_{j-1}, say so that the

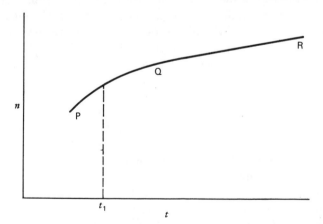

Fig. 47.   Form of *t*-plot for a vapor composed of microporous particles.

difference $\Delta S_j = S_j - S_{j-1}$ was equal to the surface of those pores having lying between $2(t_j - \Delta t)$ and $2(t_j + \Delta t)$. The pores being slit-shaped, the volume of these pores is $\Delta v_j = t_j \Delta S_j$, and the total volume of all the micropores is accordingly

$$\Sigma(\Delta v_j) = \sum_1^j t_j \Delta S$$

The curve of $\Delta v_j/2t_j$ against $2t_j$ would then directly express the distribution of pore volume with pore width. For other shapes of pores it would of course be necessary to work out the relationship between ΔS_j and Δv_j from a knowledge of the pore model.

The method is open to criticism inasmuch as it involves the tacit assumption that the t-curve—and therefore the standard isotherm from which it was derived—remains exactly the same for adsorption within micropores as on an open surface, up to the point where further thickening is prevented by the narrowness of the pore; by implication, the heat of adsorption within micropores is thus assumed to be identical with that for an open surface, whereas it is almost certainly higher (Section IV.C). The consequent upward distortion of the isotherm at low pressures, as compared with its course on an open surface, must mean that one is no longer justified in applying the standard t-curve to adsorption in micropores.

An attempt to circumvent this objection has been made by Mikhail, Brunauer, and Bodor (197), who recall that the shape of the isotherm in the region around point B (often located around $0.05 < P/P° < 0.30$) may be expressed through the BET c constant (Section II.B). They accordingly argue that when micropores are present the standard t-curve should be replaced by one derived from a reference isotherm which has the same c value as the isotherm under study. The method was illustrated by results for certain isotherms of nitrogen on silica; for one microporous gel the c value was found to be $c = 130$ and the de Boer–Linsen t-curve (which corresponds very nearly to this value of c) was used; for four other gels where the value of c was high the Cranston–Inkley isotherm (66) was taken as the best available.

A special study of the case of water vapor adsorbed on silica has been made by Hagymassy, Brunauer, and Mikhail (198). In view of the considerable variety of behavior displayed by this system, these workers found it necessary to allow for a wide range of c values; accordingly they collected isotherms of water vapor adsorbed on a variety of solids (including anatase, barium sulphate, quartz, and calcite) covering the following c values: 50–200, 23, 10–14.5, 5.2, and (for pressures above $0.5P°$) 10–200. The five corresponding t-curves were constructed and were applied (199) to the calculation of the pore size distribution of five silica gels; the results were then compared with those from the nitrogen isotherms using the "modelless" method (Section III) incorporating the hydraulic radius. Significant differences between the results obtained with the respective adsorbates led the authors to suggest that the choice of the pore size distribution curve must depend on the use to which the distribution curve was to be put.

The soundness of this revised t-method is still open to question, however. Because of the enhancement of the heat of adsorption within a micropore, there can be no single value of this heat, and therefore of c, which applies equally to adsorption within micropores on the one hand, and on an open surface or the wall of a mesopore on the other. When c is evaluated from an isotherm the linear part of the BET plot is used (Section II.B), and this corresponds not to the steeply rising part of the isotherm (cf. AB of Figure 39), but to the region around point B where the monolayer is being completed and the multilayer is beginning to form on the open surface and the walls of the mesopores: the value of c in effect gives numerical expression to the sharpness of the knee of the isotherm.

This point is brought out clearly in the isotherms, already referred to, obtained by Gregg and Langford for carbon black in which the micropores had been progressively emptied of pre-adsorbed nonane. The progressive increase in c (Table 13) as the micropores are emptied in stages is obvious, and from Figure 34a is seen to correspond to increasing sharpness of the isotherm knee. Clearly the value of c which corresponds to adsorption on the open surface is that corresponding to isotherm A, viz. $c = 59$; and this is far lower than the value, $c = 1940$, which according to the procedure of Brunauer and Mikhail would be used to select the appropriate t-curve.

Finally it should be realized that the *mechanism* of adsorption within micropores is most probably different from that on an open surface. Adsorption on an open surface is the result of the interaction of adsorbate molecules with one surface only, and leads to the formation of a multilayer; but within micropores a cooperative process probably occurs such that the pores fill completely at a low relative pressure. The process thus superficially resembles capillary condensation in mesopores and was indeed so termed by Pierce, one of the pioneers in the field; however, since the density and mode of packing of the adsorbate in very narrow pores cannot be identical with that of the bulk liquid (cf. Section IV.A), and since there is no meniscus (Section III.B) the designation "pore filling," is preferred. Capillary condensation is to be regarded as a secondary process as compared with pore filling, a primary process.

Regrettably, then, it must be concluded that at present we have no means of calculating the pore size distribution of micropores, though by use of a sufficiently wide range of adsorbates it is possible to gain some information as to the size distribution of the pore *entrances*.

F. An Extreme Case of the Micropore Effect—Carbon Tetrachloride on Silica

It has already been emphasized that the critical value of width w at which micropore effects begin to appear will increase with increasing diameter σ of the adsorbate molecule, since the relevant parameter is the ratio w/σ rather than w itself. The quantity σ is involved, not only for steric reasons, but also because the magnitude of the dispersion interaction increases as the polarizability increases.

A molecule of particularly high polarizability is carbon tetrachloride (201) ($\alpha(CCl_4) = 10.1 \times 10^{-24}$ as compared with $\alpha(N_2) = 1.73 \times 10^{-24}$ or $\alpha(CO_2) = 2.59 \times 10^{24}$ cm^3/molecule); consequently one might expect this adsorptive to be particularly sensitive probe for the micropore effect. Results obtained by Cutting and Sing (202), for the adsorption of carbon tetrachloride on silica gel, bear this out (Figure 48). The isotherms of carbon tetrachloride on two nonporous silicas, Fransil and TK 800, (Curves A and B) were close to type III, the value of c being $c \sim 3$. (The theoretical maximum for a type III isotherm is $c = 2$). On a silica gel characterized by nitrogen adsorption as mesoporous, the isotherm of carbon tetrachloride was also of type III (curve C). Now, as explained in Section II, the emergence of a type III isotherm is indicative of relatively weak adsorbent–adsorbate interactions; it is particularly striking, therefore, that on a silica that was microporous (as

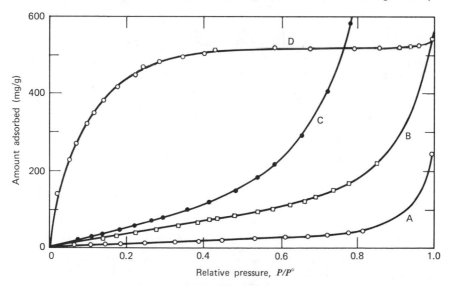

Fig. 48. Adsorption isotherms of carbon tetrachloride vapor at 293 °K on various samples of silica (202). A, "Fransil" (nonporous particles); B, "TK 800" (nonporous particles); C, a mesoprous gel; D, a microporous gel.

demonstrated by the nitrogen α_s- plot) the carbon tetrachloride isotherm was actually of type I (curve D). The presence of micropores has thus actually been able to convert an isotherm that was convex to the pressure axis into one concave to the axis, that is, to change a type III into a type I isotherm.

G. Low Pressure Hysteresis

The type of hysteresis described in Section III, with its well-defined loop, is readily interpreted in terms of capillary condensation. In 1957 Arnell and McDermott (203) drew attention to a second kind of hysteresis which persists to the lowest pressures; some adsorbate is retained even on long pumping ($\sim 10^{-4}$ torr) at the temperature of the isotherm run, and can only be removed completely by pumping at the (elevated) temperature of the original outgassing. Many other examples of the phenomenon are now known (94, 204–210), the majority of detailed measurements having been made with porous glass

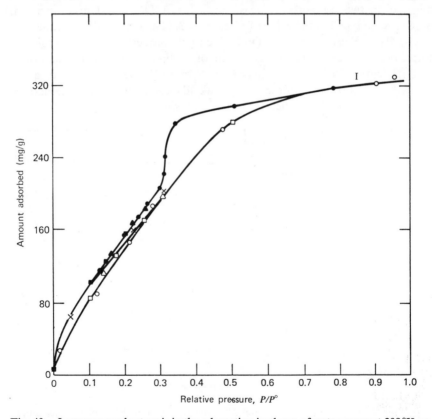

Fig. 49.   Low pressure hysteresis in the adsorption isotherm of water vapor at 298 °K on a silica gel (211). The different symbols for experimental points refer to two successive runs.

(205) and various types of carbon (209) as adsorbents. Frequently the low pressure hysteresis is superimposed on the conventional hysteresis loop, the desorption branch in the region below the shoulder running parallel to the adsorption branch (Figure 49). In many cases the effect is subject to a pressure threshold. McDermott and Arnell for example, found that provided the relative pressure did not exceed a critical value, the isotherm was reversible. Similarly in the experiments of Hickman (138) referred to in Section III (III.D), it was found that so long as the highest pressure reached was below the shoulder F of Figure 24 there was no low pressure hysteresis.

To explain their results (obtained with nitrogen and other adsorptives on graphite) McDermott and Arnell suggested that at the higher relative pressures the degree of intercrystalline swelling becomes sufficient to open up spaces, until then inaccessible, to adsorbate molecules; and that once adsorbed these molecules become effectively trapped (unless the temperature is raised), so that hysteresis extending to the lowest pressures is produed.

A recent comprehensive paper on low pressure hysteresis by Everett and his co-workers (212) is based on the work of the Bristol school extending over 15 years or so, with various forms of active carbon as adsorbent. Many experimental results, which at first sight seem puzzling and even contradictory, are collected together, and are shown to be capable of a rational and plausible explanation.

In one study the adsorbent was a carbon that had been prepared by heating polyvinyl chloride (previously pelletted at 50,000 1b/in.^{2}) at 180°C until one-half its HCl had been driven off, then raising the temperature at a uniform rate to 700°C and holding it there until decomposition was complete.

A series of isotherms of benzene were then determined without intermediate removal of the sample form the apparatus, outgassing at the end of each

TABLE 17

Adsorption of benzene by carbon 8P[a]
Height (h) of hysteresis loop at $P/P° = 0.25$, and uptake at saturation (w_s).
Runs were carried out in the order given.

Run	Temp. (°C)	h (%)	w_s (%)
1	25	0.00	35.5
2	35	0.12	35.5
3	45	0.50	36.3
4	40	0.27	38.2
5	25	0.42	38.2
	Store *in vacuo* for 2 months at room temp.		
6	25	0.37	36.5
	Anneal at 305°C for 80 hr		
7	25	0.10	36.15

[a]Reduced from the data of Everett and co-workers (212).

run being carried out at the temperature of the isotherm. As is seen from Table 17, hysteresis was absent from isotherm 1 at 25°C, was slight in isotherm 2 at 35° and was more marked in isotherm 3 at 45°; in the next run 4 at the slightly lower temperature of 40°, the saturation uptake w_s was higher than in any previous run, and in a repeat isotherm (5) at 25°, hysteresis (absent from run 1) was now present and the saturation uptake was much higher than in run 1. Storage *in vacuo* at room temperature for 2 months led to a reduction in hysteresis and also in w_s, in the next run (6) at 25°; but even long annealing at the elevated temperature of 305° did not quite suffice to eliminate hysteresis from the 25° isotherm (7), nor did w_s quite return to its original value. Clearly, low pressure hysteresis is associated with some distortion of the carbon structure, which can persist for long periods and is of such a nature as to increase the saturation uptake. Everett and his co-workers suggest a mechanism based on an irreversible intercalation of adsorbate within pores of molecular dimensions producing an inelastic distortion of the solid which may relax only slowly after removal of the adsorbate. Experimental support for intercalation is provided, for example, by the fact that with a sample of the carbon pelleted at 25,000 lb/in.[2] hysteresis appeared in the 25° isotherms of *n*-hexane and cyclohexane, but not of benzene, which has smaller molecules; with a similar carbon pelletted at 125,000 lb/in.[2] however benzene did show hysteresis even at 25°, whereas in nine runs at 25 to 45° no hysteresis appeared if the pelleting was carried out at 25,000 lb/in.[2] —the higher pelleting pressure presumably having produced many narrow crevices.

The problem was expressed in thermodynamic terms by regarding the adsorbent (1) together with the adsorptive (2) as a two-component system. At constant temperature and external pressure, the Gibbs–Duhem relation may be written

$$d\mu_1 = -\left(\frac{n_2}{n_1}\right) d\mu_2 \tag{83}$$

where μ_1 and μ_2 are the chemical potentials, and n_1 and n_2 the amounts, of the respective components.

If the adsorbate is in the equilibrium with its own vapor, we have

$$d\mu_2 = -RT\, d \ln P_2$$

so that

$$d\mu_1 = -RT\left(\frac{n_2}{n_1}\right) d \ln P_2 \tag{84}$$

or

$$\mu_1 = \mu_1^0 - RT \int_0^P \left(\frac{n_2}{n_1}\right) d \ln P_2 \tag{85}$$

where μ_1^0 = chemical potential of the empty solid structure. Low pressure

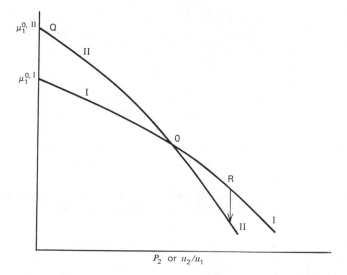

Fig. 50. Low pressure hysteresis. Chemical potential in an adsorbent–adsorbate system which can exist in two states I and II. Suffix 1 refers to the adsorbent and suffix 2 to the adsorbate. (After Everett et al. (212).)

hysteresis will occur if at some point on the adsorption isotherm the structure of the solid jumps irreversibly to a new configuration which, when empty, has a different value of chemical potential (Figure 50).

The jump could, thermodynamically, occur at Q but the presence of hysteresis suggests that in the strained solid the jump is delayed until it occurs, irreversibly, at R, say. On desorption the system follows path II; reversion to the original properties (I) occurs only at $P_2 \to 0$, and even then in general it encounters an energy barrier and so proceeds slowly at a given temperature, and faster if the temperature is raised.

If intercalation occurs by interlamellar penetration, it is appropriate to think of a "spreading pressure" that enables an adsorbed film to "prise open" the lamellae. In practice the structure of a solid is likely to consist of many domains, so that there will be many points such as R, the position of which will vary from one domain to another. Thus, in order to make low pressure hysteresis possible the solid must be such that local regions can be strained beyond the elastic limit; consequently low pressure hysteresis is not to be expected if the solid structure is strong and rigid (cross-linked) on the one hand, or if it is loosely knit so that it can expand or contract over a wide range, on the other.

The model is able to explain the existence of the molecular size effect, since low pressure hysteresis will be absent either if the adsorbate molecules are small enough to penetrate into the adsorbent when it is in state I or if they are too large to take advantage of the structural change from state I to II: for low pressure hysteresis to occur the molecules need to be close in size to the widths of the pores (or pore entrances) which are opened up by the structural change.

The existence of a pressure threshold is also explicable, since the system will behave reversibly so long as its passage along paths I stops short of the cross-over point Q. A temperature threshold is also implicit in the model, inasmuch as the relative disposition of curves I and II—both their separation and their slopes—will depend on temperature; moreover the diffusion along narrow pores which have been opened up will be an activated process (Section IV.D) so that molecules which penetrate at a negligible rate at temperature T_1, say, may pass through with reasonable rapidity at a sufficiently higher temperature T_2.

The structural change referred to in the Everett interpretation must be such as to increase the capacity for adsorption. If the solid is composed of primary particles aggregated together, then a simple picture presents itself in which the junctions between particles become prized open as adsorption progresses; this could occur either as an indirect result of the swelling of the individual particles consequent on adsorption of vapor; or directly by penetration of the adsorbed film into the junction, which would cause an expansion of the aggregate. Support for the model is provided by results of experiments on compacts of coal (213) in which simultaneous measurements of amount adsorbed, linear expansion and electrical conductivity were made for two organic vapors, ethyl cholride and n-butane. The curves for both vapors were very similar, those for ethyl chloride being shown in Figure 51, and are readily explained in terms of the model. The hysteresis in conductivity clearly points to an irreversible opening up of interparticulate junctions; since many of the passages thus produced would be narrow enough to function as constrictions in the sense of Section IV.D, adsorption would be a rate process and hysteresis would result. In other experiments, hysteresis in adsorption and swelling of the kind illustrated in Figure 51 was accompanied by a marked reduction (as much as fourfold) in the mechanical strength of the compacts (214), reinforcing the view that low pressure hysteresis in coal compacts was associated with a weakening of interparticulate bridges.

Low pressure hysteresis is especially common when water vapor is the adsorptive. Consistent with the complex properties of water, the causes of this hysteresis are diverse; in some cases it is reasonably attributed to a slow hydration of the adsorbent, but in others it results from a penetration process —often complicated by the slight solubility of the adsorbent in liquid water.

An example of the penetration into fine pores appears in the detailed investigation of Anderson and Horlock (215) of the adsorption of water vapor

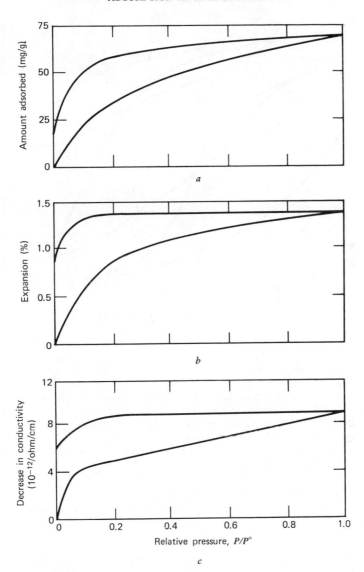

Fig. 51. Swelling and low pressure hysteresis, in the adsorption of ethyl chloride on compacts of coal (213) at 273°K. In (a) the amount adsorbed, in (b) the percentage increase in length, and in (c) the decrease in electrical conductivity, is plotted against relative pressure of the vapor. The curves for the adsorption of n-butane were very similar to the above curves.

on beryllium oxide prepared by the decomposition of the hydroxide *in vacuo* at 200°C. The state of the adsorbent was monitored by nitrogen adsorption; the amount $n_{0.06}$ of nitrogen at a relative pressure of 0.06 was taken as a

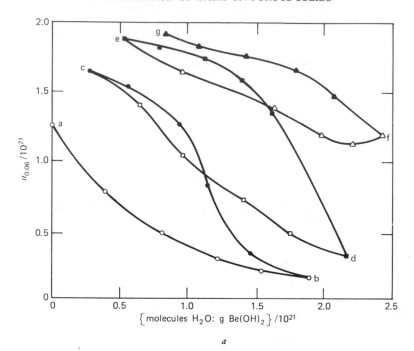

a

Fig. 52. (*a*) The adsorption of water vapor on BeO prepared by the decomposition of Be(OH)$_2$ *in vacuo* (215). $n_{0.06}$ is taken as the apparent monolayer capacity and is plotted against the water content of the sample after the alternate exposure of the sample to water vapor (b, d, f) and its outgassing at 100–200°C (c, e, g). Page 347: (*b*) Plots of amount N adsorbed against *t* (*t*-plots) constructed from adsorption isotherms of water vapor on samples of BeO which already contained amounts of pre-adsorbed water corresponding to points (a), (b), and (c), respectively, of Fig. 52*a*. (Redrawn from the original diagrams of Anderson and Horlock (215).)

measure of the apparent monolayer capacity, since with a standard, nonporous sample of BeO (prepared by calcination > 1000°C) the statistical monolayer was found to be complete at $P/P° = 0.06$. Water vapor was adsorbed in stages at 20°C till no further decrease in $n_{0.06}$ occurred (a–b of Figure 52*a*); the uptake of water occurred effectively at zero pressure and could not be removed by pumping at temperatures below 60°C. After (b) of Figure 52*a* the water was removed by pumping at elevated temperature (100–200°C) and as is seen (cf. (c)), $n_{0.06}$ increased significantly above its original value at (a). Further adsorption–desorption cycles resulted in still further increases in the value of $n_{0.06}$ ((c), (e), and (g)).

The *t*-plots corresponding to points (a), (b), and (e) are shown in Figure 52*b* (the calcined BeO being taken as internal standard for construction of the *t*-curve); they closely resemble the *t*-plots of Figure 34 and clearly point to the

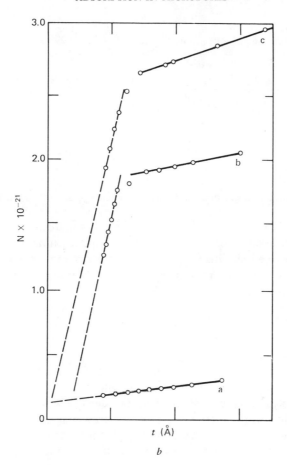

t (Å)

b

microporosity of the specimen at (a) and the filling of the micropores with adsorbed water at (b); they also show that the three adsorption–desorption cycles have led to an increase in micropore volume. This interpretation is supported by the presence of low pressure hysteresis in the isotherm corresponding to (b) and (c) and its absence from the isotherm for (a). A comparative analysis of the isotherm for water, nitrogen, and carbon tetrachloride suggested that pores in the range of 3 to 4Å have been widened to > 6Å by the adsorption and desorption of water.

Sing and his collaborators (216) have found evidence for penetration by water in their studies of chromia gels, by comparing the values of $S(H_2O)$ and $S(N_2)$, calculated by the BET procedure, from the corresponding isotherms, taking $a_m(H_2O) = 10.6$Å^2 and $a_m(N_2) = 16.2$Å^2. For certain samples the two values of S agreed reasonably closely, but for others $S(N_2)$ was close to

TABLE 18

Adsorption of nitrogen at 77°K and of water at 293°K.[a]

Gel	$S(N_2)$[b] (m²/g)	$S(H_2O)$[b] (m²/g)	$v_s(N_2)$ (cm³/g)	$v_s(H_2O)$[c] (cm³/g)	Isotherm Type N_2	Isotherm Type H_2O
A	150	174	—	—	II	II
B	49	172	—	—	II	II
C	1	267	0	0.096		I
D	1	232	0	0.089		I
E	165	94	0.145	0.146	IV	IV

[a]Sing and c-workers (216).
[b]S = BET specific surface.
[c]v_s = uptake at saturation, calculated as volume of liquid ("Gurvitsch volume").

zero whereas $S(H_2O)$ amounted to many tens of square meters per gram (Table 18); in these samples, therefore, water was penetrating into pores that were too narrow to admit nitrogen molecules. The driving force was believed to be the rehydration of the Cr^{3+} ion by completion of its coordination sphere.

Another illustration of low pressure hysteresis caused by penetration is provided by the measurements by Gammage (217) of the adsorption of water vapor on samples of calcite which had been ball-milled for various lengths of time; the isotherms were plotted as adsorption per unit area (as determined by nitrogen adsorption).

In the region below point B the water isotherms (apart from that for the 1000-hr sample) agreed reasonably closely with each other and with the isotherm of precipitated calcite taken as standard; but at higher relative pressures they diverged in an upward direction from the standard, and to an extent increasing with the duration of milling; for the 1000-hr sample, indeed, the divergence extended over the whole range of the isotherm. Low pressure hysteresis was present in all except the reference isotherm, and it became more marked as time of grinding increased.

All these results are consistent with the hypothesis that the grinding introduces fault lines and cleavage cracks into the solid, into which molecules can penetrate once the pressure threshold has been mounted; it is interesting that this threshold, for all except the 1000-hr sample, corresponds to a completed monolayer, and this suggests that the more liquidlike properties of the multilayer play a role in the penetration. Presumably the 1000-hr sample is so badly damaged that penetration can occur readily even at the lowest pressures. It seems that the penetration of water causes irreversible changes (probably arising from the slight solubility of calcite in water, so that leaching is possible) since the BET–nitrogen area of all the samples was higher after a water adsorption run than before, and in the case of the 1000-hr sample, much higher (28.6 as compared with 7.8 m²/g).

V. CONCLUSION

The material presented and discussed in the foregoing pages has been selected from a vast literature in order to provide a reasonably balanced picture of progress made in the study of the adsorption of vapors by porous solids. It has brought to light a number of areas in which further research effort should be especially profitable. Since the problems vary in nature according to the range of pore size involved, they may appropriately be considered under the respective headings mesoporosity, microporosity, and the region between the two.

A. Mesoporosity

In this range the major pore-filling process is capillary condensation. Since virtually all the theoretical treatments are based on the Kelvin equation, the extent of the practical applicability of the equation needs to be thoroughly examined by measurements of vapor adsorption on a variety of model solids, using a representative selection of vapors so as to remedy the present overdependence on the one vapor, nitrogen.
This work would include:

1. The preparation of adsorbents having well-defined pore structures, by methods such as compaction under standard conditions of powders comprising particles of definite size and shape, or selective leaching of composite substances. The resultant solids would need to be characterized by independent methods, especially high-resolution electron microscopy.

2. The use of these model solids to measure the adsorption isotherms of a variety of vapors, so chosen as to cover a wide range of isotherm temperature from, say 60 to 350°K.

3. Detailed study of the lower closure point of the hysteresis loop, with a view to obtaining further evidence as to the validity of the "tensile strength" hypothesis. In addition, critical assessment is required of the significance of hysteresis in the interpretation of results of pore size distribution calculations.

B. Microporosity

The major mechanism in this region is "micropore filling" rather than capillary condensation, and the problems tend to be more intractable, especially when—as is often the case—nonequilibrium effects are present. The use of model solids is again highly desirable, though in practice the choice would have to be restricted to materials in which the pores form part of the crystalline structure. On the technique side, heat of immersion measurements can often play a useful supporting role. Possible lines of work include:

1. Measurements of adsorption isotherms and of heats of immersion, using "molecular probes," that is, a series of adsorptives of gradually increasing molecular size and polarizability, and having various molecular shapes.

2. The study of activated diffusion through fine constrictions and within layer structures, by adsorption and heat measurements conducted at a series of temperatures.

3. The pre-adsorption technique for micropore evaluation extended to a variety of adsorptives and pre-adsorptives, both in order to test the general validity of the method, and to glean information as to the size distribution of pores and pore constrictions.

C. The Intermediate Region between Mesopores and Micropores

Very little is known about this region, which corresponds to the part of the type IV isotherm below the hysteresis loop. A useful approach would be through the systematic study of low pressure hysteresis, especially with non-rigid adsorbents, notably those having layer structures. Further study is also required of the dependence of *reversible* capillary condensation on particle size, shape, and packing.

SYMBOLS

A	Affinity of adsorption; constant of the Lennard–Jones expression.
a_m	Average area occupied per molecule in the completed monolayer
B	Constant of the Lennard–Jones expression; number of adsorption sites
b_t	Slope of the t-plot
b_s	Slope of the α_s-plot
C	Constant of the expression for potential energy of adsorption
C_1, C_2, C_3	Constants of the virial equation for the adsorption isotherm
c	Constant of the BET equation; velocity of light
D	Diameter of cylindrical micropore; constant of the Dubinin–Radushkevich equation
d_0	Molecular separation in the liquid adsorptive
d_k	Width of slit-shaped core
d_p	Width of slit-shaped pore
E_1	Heat of adsorption in the first molecular layer
ΔG	Free energy of adsorption
ΔH	Enthalpy of adsorption

h	Height of hysteresis loop
$j(T)$	Partition function
L	Latent heat of evaporation of adsorptive
$L(r)$	Length of pore having radius r
M	Molecular weight of adsorptive
m	Mass of adsorbent; mass of electron
N	Avogadro constant
N	Number of molecules
n	Amount adsorbed (moles)
n_m	Monolayer capacity (moles)
P	Equilibrium pressure
$P°$	Saturated vapor pressure of adsorptive
P_c	Critical pressure of adsorptive
Q	Factor to convert core volume into pore volume
q_{st}	Isosteric heat of adsorption
r	Exponent of Halsey equation for the isotherm
r_1, r_2	Principal radii of curvature of the meniscus
r_H	Hydraulic radius of core
r_k, r^k	Core radius
r_m	Mean radius of curvature of meniscus
r_n	Core radius of narrowest part of pore
r_p, r^p	Pore radius
r_w	Core radius of widest part of pore
S	Specific surface area
t	Statistical thickness of adsorbed film
V	Molar volume of adsorptive
v_k, v^k	Volume of core
v^f	Volume of adsorbed film
v_p, v^p	Volume of pore
v^s	Amount adsorbed at saturation pressure, calculated as volume of liquid
W_0	Micropore volume
w	Width of constriction in micropore; width of pore
z	Distance from isolated molecule to adsorbent surface
α	Polarizability of adsorptive
α_s	The ratio n/n_s where n_s = amount adsorbed when $P/P° = s$
Γ	Amount (moles) adsorbed per unit surface area of adsorbent
γ	Surface tension of liquid adsorptive
ε	Energy of interaction of two isolated atoms
θ	Angle of contact; surface coverage ($= n/n_m$)
μ	Dipole moment; chemical potential
ρ	Density

σ Statistical thickness of a single molecular layer
τ_0 Tensile strength of liquid adsorptive
τ Tension in the capillary-condensed adsorbate
ϕ_0 Potential energy of adsorption
χ Magnetic susceptibility

REFERENCES

1. F. Fontana, *Mem. Mat. Fis. Ital.*, **1**, 679 (1777).
2. C.W. Scheele, *Chemical Observations and Experiments on Air and Fire*, transl. by J. R. Forster, 1780
3. H. Kayser, *Wied. Ann.,* **14**, 451 (1881)
4. W. Ostwald, *Lehrbuch der allgem. Chemie*, Zweite Auflage, 1906, p. 242.
5. J.W. McBain, *Z. Phys. Chem.,* **68**, 471 (1909); *Phil. Mag.,* **18** (6), 916 (1909).
6. W. Thomson, *Phil. Mag.,* **42** (4), 448 (1871).
7. R. Van Helmholtz, *Wied. Ann.,* **27**, 508 (1886).
8. J.M. Van Bemmelen, *Z. Anorg. Chem.,* **13**, 233 (1897); **59**, 225 (1908); **62**, 1 (1909)
9. H. Freudlich, *Capillary and Colloid Chemistry*, Methuen, London, 1926, p. 44.
10. R. Zsigmondy, *Z. Anorg. Chem.,* **71**, 356 (1911).
11. A.G. Foster, *Trans. Faraday Soc.,* **28**, 645 (1932).
12. S. Brunauer, P.H. Emmett, and E. Teller, *J. Amer. Chem. Soc.,* **60**, 309 (1938). S. Brunauer, *The Physical Adsorption of Gases*, Princeton Univ. Press, Princeton, N.Y. 1945.
13. M. Dubinin, *Quart. Rev. Chem. Soc.,* **9**, 101 (1955).
14. C. Pierce and R.N. Smith, *J. Phys. Chem.,* **57**, 64 (1953).
15. *I.U.P.A.C. Manual of Symbols and Terminology*, Appendix 2, Part I, Colloid and Surface Chemistry, *Pure and Applied Chem.,* **31**, 578 (1972).
16. S. Brunauer, L.S. Deming, W.S. Deming, and E. Teller, *J. Amer. Chem. Soc.,* **62**, 1723 (1940).
17. A.D. Crowell, in *The Solid-Gas Interface*, E.A. Flood, Ed., Arnold, London, 1967. R.H. Dongen, *Physical Adsorption on Ionic Solids*, Uitgeverij Waltman, Delft, 1972. K.S.W. Sing, *Colloid Science*, **1**, 1, Specialist Periodical Reports, Chem. Soc., London, 1973. H. Krupp, *Advan. Colloid. Interface Sci.,* **1**, 111 (1967).
18. F. London, *Z. Physik*, **63**, 245 (1930).
19. J.O. Hirschfelder, C.F. Curtis, and R.B. Bird, *Molecular Theory of Gases and Liquids* Wiley, New York, 1954.
20. M. Born and J.E. Mayer, *Z. Phys.,* **75**, 1 (1932).
21. F. Ricca and E. Garrone, *Trans. Faraday Soc.,* **66**, 959 (1970)
22. J.E. Lennard-Jones, *Proc. Phys. Soc.,* **43**, 461 (1931).
23. J.E. Lennard-Jones, *Physica*, **4**, 941 (1937); cf. R.H. Fowler and E.A. Guggenheim, *Statistical Thermodynamics*, Cambridge, 1939, p. 280. J.C. Slater and J.G. Kirkwood, *Phys. Rev.,* **37**, 682 (1931). J.G. Kirkwood, *Phys. Z.* **33**, 57 (1932).

24. A. Muller, *Proc. Roy. Soc.*, **A154,** 624 (1936).

25. F. Ricca, C. Pisani, and E. Garrone, *Adsorption–Desorption Phenomena*, Proc. 2nd Intern. Conf., 1971, p. 111 (1972).

26. W.J.C. Orr, *Trans. Faraday Soc.*, **35,** 1247 (1939).

27. C. Pisani, F. Ricca, and C. Roetti, *J. Phys. Chem.*, **77,** 657 (1973).

28. F. London, *Z. Phys. Chem.*, **11,** 222 (1930).

29. J.H. de Boer, *Advances in Catalysis* **8,** 33, Academic Press, New York, 1956.

30. A.D. Buckingham, *Quart. Rev. Chem. Soc.*, **13,** 183 (1959).

31. L.E. Drain and J.L. Morrison, *Trans. Faraday Soc.*, **49,** 654 (1953).

32. R.W. Barrer, *J. Colloid Interface Sci.*, **21,** 415 (1966).

33. J.M. Honig, *Ann. N.Y. Acad. Sci.*, **58,** 749 (1954).

34. P.J. Anderson and R.F. Horlock, *Trans. Faraday Soc.*, **65,** 251 (1969).

35. P.R. Anderson, *Surface Sci.*, **27,** 60 (1971).

36. A.V. Kiselev, A.A. Lopatkin, and E.R. Razumova, *Zhur. Fiz. Khim.*, **44,** 150 (1970); *Russ. J. Phys. Chem.*, **44,** 82 (1970).

37. M. Leard and A. Mellier, *Compt. Rend.*, **B272,** 1477 (1971).

38. A.V. Kiselev, *Discuss. Faraday Soc.*, **40,** 205 (1965).

39. A.V. Kiselev, N.V. Kovaleva, and Yu. S. Nikitin, *J. Chromatog.*, **58,** 19 (1971).

40. R.M. Barrer, *J. Colloid Interface Sci.*, **21,** 415 (1966).

41. K.S.W. Sing and V.R. Ramakrishna, *Colloques Internationaux de Centre National de la Recherche Scientifique*, No. 201 435 (1971).

42. J. Frenkel, *Kinetic Theory of Liquids*, Clarendon Press, Oxford 1947.

43. T.L. Hill, *Advan. Catal.* **4,** 211, 1952.

44. G.D. Halsey, *J. Chem. Phys.*, **16,** 931 (1948).

45. N.N. Avgul and A.V. Kiselev, *Chem. Phys. Carbon*, **6,** 1 (1970). N.N. Avgul, A.G. Benzus, E.S. Debrova, and A.V. Kiselev, *J. Colloid Interface Sci.*, **42,** 486 (1973).

46. T.L. Hill, *J. Chem. Phys.*, **14,** 263 (1946).

47. C. Kemball and G.D.L. Schreiner, *J. Amer. Chem. Soc.*, **72,** 5605 (1950).

48. S.J. Gregg and J.J. Jacobs, *Trans. Faraday Soc.*, **44,** 574 (1948).

49. D.C. Jones, *J. Chem. Soc.*, **1951,** 126.

50. A.C. Zettlemoyer, *J. Colloid Interface Sci.*, **28,** 343 (1968).

51. D.M. Young and A.D. Crowell, *Physical Adsorption of Gases*, Butterworths, London, 1962, p. 170.

52. C.F. Prenzlow and G.D. Halsey, *J. Phys. Chem.*, **61,** 1158 (1957).

53. J.P. Freeman, *J. Phys. Chem.*, **62,** 723, 729 (1958).

54. K.S. Pitzer and O. Sinanglu, *J. Chem. Phys.*, **32,** 1279 (1960).

55. A.G. Bezus, A.V. Kiselev, and Pham Quang Du, *J. Colloid Interface Sci.*, **40,** 223 (1972).

56. R.M. Barrer and J.A. Davis, *Proc. Roy. Soc.*, **A320,** 289 (1970).

57. R.A. Pierotti and H.E. Thomas, *Surface and Colloid Science*, E. Matijević, Ed., Wiley, New York, 1971, p. 93.

58. P.H. Emmett and S. Brunauer, *J. Amer. Chem. Soc.*, **59,** 1553 (1937).

59. A.L. McClellan and H.F. Harnsberger, *J. Colloid Interface Sci.*, **23**, 577 (1967).

60. S.J. Gregg and K.S.W. Sing, *Adsorption, Surface Area, and Porosity*, Academic Press, London, 1967.

61. D.M. Young and A.D. Crowell, *Physical Adsorption of Gases*, Butterworths, London, 1962, p. 198.

62. A.V. Kiselev, *Proc. 2nd Int. Congr. Surface Activity*, **2**, 179 (1957).

63. C. Pierce, *J. Phys. Chem.*, **62**, 1076 (1958); **63**, 1077 (1959).

64. J.W. Lucas, G.W. Newton, and K.S.W. Sing *J. Appl. Chem.*, **13**, 265 (1963). M.R. Harris and K.S.W. Sing, *Proc. 3rd Int. Congr. Surface Activity*, Cologne, 1960, **2**, p. 42.

65. C.G. Shull, *J. Amer. Chem. Soc.*, **70**, 1410 (1948).

66. R.W. Cranston and F.A. Inkley, *Advan. Catal.*, **9**, 143 (1957).

67. C. Pierce, *J. Phys. Chem.*, **72**, 3673 (1968).

68. B.C. Lippens, B.G. Linsen, and J.H. de Boer, *J. Catal.*, **3**, 32 (1964).

69. J.D. Carruther, P.A. Cutting, R.E. Day, M.R. Harris, S.A. Mitchell, and K.S.W. Sing, *Chem. Ind.*, 1772 (1968).

70. D.A. Payne and K.S.W. Sing, *Chem. Ind.*, 918 (1969).

71. D.A. Payne, K.S.W. Sing, and D.H. Turk, *J. Colloid Interface Sci.*, **43**, 287, 1973.

72. G.A. Nicolaon and S.J. Teichner, *J. Colloid Interface Sci.*, **38**, 172 (1972).

73. M.H. Polley, W.D. Schaeffer, and W.R. Smith, *J. Phys. Chem.*, **57**, 469 (1953).

74. J.M. Holmes and R.A. Beebe, *J. Phys. Chem.*, **61**, 1684 (1957).

75. A.A. Isirikyan and A.V. Kiselev, *J. Phys. Chem.*, **65**, 601 (1961).

76. C. Pierce, *J. Phys. Chem.*, **68**, 2562 (1964).

77. K.M. Hanna, I. Odler, S. Brunauer, J. Hagymassy, Jr., and E.E. Bodor, *J. Colloid Interface Sci.*, **45**, 27 (1973).

78. B.C. Lippens and J.H. de Boer, *J. Catal.* **4**, 319 (1965).

79. B.C. Lippens, B.G. Linsen, and J.H. de Boer, *J. Catal.*, **3**, 32 (1964).

80. K.S.W. Sing, *Chem. Ind.*, 1520 (1968).

81. K.S.W. Sing, *Surface Area Determination*, D.H. Everett and R.H. Ottewill, Eds., Butterworths, London, 1970, p. 25.

82. P.C. Carman and F.A. Raal, *Proc. Roy. Soc.*, **209A**, 59 (1951).

83. A.V. Kiselev, in *The Structure and Properties of Porous Materials*, Butterworths, London, 1958, p.195. B.G. Aristov, V. Yu. Davydov, A.P. Karnaukhov, and A.V. Kiselev, *Russ. J. Phys. Chem.*, **36**, 1497 (1962).

84. P. Zwietering, in *The Structure and Properties of Porous Materials*, Butterworths, London, 1958, p. 68.

85. J.F. Langford, Ph.D., Thesis, Exeter University, 1967, S.J. Gregg and J.F. Langford, *Trans. Faraday Soc.*, **65**, 1394 (1969).

86. W.H. Wade, *J. Phys. Chem.*, **68**, 1029 (1964); **69**, 322, 1395 (1965).

87. B.G. Aristov, A.P. Karnaukhov, and A.V. Kiselev, *Russ. J. Phys. Chem.*, **36**, 1159 (1962).

88. R.G. Avery and J.D.F. Ramsay, *J. Colloid Interface Sci.*, **42**, 597 (1973).

89. J.H. de Boer, in *The Structure and Properties of Porous Materials*, Butterworths, London, 1958, p. 68.

90. D.H. Everett, in *The Solid-Gas Interface*, E.A. Flood, Ed., Arnold, London, 1967, p. 1055.

91. R. Defay, I. Prigonine, A. Bellemans, and D.H. Everett, *Surface Tension and Adsorption*, Longmans, London, 1966, p. 218.

92. T. Young *Miscellaneous Works*, Vol. 1, Murray, London, 1855, p. 418.

93. P.S. de Laplace, *Mechanigne Celeste*, Supplement to Book 10, 1806.

94. E.W. Sidebottom and G.G. Litvan, *Trans. Faraday Soc.*, **67**, 2726 (1971).

95. L.H. Cohan, *J. Amer. Chem. Soc.*, **60**, 433 (1938).

96. B.V. Derjaguin, *Acta Phys. Chem., U.S.S.R.*, **12**, 181 (1940). B.V. Derjaguin *Proc. 2nd Int. Congr. Surface Activity*, **2**, 153 Butterworths, London, 1957.

97. D.H. Everett and J.M. Haynes, *Specialist Periodical Reports in Colloid Science*, **1**, 147, Chem. Soc., London, 1973.

98. J.H. de Boer and J.C.P. Broekhoff *J. Catal.*, **10**, 391 (1968).

99. D.H. Everett, in *The Solid-Gas Interface*, E.A. Flood, Ed., p. 1055, Dekker, New York, 1967.

100. T.D. Blake and J.M. Haynes, *Prog. Surface Membrane Sci.*, **6**, 125 (1973).

101. E.O. Kraemer, in *Treatise on Physical Chemistry*, H.S. Taylor, Ed., Van Nostrand, 1931, p. 1663.

102. J.W. McBain, *J. Amer. Chem. Soc.*, **57**, 699 (1935).

103. D.H. Everett, in *The Solid-Gas Interface,* E.A. Flood, Ed., Dekker, New York, 1967, p. 1081.

104. A.P. Karnaukhov, *Kinetika i Kataliz*, **8**, 172 (1967) (Transl.)

105. D.H. Everett, *Trans. Faraday Soc.*, **48**, 749 (1952); **50**, 187, 1077 (1954); **51**, 835 (1955); **52**, 106 (1956).

106. D.H. Everett, in *The Solid-Gas Interface*, E.A. Flood, Ed., 1095, Dekker, New York, 1967, p. 1073.

107. A. Wheeler, *Catalysis*, Vol. 2, Reinhold, New York, 1955, p. 116; *Advan. Catal.*, **3**, 250 (1950).

108. C. Orr and J.M. Dalla Valle, *Fine Particle Measurement*, MacMillan, London, 1959, p. 271.

109. C. Pierce, *J. Phys. Chem.*, **57**, 149 (1953).

110. D. Dollimore and G.R. Heal, *J. Appl. Chem.*, **14**, 109 (1964).

111. D. Dollimore and G.R. Heal, *J. Colloid Interface Sci.*, **33**, 508 (1970).

112. B.F. Roberts, *J. Colloid Interface Sci.*, **23**, 266 (1967).

113. J.J. Steggerda, Thesis, Delft University, 1955.

114. E.P. Barrett, L.F. Joyner, and P.H. Halenda, *J. Amer. Chem. Soc.*, **73**, 373 (1951).

115. S. Brunauer, R.Sh. Mikhail, and E.E. Bodor, *J. Colloid Interface Sci.*, **24**, 451 (1967).

116. A.V. Kiselev, *Usp. Khim.*, **14**, 367 (1945). *Proc. 2nd Int. Congr. Surface Activity*, **2**, 189 (1957).

117. S. Brunauer, *Chem. Eng. Progr. Symp.*, No. 96, 1 (1969).

118. C.F. Gausss, in *Comment. Soc. Reg. Göttingen. Rec.*, **7** (1830), trans. by R.H. Weber in *Ostwald's Klassiker der Exakten Wissenschaften*, No. 135, Engelmann, Leipzig, 1903.

119. K.S. Rao, *J. Phys. Chem.*, **45**, 506 517 (1941).

120. G.L. Kington and P.S. Smith, *Trans. Faraday Soc.*, **60**, 705, 721 (1964).

121. S.P. Zhdanov, *Coll. Methods Investigating the Structure of Highly Disperse Porous Solids*, Izd. AN SSSR, No. 2, 71 (1958). (In Russian.)

122. A.V. Kiselev, *Proc. 2nd Int. Congr. Surface Activity*, **2**, 168, Butterworths, London, 1957.

123. P.A. Cutting, Ph.D. Thesis, Brunel University, 1969; *Vacuum Microbalance Techniques*, Vol. 7, Plenum Press, New York, 1970, p.71.

124. G. Nicolaon and S.J. Teichner, *J. Chim. Phys.*, **65**, 871 (1968).

125. D.H. Everett, in *The Structure and Properties of Porous Materials*, Butterworths, London, 1958, p. 95.

126. M.R. Harris, *Chem. Ind.*, 269 (1965).

127. M.M. Dubinin, *Russ. J. Phys. Chem.*, **34**, 959 (1960).

128. S.J. Gregg and K.H. Wheatley, *Proc. 2nd Int. Congr. Surface Activity*, **2**, 102 (1957). K.H. Wheatley, Thesis, London University, 1953.

129. M.I. Pope, Ph.D. Thesis, Exeter University, 1957. S.J. Gregg and M.I. Pope, *J. Chem. Soc.* 1252 (1961).

130. R.C. Asher, Ph.D. Thesis, London University, 1955. R.C. Asher and S.J. Gregg, *J. Chem. Soc.*, 5057 (1960).

131. R. Stock, Ph.D. Thesis, London University, 1955.

132. S.J. Gregg and K.S.W. Sing, *J. Phys. Chem.*, **55**, 592 (1951).

133. E.G.J. Willing, Ph.D. Thesis, London University, 1952.

134. J.F. Goodman, Ph.D. Thesis, Exeter University, 1955.

135. K.J. Hill, Ph.D. Thesis, London University, 1950.

136. P.K. Packer, Ph.D. Thesis, London University, 1952.

137. J. Hickman, Ph.D. Thesis, Exeter University, 1958.

138. S.J. Gregg and J. Hickman, *Third Conf. on Industrial Carbon and Graphite of the Soc. of Chem. Ind., 1970*, Academic, London, 1970, p. 146.

139. R.K. Schofield, *Discuss. Faraday Soc.*, **3**, 105 (1948).

140. E.A. Flood, in *The Solid-Gas Interface*, E.A. Flood, Ed., **1**, 54. Dekker, New York, 1967.

141. C.G V Burgess and D.H. Everett, *J. Colloid Interface Sci.*, **33**, 611 (1970).

142. O. Kadlec and M.M. Dubinin, *J. Colloid Interface Sci.*, **31**, 479 (1969).

143. J.C. Melrose, *A.I. Ch. E.J.*, **12**, 986 (1966).

144. W. Benson, *J. Chem. Phys.*, **17**, 914 (1949).

145. L.J. Briggs, *J. Appl. Phys.* **21**, 721 (1950).

146. J.L. Gardon, in *Treatise on Adhesion and Adhesives*, Vol. 1, R.L. Patrick, Ed., Dekker, New York, 1967, p. 219.

147. E.A. Guggenheim, *Trans. Faraday Soc.*, **36**, 407 (1940).

148. N. Blackman, N.D. Lisgarten, and L.M. Skinner, *Nature*, **217**, 1245 (1968).

149. J.C. Melrose, *Pure Appl. Chem.*, **22**, 273 (1970); *Ind. Eng. Chem.*, **60**(3), 53 (1968).

150. W.S. Ahn, M.S. Jhon, H. Pak, and S. Chang, *J. Colloid Interface Sci.*, **38**, 605 (1972).

151. M.M. Dubinin, in *Surface Area Determination*, Proc. Intern. Symp., 1969, Butterworths, London, 1970, p. 131.

152. B.P. Bering, M.M. Dubinin, and V.V. Serpinsky, *J. Colloid Interface Sci.*, **21**, 378 (1968).

153. M. Polanyi, *Trans. Faraday Soc.*, **28**, 316 (1932).

154. J.H. De Boer and J.F.H. Custers, *Z. Phys. Chem.*, **25B**, 225 (1934).

155. R.M. Barrer, *Proc. Roy. Soc.*, **A161**, 476 (1934); *Nature*, **181**, 176 (1958).

156. J.M. Barrer, *J. Colloid Interface Sci.*, **21**, 415 (1966).

157. D.A. Cadenhead and D.H. Everett, *Industrial Carbon and Graphite*, Soc. Chem. Ind., London, 1958, p. 272.

158. M.M. Dubinin, *J. Colloid Interface Sci.*, **23**, 487 (1967).

159. W.A. Steele, *Advan. Colloid Interface Sci.*, **1**, 3 (1967).

160. W.A. Steele and G.D. Halsey, *J. Phys. Chem.*, **59**, 57 (1955).

161. N.S. Gurfein, D.P. Dobychin, and L.S. Koplienko, *Zhur. Fiz. Khim.*, **44**, 741 (1970).

162. C. Pierce, J.W. Wiley, and R.N. Smith, *J. Phys. Chem.*, **53**, 669 (1949).

163. M.M. Dubinin, E.D. Zaverina, and L.V. Radushkevich, *Zhur. Fiz. Khim.*, **21**, 1351 (1947).

164. A.V. Kiselev, in *The Structure and Properties of Porous Materials*, Butterworths, London 1958, p. 51.

165. L. Gurvitsch, *J. Phys. Chem. Soc., Russ.*, **47**, 805 (1915).

166. R.U. Culver and N.S. Heath, *Trans. Faraday Soc.*, **51**, 1569 (1955).

167. K.S.W. Sing, *Chem. Ind.*, 829 (1967).

168. R.L. Mieville, *J. Coll. Interface Sci.*, **41**, 37 (1972).

169. K.S.W. Sing, *Chem. Ind.*, 1520 (1968); *Surface Area Determination Proc. Int. Symp.*, 1969, Butterworths, London, 1970, p. 25.

170. D.H. Everett, G.D. Parfitt, K.S.W. Sing, and R. Wilson, SCI/IUPAC Project on Surface Area Standards, *J. Appl. Chem. Biotechnol.*, **24**, 199 (1974).

171. A.V. Kiselev, V.N. Novikova, and Yu. A. Eltekov, *Dokl. Acad. Nauk*, **149**, 131 (1963). Yu. A. Eltekov, *Surface Area Determination*, Proc. Int. Symp., 1969. Butterworths, London, 1970, p. 295.

172. R. M. Barrer, *Surface Area Determination*, Proc. Int. Symp, 1969, Butterworths, London, 1970, p. 90.

173. A. J. Juhola and E. O. Wiig, *J. Amer. Chem. Soc.*, **71**, 2069 (1949).

174. Yu. F. Berezkina, M. M. Dubinin, and A.I. Sarakhov, *Bull. Acad. Sci. U.S.S.R., Chem. Sci.*, 2495 (1969).

175. M. M. Tayyab, Ph. D. Thesis, Brunel University, 1971.

176. M. M. Dubinin and E. D. Zaverina, *Zhur. Fiz. Khim.*, **23**, 1129 (1949). M. M. Dubinin, *Russ. J Phys. Chem*, **39**, 697 (1965).

177. M. M. Dubinin and L. V. Radushkevich, *Proc. Acad. Sci. U.S.S.R.*, **55**, 331 (1947). L. V. Radushkevich, *Zhur. Fiz. Khim*, **23**, 1410 (1949).

178. M. Polanyi, *Verh. Deut. Phys. Ges.*, **16**, 1012 (1914).

179. E. M. Freeman, T. Siemieniewska, H. Marsh, and B. Rand, *Carbon*, **8**, 7 (1970).

180. M. M. Dubinin, *Industrial Carbon and Graphite*, Soc. Chem. Ind., London, 1958, p. 219.

181. H. Marsh and B. Rand, *Third Conference on Industrial Carbon and Graphite of the Society of Chemical Industry*, Academic Press, London 1970, p. 93.

182. M. G. Dovaston, B. McEnaney, and C. J. Weedon, *Carbon*, **10**, 277 (1972).
183. B. A. Gottwald, *Surface Area Determination*, Proc. Intern. Symp., 1969, Butterworths, London, 1970, p. 59.
184. A.V. Kiselev, A. A. Lopatkin, and L. G. Ryabukina, *Bull. Soc. Chim.*, 1324, 1972.
185. A.V. Kiselev, *Advances in Chromatography*, J. C. Giddings and R.A. Keller, Eds., Dekker, New York, 1967, p. 113.
186. M.R. Bhambhani, P.A. Cutting, K.S.W. Sing, and D.H. Turk, *J. Colloid Interface Sci.*, **38**, 109 (1972).
187. W. Diano, Ph. D. Thesis, Exeter University, 1969. W. Diano and S. J. Gregg, *Colloques Internationaux de Centre National de la Recherche Scientifique*, No. 201 (1971).
188. F.A.P. Maggs, *Research*, **6**, S.13 (1953).
189. P. Zwietering and D.W. van Krevelin, *Fuel*, **33**, 331 (1954).
190. S. J. Gregg, F.M. Olds, and R.F.S. Tyson, *Third Conference on Industrial Carbon and Graphite,* Soc. Chem. Ind., p. 184 (1970).
191. R. M. Barrer, *Quart. Rev. Chem. Soc.*, **3**, 293 (1949).
192. H. Marsh and W.F.K. Wynne-Jones, *Carbon*, **1**, 269 (1964).
193. S.S. Barton, F.G. Beswick, and B.H. Harrison, *Chem. Soc. Faraday Trans. I*, **68**, 1647 (1972).
194. P.L. Walker, L.G. Austin, and S.P. *Chem. Phys. Carbon*, **2**, 257 (1966).
195. V.L. Keibal, A.V. Kislev, I.M. Savinov, V.L. Khudyakov, K.D. Shcherbakova, and Ya. I. Yashin, *Russ. J. Phys. Chem.*, **41**, 1203 (1967).
196. J.H. De Boer, B.G. Linsen, Th. van der Plas, and G.J. Zondervan, *J. Catal.*, **4**, 649 (1965).
197. R. Sh. Mikhail, S. Brunauer, and E. E. Bodor, *J. Colloid Interface Sci.*, **26**, 45 (1968).
198. J. Hagymassy, S. Brunauer, and R.Sh. Mikhail, *J. Colloid Interface Sci.*, **29**, 485 (1969).
199. J. Hagymassy and S. Brunauer, *J. Colloid Inteface Sci.*, **33**, 317 (1970).
200. M.M. Dubinin, *J. Colloid Interface Sci.*, **21**, 378 (1966).
201. E.A. Moelwyn-Hughes, *Physical Chemistry*, Cambridge, 1957, p. 373.
202. P.A. Cutting, Ph.D. Thesis, Brunel University, p. 98 (1970).
203. J.C. Arnell and H.L. McDermott, *Proc. 2nd Int. Congr. Surface Activity*, **2**, 113, Butterworths, London, 1957.
204. H. W. Quinn and R. McIntosh, *Proc. 2nd Int. Congr. Surface Activity*, **2**, 122, Butterworths, London, 1957; *Can. J. Chem.*, **35**, 745 (1957). C.H. Amberg and R. McIntosh, *Can. J. Chem.*, **30**, 1012 (1952).
205. P.H. Emmett and Th. DeWitt, *J. Amer. Chem. Soc.*, **65**, 1253 (1943).
206. R.M. Barrer and J.A. Barrie, *Proc. Roy. Soc.*, **213A**, 250 (1952).
207. D.A. Cadenhead and D. H. Everett, *J. Phys. Chem.* **72**, 3201 (1968).
208. R.B. Jones and W.H. Wade, *J. Colloid Inteface Sci*, **28**, 415 (1968). C.B. Ferguson and W.H. Wade, *J. Colloid Interface Sci.*, **24**, 366 (1967).
209. B. McEnaney, *Chem. Soc. Faraday Trans.*, *I*, **70**, 84 (1974).
210. H.L. McDermott and J. C. Arnell, *Can. J. Chem.*, **33**, 915 (1955).
211. K.S.W. Sing and J.D. Madeley, *J. Appl. Chem.*, **4**, 365 (1954).

212. A. Bailey, D.A. Cadenhead, D.H. Davies, D.H. Everett, and A.J. Miles, *Trans. Faraday Soc.*, **67,** 231 (1971).

213. S.J. Gregg and M.I. Pope, *Fuel*, **39,** 301 (1960).

214. D. Dollimore and S.J. Gregg, *Research,* **11,** 183 (1958). N. Desai and S.J. Gregg, *Conference on Science in the Use of Fuel*, Institute of Fuel, A.56, 1958.

215. P.J. Anderson and R.F. Horlock, *Trans. Faraday Soc.*, **63,** 717 (1967).

216. F.S. Baker, K.S.W. Sing, and L.J. Stryker, *Chem. Ind.*, 718 (1970).

217. R.B. Gammage and S. J. Gregg, *J. Colloid Interface Sci.*, **38,** 118 (1972). R.B. Gammage, Ph.D. Thesis, Exeter University (1964).

218. T.G. Lamond and H. Marsh, *Carbon*, **1,** 281, 293 (1964).

219. S.J. Gregg, *J. Chem. Soc.*, 351 (1943).

220. S. Brunauer, J. S. Skalny, and E. E. Bodor, *J. Colloid Interface Sci.*, **30,** 546 (1969).

221. E.A. Guggenheim, *Applications of Statistical Mechanics*, Clarendon Press, Oxford, 1966.

222. C.F. Prenzlow, H.R. Beard, and R.S. Brundage, *J. Phys. Chem.*, **73,** 969 (1969). C.F. Prenzlow, *J. Colloid Interface Sci.,* **37,** 849 (1971).

Author Index

Numbers in parentheses are reference numbers and show that an author's work is referred to although his name is not mentioned in the text. Numbers in *italics* indicate the pages on which the full reference appears.

Subject Index

Cumulative Index